FLORIDA STATE
UNIVERSITY LIBRARIES

DEC 28 1995

TALLAHASSEE, FLORIDA

Technological Innovation and the Great Depression

Technological Innovation and the Great Depression

Rick Szostak

WestviewPress
A Division of HarperCollinsPublishers

HC
110
T4
S96
1995

All rights reserved. Printed in the United States of America. No part of this publication may be reproduced or transmitted in any form or by any means, electronic or mechanical, including photocopy, recording, or any information storage and retrieval system, without permission in writing from the publisher.

Copyright © 1995 by Westview Press, Inc., A Division of HarperCollins Publishers, Inc.

Published in 1995 in the United States of America by Westview Press, Inc., 5500 Central Avenue, Boulder, Colorado 80301-2877, and in the United Kingdom by Westview Press, 12 Hid's Copse Road, Cumnor Hill, Oxford OX2 9JJ

A CIP catalog record for this book is available from the Library of Congress.
ISBN 0-8133-8941-0

The paper used in this publication meets the requirements of the American National Standard for Permanence of Paper for Printed Library Materials Z39.48-1984.

10   9   8   7   6   5   4   3   2   1

Contents

List of Tables ix
Acknowledgements xi

PART ONE
THEORETICAL CONSIDERATIONS

1. **The Great Depression Revisited**   3

    A Synopsis of the Theory Employed in This Work, 4
    Outline of Work, 10
    What Happened? 13
    Notes, 23

2. **Questions Unanswered**   27

    The Monetary Approach, 27
    The Keynesian Approach, 34
    Fisher's Debt-Deflation Theory, 38
    Labor Markets, 39
    Other Market Imperfections, 44
    The Role of the Stock Market Crash, 46
    Real Business Cycles, 48
    Underconsumptionist and Marxian Theories, 51
    Long Waves, 55
    The Resurgence of Structural Approaches, 57
    Questions Unanswered, 60
    Notes, 61

3. **Building a New Theoretical Approach: Consumption and Investment**   67

    Consumption Demand, 68
    The Decline in Investment, 81
    A Note on Fiscal Policy and Export Demand, 93
    Notes, 95

4. The New Theory: Technological Change — 101

Product Versus Process Technology, 102
Process Innovation in the 1920s and 1930s, 103
Why This Time Path of Innovation? 106
Solving the Technological Riddle, 110
A Further Step Back, 115
Technological Determinism? The Role of
  Industrial Research, 117
Echoes of Schumpeter? 122
Notes, 124

5. The Labor Market — 129

Technological Unemployment, 130
Structural Unemployment, 132
Clearing the Labor Market, 134
Second and Third Degree Unemployment, 136
Price Rigidity Revisited, 137
Notes, 138

PART TWO
SECTORAL ANALYSIS

6. The Procedure for Sectoral Analysis — 143

Extent of Sectoral Survey, 144
Outline of Sectoral Studies, 147
A Note on Methodology, 153
Notes, 153

7. Chemicals — 155

Electrochemicals, 159
Petrochemicals, 161
Plastics and Synthetic Fibers, 161
Photography and Motion Pictures, 166
Pharmaceuticals, 170
Concluding Remarks, 173
Notes, 173

8. Electrical Products — 175

Interwar Employment Trends, 178
Telephone, 178
Radio, 179

Television, 185
Appliances, 188
Machinery, 192
Computers, 196
Transistors, 198
Notes, 200

9. **Internal Combustion** 205

   Automobiles, 205
   The Rubber Industry, 216
   The Glass Industry, 217
   Iron and Steel, 218
   Airplanes, 221
   Notes, 226

10. **Selected Other Industries** 231

    Energy, 231
    Food Processing, 239
    Lumber, 242
    Textiles, 244
    Notes, 248

11. **Non-Manufacturing Sectors** 253

    Construction, 253
    Transport, 258
    Agriculture, 264
    Services, 269
    Notes, 282

12. **Summary of Sectoral Studies** 287

    Establishing Plausibility: An Exercise in Addition, 291
    Recovery, 296
    Notes, 297

13. **International and Regional Comparisons** 299

    The Depression Abroad, 299
    The Regional Distribution of Unemployment in the
      United States, 307
    Unemployment by Race and Gender, 308
    Notes, 309

14. Conclusion 313

A New View of the Economic Process, 315
The Role of Economic History, 317
Will It Happen Again? 318
Public Research Policy, 322
Growth Is the Only Answer? 324
Unemployment, 326
What Should Hoover and Roosevelt Have Done? 330
Trade Theory, 331
Further Research, 332
Notes, 333

Bibliography 337
Index 357
About the Book and Author 367

Tables

1.1	Indexes of Employment, Hours, and Unemployment, 1920-1945	16
1.2	Indexes of Output and Productivity, 1920-1945	19
1.3	Labor Turnover in Manufacturing, 1920-1945	22
6.1	Sectoral Employment Changes 1920-1929 and 1929-1933	145
6.2	Major Employment Losses Within Manufacturing, 1929-1933	145
6.3	Manufacturing Employment: Selected Years	146
6.4	Labor Productivity Growth in Manufacturing, 1919-1929	148
13.1	Average Unemployment Rates, Selected Countries	302

Acknowledgements

I have always been intrigued by the enigma of the Great Depression, perhaps because my parents often told tales of their experience of the 1930s in drought-stricken western Canada. Only in the late 1980s, though, did I begin to develop the theoretical approach of this book. I was helped at an early stage by a paper presentation at Northwestern University, and the lengthy conversations with Joel Mokyr and Lou Cain which that entailed.

Much of the first draft of this book was completed while I was on sabbatical at the University of New South Wales in 1991-2. John Perkins organized a one-day workshop on the Depression at which I was one of the speakers. I received a great deal of helpful advice that day and during my entire stay from many people, including Alex Blair, Craig Freedman, Peter Kriesler, Ian Inkster, John Lodewijks, and John Perkins. I also benefitted from being able to present papers at the University of Newcastle, Australian National University, the University of Western Australia, and Curtin University. I gained insights of great value at each of those places, and thank John Fisher, Mac Boot, Graeme Snooks, and Malcolm Tull for having made those visits possible.

I would like to take this opportunity to thank the entire school of economics at the University of New South Wales for the tremendous hospitality which they bestowed upon me. Special thanks must go to David Meredith, Ross Millbourne, and Bob Conlon, who occupied the key administrative posts while I was there. The School of Economics provided, at the time, a wonderful melange of respected scholars representing varied viewpoints. I hope that those in charge can overcome the temptation to turn it into just another mainstream economics department, for surely the world already has enough of these. I was blessed not only with the friendship and intellectual insights of all those above, and others such as Lance Fisher, Frank Barry, and Barrie Dyster, but also the warm smile, masterful typing, and encouragement of Charleen Borlase. And even this, let me be clear, is only a partial list.

Since my return to North America, I have had the opportunity to present some of this material at the meetings of the Economic History Association and the Nineteenth Conference on the Use of Quantitative Methods in Canadian Economic History. Both of these occasions aided me considerably as I revised the manuscript. I would like to thank Morris Altman in particular for some sage advice.

I have been aided over the years by two grants from the Central Research Fund of the University of Alberta related to this book. As always, I have benefitted greatly from the advice and encouragement of Mike Percy, Ken Norrie, and Melville McMillan. Charlene Hill has finished typing that which Charleen Borlase had begun. One can only hope that shortsighted government budget cuts will not destroy the positive environment which made this book possible.

Finally, let me thank Alison Auch and the staff at Westview Press. It has been a pleasure dealing with them.

Rick Szostak

PART ONE

Theoretical Considerations

1

The Great Depression Revisited

Most economists would agree that we have as yet no widely accepted explanation of the Great Depression. Various theories have been put forward, and some are felt to be capable of explaining some, though far from all, of the calamitous events of the 1930s (as we will see in Chapter 2). The situation is ripe for a new approach. The purpose of this work is to proffer a technological explanation of much of the Depression experience. While Schumpeter (1939) and others, especially at the time of the Depression, have suggested that technology played a substantial role in causing the Depression, the approach pursued here is quite novel. I will detail in the next five chapters a theoretical explanation of how an abundance of labor-saving production technology coupled with a virtual absence of new product innovation could affect consumption, investment, and the functioning of the labor market in such a way that a large and sustained contraction in employment would result. I will also describe how such a technological situation could have arisen historically. In the succeeding several chapters, I will provide evidence at the industry level that these technological forces were at work, and that they were powerful enough to generate most of the observed unemployment in the 1930s.

The fundamental premise of this work is that the key to understanding the Great Depression is to be found through analysis at the industry level rather than through the highly aggregated research that has dominated the field in the past. This does not mean, of course, that we can turn our back on the concepts of macroeconomics. We must, indeed, couch our analysis carefully in a general equilibrium framework. It is too easy to describe why some sectors of the economy went into decline in or near 1929. This may provide limited insight into why the Depression began, but can tell us little about why recovery was so slow. We must understand why the resources released by some sectors did not — indeed, could not — flow smoothly into other sectors. Why didn't the labor market clear? Once we remove some simple misconceptions about the functioning of the labor market, and the determinants of consumption demand, we will find that the answer to this question is also to be found at the industry level.

We can already see that there is not just one Depression question. Why was the initial turndown so severe? Why was recovery so slow? We can add others.[1] Why was the recovery to 1936 so feeble and so overwhelmed by the downturn of 1937? How was the Depression transmitted around the world at a time when links between nations were much weaker than today? On the other hand, why did some countries, such as Britain, fare so much better than others in the 1930s? Most previous works have tended to focus on only one of these questions. There is, naturally, no logical necessity that the same factors provide the answer to all. It would, though, be an awful coincidence for a number of horrible events to occur in a short time period for quite different reasons. I will, in what follows, outline an approach which can deal with all of these questions.

This does not mean that I am pursuing some simple monocausal explanation of the entire Depression experience (though I must inevitably focus on my own theoretical approach to the exclusion of others in much of what follows). As Scitovsky (1986) so aptly put it, "Nothing is ever so simple that a single explanation will adequately explain it." Certainly the Great Depression is not so simple. As noted above, the existing literature does adequately explain some of what happened, but is unconvincing in its attempt to explain the entire Depression. My purpose is to fill much of that gap.

A Synopsis of the Theory Employed in This Work

What is the connection between new product innovation and economic performance? Let us start by asking a question which never arises in standard attempts to comprehend deficient aggregate demand; what goods or services should the mass of unemployed have produced in the 1930s? Full employment would only have been possible if the goods and services produced could be sold profitably. It would seem natural to look ahead to the next period of peacetime full employment and see what workers were producing during the phenomenal postwar economic expansion.

The sectors which provided employment growth to power postwar economic expansion could not have done so in the 1930s. In most cases, this was due to the timing of technological innovation. Television, the modern airplane (the DC-3 of 1935), sulfanomides and vitamins, synthetic fibres, and improved plastics were developed only in the late 1930s. With the exception of the electric refrigerator — where sales expanded steadily through the 1930s — there was quite simply a dearth of new product technology in 1929. Moreover, the "new" products of the 1920s, notably the automobile and the radio, had largely saturated their markets, and thus investment and employment in these sectors were destined to fall in the early 1930s. Sluggish sales during the 1930s and World War Two, along with product improvements, provided the auto industry with much room to expand postwar.

Is it simply a coincidence that none of the sectors which powered postwar growth could possibly have done so in 1929 for technological reasons? To be more modest, we could pose a counterfactual; if this technology had been available in 1929, could the Depression have been avoided? The answer must surely be that if TV, the DC-3, and vitamins were ready to go in 1929, investment and consumption would have been significantly higher, and the Depression much less severe. To recognize that a shortage of growing sectors may be responsible for the depth and length of the Depression is to suggest that perhaps declining sectors — in the absence of counteracting growth potential — may have caused the downturn. The Keynesian multiplier would have time to operate at full force.

What caused consumption expenditure to fall from 1929? To answer this question, we need to take a longer view than is common in economic theory. Our propensity to consume naturally depends on the range of goods available for consumption. Much of the increase in incomes over the last century has been spent on goods and services our great-grandparents had never heard of or could not dream of affording: TVs, VCRs, cars, pharmaceuticals, radios, movies etc. It is surprising, then, that the discussion of particular episodes in the behavior of aggregate consumption demand is never related to the particular bundle of goods available at that time and place. The simple fact is that, in the absence of the creation of new products, aggregate consumption demand can be highly inelastic. Increases in incomes will result in higher savings rates. Moreover, as we enter a world of consumer durables, it not only becomes potentially difficult to extricate an economy from a situation of deficient demand, but quite straightforward to enter such a world. Purchases tend naturally to bunch as a new product is introduced. Automakers and other durable producers, fighting for market share, could not even out the time path of production. Consumption expenditure thus contracted in 1929.

Why did investment not absorb the increased savings, but rather fall sharply from 1929? We can not answer this question if we examine aggregate data alone. Trends in investment can only be understood at the sectoral level. Demographic variables limited the scope for expansion within the traditional necessities. Saturation in durables naturally discouraged investment there. While the rise and decline of industries is a natural characteristic of modern economies, in the late 1920s there were no new products to pick up the slack. Whatever limited investment potential there was in "old" industries had largely been achieved earlier in the decade. An automobile-induced construction boom led to overbuilding of all types of structures in the mid-1920s. A new generation of machine tools appeared at the beginning of that decade which would not be rendered obsolete for decades.

Why was unemployment so high? As Keynesians have long argued, the fall in consumption and investment induced a dramatic fall in output and employment through the multiplier effect (though they could not explain why consumption and/or investment had fallen). A third source of unemployment was labor-saving technology. At the same time that no new products were being developed, it became possible to produce the existing range of goods with a much smaller labor force. Electrification, assembly lines and continuous processing induced the largest decadal increase in labor productivity the country had ever seen in the 1920s. This had contributed to chronic unemployment through that decade. However, many redundant workers were only let go in the downturn of the 1930s. This dishoarding of labor swelled the unemployment rolls.

The idea that labor-saving technology must reduce employment has been around since at least the Industrial Revolution. At the industry level, depending on the division of gains between employer, employee, and consumer, and on the elasticity of demand, this may be true. At the economy-wide level and in the long run we would expect displaced workers to be absorbed elsewhere. This, though, depends on there being expanding sectors. The phrase "technological unemployment" emerged in the 1920s, not because the problem was new but because it involved much greater numbers than before.

In the absence of growing sectors, the multiplier mechanism was able to operate at full force. Those who lost their jobs reduced their expenditures and induced further unemployment. The unemployed were, unless coupled with capital, wholesale raw materials, and access to wide markets, inherently unproductive. For full employment to be restored, entrepreneurs had to come forth who saw profitable opportunities in particular product lines. The multiplier mechanism might then be put to work in the opposite direction. Aggregate expectations remained high until 1932. Never having seen a depression so severe, the natural supposition was that recovery must be imminent. Market saturation and a lack of new products prevented such expectations from being translated into investment and job creation.

Keynesian explanations of the Depression have been justly criticized for relying on coincidental causes of declines in consumption and/or investment. If a surfeit of labor-saving process innovation and deficiency of new product innovation are to be credited with causing the Depression, the timing of innovation must be explained. Economic historians often speak of a Second Industrial Revolution of the late nineteenth century, and identify three main elements: the internal-combustion engine, economical electric power production, and a revolution in chemical understanding and application. It is widely recognized that much of the technological — and economic — history of the twentieth century is derivative of these three breakthroughs. A perusal of a list of key innovations of the first half of the twentieth century, indeed, indicates that the most important innovation which can not be traced to any of these three is the zipper.

If these three new sources of technological potential had spawned a steady stream of new products over the next decades, economic stability might have been ensured. This was not to be the case. The automobile was a (temporarily) spent force in 1929, but the modern aircraft was not yet ready. The most far-reaching product breakthroughs in chemistry — in plastics, synthetic fibers, and pharmaceuticals — occurred in the 1930s. The radio expanded and declined somewhat less spectacularly than the automobile in the 1920s; the television was not perfected for another decade. Each of the three encouraged widespread process innovation in the 1920s. The automobile spawned the assembly line in 1913. The counterpart to the assembly line for homogeneous output is continuous processing; this was developed in chemicals and spread through such industries as food processing and oil refining in the 1920s. Electrification replaced steam engine belting, allowed improved factory layout, facilitated flexibility in machine operation, and encouraged a 50% increase in horsepower per worker in that one decade. Manufacturing output rose 64% in the 1920s, with virtually no change in employment.

The time-path of innovation is a function of both technical and social considerations. The unfolding of the potential of the Second Industrial Revolution could not be unaffected by the rise of the industrial research laboratory (itself largely a creature at first of the chemical and electrical industries). These, emerging shortly after the turn of the century in the United States, emphasized process improvement and minor product development almost exclusively in their first years of existence. While these have remained their foci, a greater role has been accorded radical product innovation in later decades. Since these labs attracted scarce human resources from other settings and competed with independent innovators in both marketplace and patent office they could skew inventive effort away from new product development through the first decades of this century.

Major innovations generally evolve over a period of decades. It is incorrect, therefore, to think that innovation can quickly respond to economic conditions. Schumpeterian entrepreneurs do not, in fact, have access to a pool of dormant product technology when they feel the time is right. While the Depression itself *might* encourage some types of research, such investigations could only build on the existing technical base.

Can secular (i.e. slowly acting) forces explain the Great Depression? Gordon (1961: 449) raised this question, and replied that, "It is not sufficient to reply that secular stagnation was also operating in the 1920s but was temporarily offset by a speculative boom and by the investment opportunities created by the spread of the automobile and electric power." Yet surely he overstates the case. Fleisig (1974) is one who has argued that secular forces could be causes of the Depression. Boulding has warned that we must be aware of the great discontinuities in the world; we live on the tablelands and must not fall off the cliffs. Fisher (1933) characterized the economy as a ship on the ocean; normally it naturally rights itself but in the Depression it tipped too far and thus disequilibrating forces came to dominate. Depressing influences can thus exist for some time largely unnoticed, and then be triggered to full effect. We argue, for example, that much redundant labor was hoarded in the 1920s only to be summarily dishoarded in the 1930s. Note also that, as we will see below, the 1920s were not a decade of prosperity for all, and thus the Depression begins from a position of weakness.

Still, the very notion of a trigger mechanism implies that secular forces must be coupled with some that act more suddenly. Our approach contains many of these. We could note in particular that without the spectacular rise and fall of the automobile industry the American interwar period would have looked like the British: no spectacular decline but two decades of lackluster performance.

We should briefly discuss how this theoretical approach relates to the existing literature on the Great Depression (the subject of Chapter 2). Much of that literature has been cast in terms of a debate between two macroeconomic views: the monetary approach and the Keynesian approach. Whereas decades ago adherents of each may have claimed that they had *the* answer, it is now widely recognized that no one theory adequately explains the entire Depression experience. The major problems with the Keynesian approach are that it has not been able to tell us why consumption and/or investment should have fallen so precipitously from 1929, and that it can not explain why recovery was so sluggish. As this book's approach can answer these questions, it can be viewed as complementary to the Keynesian position. Our discussion of labor markets in Chapter 5 and of structural elements throughout should firmly establish, though, that this book does not strictly adhere to a Keynesian view.

Some Keynesians have argued that the cause of the downturn is to be found in expectations; people became pessimistic about the future and thus postponed investment (and consumption) decisions. It is always difficult to ascertain what people were thinking at any point in time. The evidence we have, though, suggests strongly that expectations were high well into the 1930s. Further, I argue that an entrepreneur considering an investment project will be much more concerned with his/her view of the potential of a particular product line then with guessing future aggregate economic performance. Refrigerator sales, after all, did expand through the 1930s. We can only fully understand investment behavior by looking at the lack of investment opportunities sector by sector. Since technology explains the lack of investment, we need not fall back on expectational arguments.

I accept the premise that the Federal Reserve Board behaved suboptimally in the 1930s. I also accept the majority view of the profession that this behavior was responsible for some but far from all of the poor economic performance of that decade. My argument here is that the timing of technological innovation *on its own* would have led to a very severe downturn. Financial disturbances exacerbated this downturn, but these two forces acted for the most part independently.[2] Financial disturbances will rarely be mentioned after Chapter 2 because they are not essential to this technological argument. This does not mean they were necessarily unimportant in their own right.

Both monetary and Keynesian approaches must cope with the fact that real wages actually rise during the 1930s, as opposed to falling so that labor markets could clear. Various authors have attempted to explain this phenomenon with appeals to unanticipated inflation, efficiency wages, government interference, personnel policies, decreased working hours, insider-outsider theory, and imperfections in other markets. All of these at best provide only a partial explanation of massive and longlasting unemployment. This can only be fully comprehended in the context of little product and much process innovation.

Recent research does not support the popular view that the stock market crash caused the Depression. Still, people can be forgiven for doubting that these events were entirely coincidental. I argue that stock market speculation reflects a lack of real investment opportunities in the late 1920s, and that this helps us to understand how secular forces led to such a sharp downturn.

Real Business Cycle theory, underconsumptionism, Marxian approaches, and Long Wave theories each suffer from severe theoretical and empirical difficulties. Yet they each point us in the direction of looking at long term and especially technological forces in order to understand a cataclysm such as the Great Depression. Structural views have increased in popularity over the last decades as the western world has again experienced a lengthy period of high unemployment. Phelps (1993) argues forcefully that mainstream macroeconomic theory is poorly equipped to deal with the booms or slumps of more than a decade in length which characterize modern economies, and suggests technology and demography as among the key structural forces on which we should focus our attention.

Bernstein (1987) has attempted to rehabilitate a technological view of the Depression. Unfortunately, his approach has many defects: he relies on the stock market crash to cause the initial downturn. He feels that sluggish recovery was due to a combination of technological and financial factors but does not attempt to explain why the time path of innovation was uneven or provide evidence for the contention that there were industries with growth potential that could not borrow money (theory and observation suggest otherwise, e.g. the refrigerator). Finally, he focuses only on manufacturing. Like Schumpeter before him, these defects detract from the power of a technological explanation.

I defer my discussion of Schumpeter to the end of Chapter 4, so that I can more clearly distinguish his argument from my own. Difficulties with Schumpeter include his strict adherence to the idea of long waves (indeed Schumpeter felt the Depression was a coincidental result of downturns in three regular cycles of different length). He posited an artificial separation between invention and innovation; this allowed him to attribute changing rates of innovation to entrepreneurial whim. His technological explanation is thus entirely disconnected from the actual course of technological change. Schumpeter does not pay sufficient attention to the distinction between product and process innovation, and therefore ignores the possibility of market saturation. Thus, he is unable to explain severe downturns without reference to weak expectational arguments. His case, then, is empirically much weaker than mine.

I discuss in Chapter 5 the possibility of structural unemployment. An alternative technological argument would be that the skills required of the workforce changed more rapidly in the interwar period than did the skills possessed by the workforce. Thus, there were enough jobs to go around; workers simply were not suited to them, and a painful decade of adjustment was required. It is not clear how riding the rails[3], waiting in soup lines, or even working on government relief projects (which deliberately did not seek to train workers) could have addressed such an imbalance. I argue that in fact there simply were not enough jobs of any kind available, and believe that the empirical evidence from my sectoral studies supports that contention. As well, my description of the evolution of process technology provides little support for the view that radically different skills were required.

Outline of Work

The foregoing has provided a mere sketch of a complete theoretical argument which will be fleshed out in the following chapters. It is necessary first to review the experience of aggregate variables, and especially of unemployment, the key variable we will focus on in what follows. We must know what happened before we can try to explain it.

Chapter 2 will discuss existing theoretical explanations of the Depression. This serves two purposes. We must recognize the insights of existing theory in order to know what still requires explanation. Further, the shortcomings of existing theories will point us in the direction of our new theoretical approach.

Chapters 3, 4, and 5 develop the theory from the previous section in much greater detail. Chapter 3 looks at the behavior of consumption and investment, and recognizes in particular the role technology plays (in conjunction with other factors) in determining both. Chapter 4 then focuses on technology; the experience of process innovation is discussed, and then reasons for the abundance of process and shortage of product innovation are explored. Chapter 5 looks at how the forces discussed above would operate in the labor market. It would be counterproductive to strictly delineate theory and evidence; thus in Chapters 3 through 5 we will also discuss aggregative evidence which supports our theory.

Innovation and market saturation can only be analyzed at the sectoral level. In Part Two, case studies are thus undertaken of most manufacturing sectors as well as mining, agriculture, construction, transport, and services. The evidence for market saturation is examined. Investment opportunities are surveyed sector by sector. Estimates are provided of employment losses by sector. It is seen that there was virtually no place for the unemployed to go. Technological change, for example, severely limited the absorptive capacity of the service sector. A special focus of the case studies is the course of innovation. Particular innovations are seen as the result of lengthy research efforts which can be traced, in logical steps, to at least one element of the Second Industrial Revolution. In some but not all cases industrial research laboratories are seen to play an important role in determining the timing of innovation. In Chapter 6, after we have discussed our theory in more detail, we provide a more detailed overview of Chapters 7 through 11.

Chapter 12 opens with a review of the sectoral evidence for market saturation, limited investment opportunity due primarily to the lack of product innovation, widespread process innovation, many job losses but few job openings, and the role of the Second Industrial Revolution and industrial research labs in determining the time path of innovation. Much of the chapter, however, is devoted to an estimation of the total number of unemployed who can be accounted for by the structural forces we have outlined. Sector by sector, the evidence is weighed in order to create plausible estimates of the net job loss. When labor dishoarding is included, the total is very nearly three million workers. This figure on its own implies that technology must bear the blame for a substantial portion of the Depression-era pain and suffering. Moreover, macroeconomic theory clearly indicates that employment losses of this magnitude will have a further multiplier effect on the economy as the newly unemployed reduce their consumption, and thus cause a further chain of job losses and consumption declines. A multiplier of four is not unreasonable when we recall that there were extremely few growth-generating forces at work in the early 1930s to counteract this depressive effect. Though I do not believe that technology was the only cause of the Depression, these numbers indicate that the vast bulk of the unemployment experience of the 1930s could be attributed to technological factors.

In Chapter 13 we discuss both the international and regional experience of the Depression. Also in that chapter, we discuss how sectoral analysis helps explain gender and racial differences in unemployment experience. Despite recent advances, economists have still not unearthed transmission mechanisms which convincingly explain how the American Depression could have been transmitted worldwide in an era when both trade and financial links were much weaker than today. This leads to the conclusion that similar forces were at work generating the Depression in other developed countries. Similar technological trajectories appear capable of explaining much of the international experience, while the implications of differing levels of national income for market saturation help us understand why a country such as Britain suffered much less in the 1930s despite having relatively strong ties to the United States. Within the United States, the differential impact of the Depression by region, gender, and race can *only* be fully comprehended within a disaggregated analysis. Our discussion of which industries suffered most and why sheds much light on why different sections of the American population were affected so differently during the Depression.

Chapter 14 provides concluding remarks and suggestions for further research. The primary policy implication of this research is that it provides another rationale for government support of research. As well, it is suggested that direct employment of the unemployed in public works projects is a superior method of dealing with severe recessions than attempting to stimulate private consumption. While it is believed that this work provides a firmer basis than the existing literature for believing that a depression as severe will not happen again, it is nevertheless asserted that the equilibrating mechanism of the economy is sluggish. An imbalance between rising and declining sectors may have characterized the 1970s, and could well cause widespread unemployment again.

What Happened?

We should begin by noting that the downturn does not start with the economy anywhere near full capacity in either Europe or North America. There was chronic unemployment throughout the 1920s. The economic boom of the 1920s was not as phenomenal as people have suggested; it is famous because of financial speculation and the subsequent Depression (Aldcroft, 1977: 193-4). We should view the Depression as an extreme continuance of weak economic performance. We would not be misguided, then, to suspect that some of the forces behind the Depression would already have been apparent in the 1920s.[4] "The key to the 1930s disaster may well lie in the boom years of the 1920s. If so it has never been clearly identified" (Hughes, 1987: 428).

The 1920s were at least a period of stable growth in GNP. The recessions of 1921, 1924, and 1927 were quite mild, the last providing just a slowdown in the rate of growth. Prices were also fairly stable. Industrial output increased by 45% and real GNP 40% between 1922 and 1929. A range of technological innovations allowed significant increases in both capital and labor productivity. Investment surged in some sectors. Thus, in some ways the 1920s do appear vibrant. Growth rates, though, were about the same as before the war, and much lower than they would be after the next one. Many authors have attributed stock market speculation and real estate speculation to a shortage of investment opportunities. Others have noted that the income distribution was worsening and therefore much of the population was not sharing in income gains. As well, between 367 and 976 banks failed every year during the 1920s. And we will see that unemployment remained high throughout. It should be clear that far from all the signs were propitious well before 1929.

Though the Depression is associated in the minds of most people with the stock market crash of October 1929, the downturn actually began some months earlier. Industrial production peaked in mid-1929, and would not return to the same level for years. Prices, nominal wages, and production continued to fall until 1933, making this one of the longest downturns in history. There is some dispute over unemployment rates, which we will discuss below; at least one quarter of the labor force was unemployed by 1933. On top of this, there was widespread underemployment, as work weeks were shortened by one third in some industries. Real GNP was only 71% of the 1929 level in 1933 (see Table 1.3). The downturn thus erased all of the economic growth achieved in the previous quarter century. Consumption expenditure was 18% lower. Gross investment almost ceased. Manufacturing output fell 50% 1929-1933. Since recovery began in 1933, these figures do not quite capture the magnitude of the drop to the nadir in early 1933. Net investment across industry was actually negative in both 1931 and 1932. American business incurred a net loss of $5 billion in 1932. Only 20% of industrial enterprises made a profit in that year (Jaeger, 1972). The wholesale price index fell 31% 1929-1933. Despite low nominal interest rates, this meant that real interest rates were the highest seen in the twentieth century (so far). After three years of turmoil, which through mergers and failures had eliminated almost half of American banks, a banking panic in 1933 so threatened the entire banking system that Roosevelt was compelled to declare a bank holiday.

From 1933 to 1937 the economy experienced the longest recovery phase since the National Bureau of Economic Research began keeping records. While considered by some to signify the resilience of the economy, this is interpreted by others as symptomatic of a sluggish recovery. In 1937, both employment and output were still far below their 1929 levels. Investment remained at low levels, and interest rates for risky investments remained unusually high (Hayes, 1951). Profits did not rise as they usually do during a recovery, for real wages rose faster than productivity (Hayes, 120). The economy then suffered another downturn. This one was even sharper than that of 1929, though it lasted less than a full year. The economy had more than recovered from this downturn by the end of 1939, but had still not attained either full employment or the level of per capita GDP of 1929. These achievements were to await the 1940s. Thus, many have been able to claim that only preparation for war took the economy of the United States out of the Great Depression.

Previous Depressions

Gill, writing in 1940, recognized that, "The decade that has just passed is unique in American economic history." In neither the so-called depressions of the 1870s or 1890s did output fall as far. Industrial production dropped 7% in the 1870s, 13% in the 1890s, and almost 50% in the 1930s. In both those previous cases, production was over 50% higher than the previous peak within a decade, while ten years after 1929 production still had not attained the 1929 level. Saul, in his survey of that period, concludes that it is inappropriate to apply the term 'Great Depression' to the last quarter of the nineteenth century (1985: 55).

Unemployment

The photographs of people in soup lines, or riding the rails,[5] or working on various government relief efforts, should ensure for all time that the Great Depression is recognized as a period of unusually severe unemployment. Yet, as Allen noted years ago, most of the unemployed were not so obvious (1952: 148). Thus, there is a great deal of scope for arguing that unemployment was really not all that bad in the 1930s. The source of controversy is the fact that statistics on unemployment are poor for the pre-World War Two period. Congress had passed a Bill in 1930 to have employment statistics collected, but the Hoover administration never provided funding for the endeavor. In the absence of statistics, the government (like some modern economists) could continually claim that unemployment rates were not that high and were coming down (Bernstein, 1960: 257, 269). We must, then, estimate both the labor force and total employed to arrive at unemployment figures.

The figures compiled by Baily (1983) on employment, hours, and unemployment are a good starting point (Table 1.1). They show unemployment rates rising to over 25% in 1933, and remaining above 14% through 1940. The figures stand out sharply from the experience of the 1920s and the later 1940s. Baily obtained his unemployment figures from the Historical Statistics of the United States, which uses Lebergott's estimates. Coen (1973) has re-estimated these figures for 1922-29, finding an average unemployment rate of 5.5%. He wished to account for cyclical variation in the size of the labor force due to discouraged workers. He used postwar data to estimate parameters. Smiley (1983b) has questioned whether we would expect the same parameter values to hold in the different periods, and has suggested Coen erred as well in applying his analysis to the farm sector. Still, Coen's figures for the 1920s seem more plausible in many ways; Lebergott's figure of 1.8% for 1926 is too good to be believed. Micro-level studies have often found much higher levels of unemployment than Lebergott suggests; one study found a rate of 10.4% in Philadelphia in April, 1929 (Fearon, 1987: 63). Douglas (1930: 588) found average unemployment of 9.8% in transport and manufacturing 1921-1926, though this was largely due to high levels in 1921 and 1922. Foreign observers and social workers also claimed that unemployment was high through the 1920s (Bernstein, 1960: 58-60). Mills (1932: 531) was convinced that unemployment rates had been high, noting that, "An increasing value of unemployment during an era of economic expansion was, considering its magnitude, a new phenomenon in our history."

Darby (1976) is the most well-known of those who have argued that unemployment figures such as Lebergott's exaggerate the degree of unemployment in the 1930s. He argued that the true rates were 20.9% in 1933, 16.2% in 1934 and 9.2% in 1937. Even these must be recognized as very high numbers relative to surrounding decades, but they do make the 1930s look significantly less disastrous. Darby indeed argued that with these new figures, and reference to Federal Reserve miscues in 1931-1933 and 1936-1938, the "fiction" of a slow return to full employment could be done away with. There is good reason, though, to believe that his numbers are much too low. The main contention of Darby was that workers receiving government relief should be counted among the employed rather

TABLE 1.1 Indexes of Employment, Hours, and Unemployment, 1920-1945

1929 = 100 unless otherwise specified.

	Employment			Hours			
Year	Total Civilian	Private Non-farm	Production Workers in Manufacturing	All Employees	Private Non-farm	Manufacturing	Unemployment Rate
1920	84.9	88.6	101.0	89.1	85.3	107.3	5.2
1921	80.2	78.3	77.3	80.5	76.2	75.4	11.7
1922	85.8	85.5	85.5	86.2	82.9	84.7	6.7
1923	91.8	94.5	97.9	92.9	91.7	99.3	2.4
1924	91.0	91.7	90.9	91.0	88.4	89.2	5.0
1925	94.6	95.4	94.1	94.3	92.2	93.4	3.2
1926	97.0	98.1	95.9	97.5	96.3	96.4	1.8
1927	97.1	97.4	93.8	97.1	96.6	95.2	3.3
1928	97.7	97.3	94.0	98.1	97.3	94.2	4.2
1929	100.0	100.0	100.0	100.0	100.0	100.0	3.2
1930	95.6	92.9	87.1	93.5	91.5	85.0	8.9
1931	89.4	82.7	73.5	85.5	80.9	69.3	16.3
1932	82.3	72.2	62.5	75.6	69.4	55.4	24.1
1933	82.4	72.7	69.1	76.0	68.4	59.4	25.2
1934	87.2	80.1	80.6	76.8	69.6	62.6	22.0
1935	90.2	83.4	86.1	81.1	73.8	70.4	20.3
1936	95.2	89.9	93.5	89.3	82.1	81.7	17.0
1937	99.7	96.4	102.6	93.2	86.7	88.4	14.3
1938	95.5	89.6	87.3	86.8	78.7	70.4	19.1
1939	99.0	94.2	97.1	90.6	84.1	81.5	17.2
1940	102.8	99.6	104.4	94.5	89.2	89.9	14.6
1941	109.0	112.8	128.6	104.5	99.8	115.5	9.9
1942	116.3	122.5	151.7	116.3	108.5	141.8	4.7
1943	117.9	128.6	176.8	131.4	113.6	168.9	1.9
1944	116.8	126.8	172.1	134.2	111.8	166.8	1.2
1945	114.3	121.8	151.9	125.8	106.1	142.9	1.9

Source: Baily, 1983, with permission of the Brookings Institution.

than the unemployed. Relief efforts are alleged to have inflated wage demands, lured workers away from private employment, and perhaps crowded out private investment through debt finance. Kesselman and Savin (1978) have shown that there was excess demand for both private and government jobs throughout the 1930s. Thus, relief workers were not pulled from private sector jobs. Margo (1987) finds that relief work was distributed quite differently by age, nationality and region than was unemployment. Moreover, the Works Progress Administration strongly encouraged its workers to seek private employment. As for crowding out, about half of relief expenditure was tax-financed, and there should have been a powerful multiplier effect in paying workers with a very high marginal propensity to consume. Kesselman and Savin thus conclude that Lebergott was right in treating relief workers as unemployed. Margo (1987: 340-1) finds that relief spending had no effect on the duration of employment, and it is possible that relief work increased the probability of reemployment. Further support comes from Green and MacKinnon (1990) who have compared similar regions in Canada and the United States, and found that employment recovery was no faster north of the border, despite the absence of programs like the WPA.

Given the difficulty with unemployment estimates, it is worthwhile to look at more reliable output figures, to see what they might tell us about employment. Table 1.2 shows that output fell by over one third in the private non-farm sector from 1929-1933. Even allowing for some decrease in output per man hour over this period (firms may hoard some workers with an eye to the future — we will see below that the opposite was likely the case for most industries in the early 1930s) we are left with the result that hours worked must have fallen by 30%. Again, we must remember that the labor force was growing through the 1930s (population grew by about 1% per year; while some men left the labor force because they became too discouraged to even look for work, more women entered when their husbands lost their jobs; see Margo, 1987). By 1939, non-farm output had still not returned to its 1929 level, while output/hour was 17% higher, yielding a decrease in hours worked of about 19%. The harsh tale that these figures tell is thus powerful corroboration for estimates of high levels of unemployment.

Not all of the decrease in hours was due to unemployment, of course. Many of those who kept their jobs found their hours reduced. While the average work week had hovered between 47 and 49 hours through the 1920s, it fell steadily from 48.6 in 1929 to 41.7 in 1934 and stayed between 42 and 45 until 1942. To some extent this can be attributed to a secular trend, for this figure had been 69 in 1850, 55 in 1900 (Hunnicutt, 1988), and would fall below 40 in the postwar era.[6] In Europe as well the average work week fell from 57 to 47 hours in the interwar period (Svennilson, 57). Yet surveying the inter-war experience indicates that there were more than secular forces at work. Employers responded to the downturn by cutting back on hours. The decrease in hours in the early 1930s was more dramatic in the United States than elsewhere. Nor was this decision welcomed by workers. Garraty has argued that unions, dominated by those with seniority, generally favored layoffs over reductions in hours (1986:

97). This was not always the case. The American Federation of Labor in 1932 did urge legislation of the 30-hour week. Unions in the late 1930s often favored work-sharing up to a point (24 to 32 hours per week) after which they urged reverse-seniority based layoffs (Jacoby, 246-7). This is a more generous attitude toward work-sharing than has characterized postwar union activity. Nevertheless, it is clear that workers only accepted hours reductions grudgingly at best. Moreover, the WPA estimated that as late as 1939 well over 10% of those counted as employed were working less than 30 hours per week and could thus be considered unemployed (Webb, 1939).

The hardship imposed on those who had their hours reduced was much less than for those who lost their entire livelihood. Nevertheless, given that these hours reductions were considered undesirable by workers, we should include them in our total of the employment impact of the Depression. To this we should add the underemployment which occurred in the agricultural sector. For decades, workers had been moving out of agriculture into industry and services. In the 1930s this movement was reversed. Faced with abject poverty in the cities, people moved back to the countryside to pursue a more self-sufficient way of life. We do not generally see explicit unemployment in the agricultural setting. When the number within the sector rises, while output remains stable, especially in a period in which agricultural technology continues to improve, we can be sure that there is a considerable wastage of labor.

Nor should we forget for a moment that these are not just numbers we are speaking about but people whose lives were crushed for an entire decade. The Hoover Committee on Social Trends had presciently noted in 1929 that "any considerable and sustained interruption in their money income exposes them to hardships which they were in a better position to mitigate when they were members of an agricultural or rural community" (in Bernstein, 1960). For the most part, they were thoroughly unprepared for the calamity which befell them. A 1930 Philadelphia survey found that 60% of the unemployed had no savings, and the rest only had enough for six weeks on average (already in 1928 one writer had observed that the incidence of part-time work in previous years had severely limited worker's financial reserves — Stricker, 1983: 19). As well, private charities were generally exhausted by late 1930, and municipal relief by late 1931 (Jacoby, 207). Social workers estimated in 1933 that less than one third of the unemployed in major cities received relief, and the picture was much worse in smaller centers (in Shannon, 1960: 38-40). There was very little actual starvation in the United States,[7] but much undernourishment. While death rates do not change, there is much evidence, including height data on people maturing in that decade, that standards of health were seriously impaired. Hospitals reported increased rates of rickets among children, and

TABLE 1.2 Indexes of Output and Productivity, 1920-1945

1929 = 100 unless otherwise specified

Year	Real GNP (billions of 1958 dollars)	Output		Output per worker-hour	
		Private Non-farm	Manufacturing	Private Non-farm	Manufacturing
1920	140.0	71.9	66.0	79.5	61.5
1921	127.8	70.6	53.5	86.1	71.0
1922	148.0	74.7	68.0	83.9	80.4
1923	165.9	85.7	77.0	87.8	77.3
1924	165.5	89.2	73.5	93.3	82.3
1925	179.4	90.7	82.0	92.6	87.7
1926	190.0	97.6	86.2	95.0	89.4
1927	189.8	98.2	87.0	95.4	91.5
1928	190.9	99.7	90.0	96.1	96.6
1929	203.6	100.0	100.0	100.0	100.0
1930	183.5	87.8	85.5	97.0	100.6
1931	169.3	78.0	72.0	98.5	103.8
1932	144.2	63.6	53.8	95.4	97.1
1933	141.5	62.0	63.7	93.2	105.6
1934	154.3	70.3	69.0	103.3	110.4
1935	169.5	77.7	83.8	106.7	117.7
1936	193.0	92.1	96.7	111.3	118.5
1937	203.2	96.2	103.2	110.4	116.7
1938	192.9	88.9	81.0	113.5	115.2
1939	209.4	98.6	102.5	117.6	125.8
1940	227.2	109.0	118.8	122.2	132.1
1941	263.7	127.2	158.0	124.2	136.9
1942	297.8	138.7	197.2	123.3	139.2
1943	337.1	149.4	238.3	124.6	141.2
1944	361.3	158.3	232.5	134.4	139.4
1945	355.2	156.4	196.5	142.0	137.5

Sources: Bureau of Economic Analysis, *Long Term Economic Growth, 1860-1970*, 185, 211; and Bureau of the Census, *Historical Statistics: Colonial Times to 1970*, 224.

Chicago schoolteachers spent from their own pockets to feed their students (Shannon, 1960: 51-2). Children from households which experienced substantial unemployment had lower educational attainment and career mobility through their lives. Localities with shrinking incomes closed schools, fired teachers, and shortened the school year. Many people were unable to maintain decent standards of clothing. Millions lost their homes (though landlords, with nobody else to rent to, generally evicted only a small proportion of those in arrears). The Southern Pacific Railroad alone ejected 683,000 hobos in 1932; it was estimated there were at least one million nationwide in 1933 (Bernstein, 1960: 325). Sociological studies of the time detail how the unemployed started out optimistic of finding a job but moved to despair after months of fruitless searching. Social workers described the hopelessness of the unemployed, their wounded pride, and their disdain of government handouts. Many studies in the 1930s and since have found that unemployment is an important contributor to mental illness. The loss of income and pride contributed to falling marriage rates. The sense of despair may explain the lack of political demonstration by the unemployed. It is almost a criminal act toward those who suffered so much to deny that the Depression was quite so severe.[8]

The 1932 report of the California Unemployment Commission captures some of the true horror of the situation in that relatively prosperous state:

> The study of the human cost of unemployment reveals that a new class of poor and dependents is rapidly rising among the ranks of young sturdy ambitious laborers, artisans, mechanics, and professionals, who until recently maintained a relatively high standard of living and were the stable self-respecting citizens and taxpayers of the state. Unemployment and loss of income have ravaged numerous homes. It has broken the spirit of their members, undermined their health, robbed them of self-respect, destroyed their efficiency and employability. Many households have been dissolved; little children parcelled out to friends, relatives, or charitable homes; husbands and wives, parents and children separated, temporarily or permanently. ... Men young and old have taken to the road. Day after day the country over they stand in the breadlines for food. ... New itinerant types develop: 'women vagrants' and 'juvenile transients'. ... The law must step in and brand as criminals those who have neither desire nor inclination to violate accepted standards of society. ... Physical privation undermines body and heart. ... Idleness destroys not only purchasing power, lowering the standards of living, but also destroys efficiency and finally breaks the spirit.

Duration of Unemployment

Beyond the aggregate figures on unemployment, we would wish to know something of the breakdown between short and long-term unemployed. Were the unemployed from one year to the next the same people, or did people tend to rotate in and out of employment through the 1930s? Table 1.3 summarizes the frequency of hires, layoffs, and quits through the interwar period. Overall labor turnover was higher in the interwar period than it has been postwar.[9] Not surprisingly, quits occurred at a much lower — almost nonexistent — level in the 1930s than in either the

1920s or the 1940s. Any attempt to attribute the Depression to an increased desire for leisure among workers would fail on this point. Layoffs occurred at a much higher rate in the 1930s. Hires did not disappear during the 1930s, though they did occur at much lower rates than in the early 1920s or mid-1940s. It is worth noting that the hiring rate was low through the late 1920s as well, indicating that the economy was having difficulty creating jobs from 1924. Throughout 1923-1929 separation rates were higher than prewar and accession rates lower. Sumner Slichter, writing in 1929, observed that the main determinant of low quit rates in the late 1920s was the lack of vacancies and widespread unemployment (Jacoby, 1985: 171). Studies in the late 1920s found that the majority of unemployed had to wait three to four months before finding a new job (Fearon, 1987: 64).

While the hiring experience was poor during the interwar period, with total hires being much less than total separations in most years in the 1930s, there was still significant hiring activity occurring. Worker turnover greatly exceeded the change in the stock of unemployed throughout the Depression (Baily, 30). This makes it likely that a substantial portion of the unemployed moved in and out of employment through the decade. In the early years of the Depression, it appears that workers often returned to the jobs they had recently been laid off from (perhaps for seasonal reasons), though this became less common later. It is important to recognize that much of the apparent turnover is illusory, reflecting the impermanent nature of labor relations at the time relative to today (Thomas, 1988: 139).

This group of workers moving in and out of employment is not the whole story, though. Indeed, Eichengreen and Hatton (1987: 42) characterize the 1930s as possessing a bifurcated labor market with one group experiencing rapid turnover and the other experiencing lengthy spells of unemployment. Determination of the quantitative importance of long versus short-term unemployment can only come from surveys of the unemployed. A survey in Philadelphia in 1936 found that 20% of those on government relief had been unemployed for more than five years (Gill, 1940: 39). A Massachusetts survey in 1934 found that 61.9% of the unemployed had been without a job for more than one year, 44.5% more than two, 24.6% more than three, and 11% more than four years (Chandler, 1970: 42). The 1940 U.S. census found that 15.1% of male unemployed had been so one to two years, 7.5% two to three years, and 11% more than three years. Similar figures are found in studies of other countries (Eichengreen and Hatton, 1987: 40). From such fragmentary evidence it appears that long-term unemployment affected millions and was more important relatively in the 1930s than it has been in the postwar era (Margo, 1987).[10] Unemployment of this sort became self-perpetuating. Workers who had been without jobs for years appeared unattractive to prospective employers. They also suffered physically and psychologically (Thomas, 133).

TABLE 1.3 Labor Turnover in Manufacturing, 1920-1945[a]

Average monthly rate per 100 employees.

Year	Accessions	Total[b]	Layoffs	Quits
		Separations		
1920	10.1	10.3	0.8	8.4
1921	2.8	4.4	1.8	2.2
1922	8.0	5.3	0.4	4.2
1923	9.0	7.5	0.3	6.2
1924	3.3	3.8	0.6	2.7
1925	5.2	4.0	0.4	3.1
1926	4.5	3.9	0.5	2.9
1927	3.3	3.3	0.7	2.1
1928	3.7	3.1	0.5	2.2
1929	5.1	3.9	0.4	3.0
1930	3.8	5.9	3.6	1.9
1931	3.7	4.8	3.5	1.1
1932	4.1	5.2	4.2	0.9
1933	6.5	4.5	3.2	1.1
1934	5.7	4.9	3.7	1.1
1935	5.1	4.3	3.0	1.1
1936	5.3	4.0	2.4	1.3
1937	4.3	5.2	3.5	1.5
1938	4.7	4.8	3.9	0.8
1939	5.0	3.7	2.6	1.0
1940	5.4	4.0	2.6	1.1
1941	6.5	4.7	1.6	2.4
1942	9.3	7.8	1.3	4.6
1943	9.1	8.6	0.7	6.3
1944	7.4	8.1	0.7	6.2
1945	7.7	9.6	2.6	6.1

a. Beginning in 1930, averages are arithmetic means; before that, they are unweighted medians. Pre-1943 data pertain to production workers only.
b. Data before 1940 include miscellaneous separations.
Source: Bureau of the Census, *Historical Statistics: Colonial Times to 1970*, 181-82.

The probability of finding employment fell very fast with increased duration of unemployment (Eichengreen and Hatton, 1987: 40). It is possible that much of the unemployment experience, especially in the later 1930s, was due to the unwillingness of employers to hire the long-term unemployed (Margo, 1987).

Unemployment data by age is not available. There are some survey results which strongly suggest that unemployment, and especially long-term unemployment, was much more prevalent among older workers (see Gill, 1940, for example). In Philadelphia in 1931, unemployment was 25%, but for those over 66 it was 34.4%; figures such as these would encourage the development of social security (Chandler, 1970: 36). The 1940 census confirms that older workers were much more likely to be unemployed, at least at that late date (Margo, 1987). One possible explanation would be that older workers had a greater probability of losing their jobs. Jacoby (1985: 219) notes that some firms did prefer to lay off older workers due to a feeling that they were inefficient. However, if we allow ourselves to rely on British statistics momentarily, it appears that older workers experienced the same likelihood of job loss as other workers (Thomas, 1988: 118-20). The reason for higher unemployment rates among older workers, then, must be a longer duration of unemployment for those who did lose their jobs.[11] Bernstein (1960: 57) felt that firms, especially those with health or pension plans, had become less willing to hire older workers because of mechanization, which caused them to emphasize energy over experience. In any case, one of the main tragedies of the period is the long-term unemployment faced by those who had previously had steady work for decades and naturally expected those days to continue.

Studies for other countries in various years often find a U-shaped unemployment pattern by age. This is the case for Australia in 1933, Belgium in 1937, Canada in 1931, the United Kingdom in 1931, and France in 1936 (Eichengreen and Hatton, 33). The very young (21-25) share with the very old (over 55) the experience of extremely-high unemployment. For Britain, at least, Hatton has found that reverse-seniority layoff rules were much less important than today. Thus, much of the poor experience of the young is to be explained by the difficulty they faced in finding their first job. In a world of much surplus labor, those without any substantial work experience could be easily passed over by potential employers. As they aged without being able to land a permanent job, they would look increasingly unattractive to employers. Government surveys in Massachusetts in 1934 found average unemployment of 24.9% but 47.6% for those 14-20; in Michigan in 1935 18.8% versus 34.3% for those 15-19; and nationally in 1937 16.4% versus 36.5% for those 20-24. The Civilian Conservation Corps was designed as an outlet for the young (male) unemployed. Still, the War would appear as salvation to many of this generation.

Notes

1. We could include the implicit question guiding much of the curiosity about the Depression, "Could it happen again?" During the heyday of Keynesian eco-

nomics in the 1960s much of the profession felt that the problems of severe economic fluctuations had largely been solved. Stimulative fiscal and/or monetary policy could quickly handle any downturn. After a lengthy period of stagflation, this optimism has faded. Many leading macroeconomists now recommend policies of fiscal restraint similar to those followed by most countries in the early 1930s. If we do not understand the sources of the last Depression, we can not be sure that there will not be another such cataclysm, nor that we would know what to do if it happened.

2. I should note that the forces outlined in this book created a situation which was very difficult for a central bank to cope with. Demand curves shifted to the left at the outset in many sectors due to market saturation; as employment fell, demand dropped across the economy. At the same time, new process technology was shifting supply curves to the right. Widespread deflation was a natural result. Bankruptcies which weakened banks and a general decline in the demand for money were also natural outcomes of the forces in this book. To be sure, cross-country comparisons do indicate that monetary expansion is associated with better economic performance in the 1930s. Nowhere, though, does it yield a miracle.

3. A WPA study in 1935 found that transients were little different from the average unemployed, just a bit younger, male, and usually single. "Migration offered an escape from inactivity." They moved in response to rumor and curiosity, but generally found no better job opportunities than at home (Webb, 1935).

4. This is easy with the advantage of hindsight. Mitchell (1929: 841) characterized the 1920s as a period in which a number of good and bad things were going on, and impressionistic writers could choose from either group. He wondered what the long-term effects of World War One might be. In 1938, Hansen recognized that the relative prosperity of the 1920s could be attributed to special temporary conditions, and thus had grave questions about future prosperity.

5. This secular trend likely increased demand for time-intensive goods like automobiles and movies, may have decreased demand for time-saving devices, and facilitated suburbanization through allowing increased commuting time.

6. Shannon (1960: 36) reprints newspaper articles which deal with unemployed workers and families facing starvation. Various other reprints display the real suffering of that era.

7. "One sometimes wryly feels that economic history is an effort to explain events shown by the next round of research not to have happened" (Jones, 1988: 138). It is perhaps natural throughout the social sciences to downplay the importance of events which existing theory has difficulty explaining. I have encountered the same phenomenon in my research on the Industrial Revolution (1991: 40-1).

8. According to census data, at least. Leonard (1988) has, by enterprise-level research, shown that rates of labor turnover in the 1980s were likely much higher than aggregate statistics would indicate. Some economic historians have questioned the accepted wisdom that worker attachment to firms has increased markedly over the last century.

9. Jensen (1989) has argued that emerging personnel policies meant that casual hiring was being replaced by hiring on merit. Such sorting meant that the lack of jobs was not as evenly distributed as before (he likely exaggerates the degree of change — see Jacoby, 1985). He feels that the problem of the late 1930s was not the absolute level of hardship but its distribution.

10. A survey of the long-term unemployed in the 1980s found that each year of age added .314 weeks to the expected duration of joblessness (Podgursky, 1988: 25). For Britain, Thomas found that older workers were both less likely to lose and to find jobs in the 1980s than the 1930s.

2

Questions Unanswered

Existing theories have — by the consensus of the profession — failed to provide a convincing explanation of the entire Depression experience. In 1933, Morris Taylor wrote; "An impartial student will observe that there is some truth in nearly every suggested cause of depression, and in most of the proposed remedies." Also in 1933, Irving Fisher recognized that existing theories all contained a grain of truth, but taken together left much unexplained. We take the same attitude here. We do not wish to demolish every theory, but merely to delineate their shortcomings and insights. From these we will be better able to develop a new theoretical structure which can fill in the gaps in our knowledge. Readers may wish to skim discussions of theories with which they are already familiar.

The Monetary Approach

It is perhaps symptomatic of the difficulty in dealing with the Depression that the vast bulk of research in macroeconomics focusses on the postwar era. Bernanke (1983) has suggested that just as seismologists can learn more from one large earthquake than from several small tremors, macroeconomists can learn more from the Great Depression than from the relatively tranquil postwar years. A similar view can be found in the philosophy of science: "It might be conjectured that in periods of transition or crisis generative structures, formerly opaque, become more visible" (Bhaskar, 1989: 84). Nevertheless, most macroeconomists have appeared to sense the intractability of the Depression question and concentrated their efforts elsewhere. Even if we question Bernanke's suggestion that study is best devoted to the 1930s, we can not reject Eichengreen's conclusion that the ultimate test of macroeconomics is "to explain satisfactorily the fluctuations in output and prices, and the dramatic rise in unemployment associated with the Great Depression" (1987: 44). Any theory which is incapable of dealing with the 1930s should be viewed sceptically. Certainly, the wider public would expect macroeconomic theory to deal with the major economic cataclysm of our time.

The monetary approach asserts that the severity of the Depression was due to a deficiency in the money supply resulting from mistakes made by the newly formed Federal Reserve System in the United States. The classic work is Friedman and Schwartz (1963). They recognize in that work the existence of a normal recession — not due to Fed bungling — in 1929-1930. From the first banking panic in 1930, banks began to increase their reserves, and depositors began to withdraw their deposits. By the quantity identity MV=PQ, we know that the total money supply multiplied by the velocity of circulation must equal the output of the economy multiplied by the price level. With a relatively constant velocity, therefore, a decrease in M must mean a decrease in P and/or Q. The Federal Reserve should have countered this tendency for M to fall, but instead exacerbated it with a tight money policy, driven by a desire to maintain exchange rate stability. Output thus fell so far and for so long because the money supply had done so. In other writings, Schwartz has extended the argument backward in time. The Federal Reserve, she argues, had actually begun to pursue restrictive policies in 1928 to reduce the degree of speculation in financial markets, and these policies largely determined the severity of the 1929 downturn.

With this extension, then, the monetary approach provides a potential explanation of both the initial severity and the extreme length of the Great Depression. Given the dominant place of the United States in the world economy, and the widespread belief that the world Depression began in the United States, reliance on an American institution could be reasonable. Much empirical work has been performed to establish the importance of the monetary explanation. However, sophisticated econometric analysis has been unable to establish convincingly the relative importance of monetary forces. Temin (1981) has questioned the validity of much of the empirical work done on the Depression. Since all aggregate variables rise together till 1929, fall together until 1933, and rise together until 1937, the result one gets depends crucially on the structural form of the model estimated. Strong results can thus be found for either monetarist of Keynesian interpretations.[1]

Many criticisms have been made of the monetary approach over the years. Some have focussed on the movements in monetary variables. Temin has argued that if monetary tightening were the true villain we should have seen an increase in nominal interest rates, but interest rates instead fell steadily through the early 1930s. Transmission of changes in the money supply to real variables through changes in the interest rate is the Keynesian mechanism, however. The monetary approach, based on the quantity identity above, posits a direct effect of monetary changes on real variables.[2]

Still, Schwartz has since argued that she feels high real interest rates were an important part of the monetary approach. In defence of this position, Cechetti (1992) has tackled the tricky issue of whether deflation was anticipated. If people did not predict substantial deflation, they would not have thought real interest rates were high. Cechetti and Hamilton (1992) — who also favors the monetary view — have recently used much the same data to argue in turn that inflation was largely anticipated and largely unanticipated. Hamilton (1987) argued that statements of contemporaries also supported the view that deflation was unanticipated. We will see below that overall expectations of economic performance remained high well into the 1930s, and might thus conclude that the burden of proof rests with those who feel deflation was anticipated. Calomiris (1993) concludes that there is widespread agreement in the profession that inflation was unanticipated. In any case, after 1933 real interest rates were very low, but recovery was sluggish.

A more telling criticism is the observation that prices in fact dropped more rapidly than the money supply in the early 1930s. Thus, within a strict interpretation of the quantity theory, it is not clear why output had to fall. It is a decrease in the real money supply, not the nominal money supply, that is needed to force output down. There are two extenuating circumstances. First, the velocity of circulation fell 9.9% during the first three quarters of the downturn (Gordon and Willcox, 1981: 55). While this downward movement might indicate that there were problems with the demand for money, not just the supply, it does restore downward pressure on output within the quantity equation. As well, since population was growing, however sluggishly, through this period, the real money stock per person did fall. Finally, we should note that while halving both money supply and prices should have no effect in a static model, it may well have an important effect in a dynamic model. This, however, leads to another inconsistency within the monetary approach. Bernanke (1983) has noted that monetarists generally assume neutrality of money in the long run, which makes it difficult to attribute a decade-long Depression to monetary forces.[3]

Temin has also questioned whether, on both theoretical and empirical grounds, changes in the money supply should be viewed as exogenous. Friedman and Schwartz assume away the possibility that changes in the money supply might themselves be due to other factors which comprise the Depression. If the fall in bank deposits actually reflects a drop in money demand, the Fed might not have been able to stimulate the increase in deposits which Friedman and Schwartz envision (Temin, 1976: 29).[4] It could be maintained that the switch from easy money to tight money in 1930 is hard to blame on the Fed unless we understand why they changed their policy. However, Wheelock (1991) has noted that the Fed had always targeted discount loan volume. Thus, by its standards, the early 1930s were a period of easy money. As banks decreased their discount borrowing, the Fed saw no need for expansion. If one believes that the appropriate policy for the Fed throughout was to ensure the steady growth of some monetary aggregate, then it is possible to see Fed policy as inappropriate from 1924, but only damaging when circumstances changed later. It could also be argued that the Fed could scarcely have done a worse job than in the pre-1913 era when it did not even exist. However, it is commonly asserted that private clearing house banks had previously prevented runs on banks much better than the Fed. Once the Fed was formed, these no longer viewed the support of other banks in times of crises as their responsibility. Alternatively, Miron (1986) has argued that the Fed, by reducing the seasonality of interest rates, had actually reduced the frequency of panics.

Other critiques of the monetary approach focus on its inability to adequately explain certain sub-periods within the Depression. Much attention has naturally focused on the initial downswing from 1929-1931. Friedman and Schwartz admitted that "Even if the contraction had come to an end in late 1930 or early 1931, as it might have done in the absence of the monetary collapse that was to ensue, it would have ranked as one of the more severe contractions on record" (1963: 303). Attempts have been made to buttress the monetarist position by ascribing the all-important initial downturn to monetary tightening in 1928. Field (1984b) has argued that stock market speculation absorbed cash balances by increasing the transactions demand for money; the Fed should have accommodated this with monetary expansion. Hamilton (1987) maintained that "in terms of the magnitudes consciously controlled by the Fed," open market operations and the discount rate, it had in fact been quite deflationary. This episode of tight money was relatively mild, however, and thus attributing a severe downturn to it would appear to be stretching the point (Aldcroft, 1977: 199). Indeed, the Federal Reserve Board endeavored not to restrict credit for legitimate business purposes in the late 1920s (Gordon, 1961: 427).

Temin has also recently redoubled his criticism of the Friedman and Schwartz interpretation of the banking crisis of 1930. He notes that if one removes the failure of the Bank of the United States, which was a special case, it was not a big event. Moreover, there was no fall in the stock of money in late 1930 (1990: 50-1). Wicker (1980) and White (1984) have looked at the experience of the banks which failed, and both noted that these bank failures were due to poor loans from the 1920s; the panic of 1930 is thus due to the same influences as bank failures in the 1920s. White (1984) suggests that weak banks were especially susceptible to monetary tightening. The Fed had felt through the 1920s that it was healthy to allow weak institutions to close; their behavior in 1930 was no different. Thus, while Temin's early assertion that the panic was a result of the economic downturn appears incorrect, it is also untrue that the panic was a major break with the past or reflected a change in Fed policy as Friedman and Schwartz suggested. Further, Wicker (1982) looks at bank debit data and finds that bank failures in 1930 had only local effects; transmission to the national economy was insignificant.

The monetary approach also appears to poorly fit the experience of the later 1930s. The correlation between money and nominal income weakens through the 1930s; the monetary approach is unable to explain the timing of the 1938-1941 recovery (Gordon, 1981). Overall, since M1 rose 49% and M2 18% over the entire period 1929-1940, it seems difficult to attribute the entire Depression to monetary forces (Gordon, 1981: 80). Romer (1992b) has suggested that the increase in money supply was an important stimulus to recovery; she recognizes that her interpretation is in opposition to those who argue that recovery was due to the natural self-correcting mechanism of the economy (not to mention those who view recovery as sluggish).

Recent developments in the literature have stressed further ways in which monetary forces could affect real variables. Temin himself has noted that while deflation may have had no real effects early in the Depression, if people came to suspect that prices might continue to fall they would have a natural tendency to postpone purchases (1990: 64). Bernanke, (1983) whose own empirical work suggests that the Friedman and Schwartz paradigm can only account for half of the decline in output 1930-1933, argues that there was a further effect through credit rationing. The waves of bank failures and mergers which began in 1930 had the eventual effect that there were only half as many banks in 1933 as there had been in 1929, and these had experienced heavy losses. Falling prices meant that both businesses and homeowners defaulted on their loans at record levels; in some cities over half of homeowners defaulted, and countrywide 45% of farmers were delinquent by 1933. This, he claims, caused banks to pressure for repayment of old loans, and to become increasingly unwilling to make new loans. Since large corporations had come to rely on retained earnings to finance investment (and banks remained willing to loan to these at very low nominal rates in any case), this credit restriction had no effect on them. Bernanke feels that large firms could substitute for small-firm investment, so that the full effect of credit rationing would show up in consumer demand (He notes that there were some increases in trade credit and consumer finance which partially offset the commercial bank decline). As the credit contraction lasted till 1935, this approach can potentially help to explain the sluggish recovery from 1933, as well as the decline from late 1930.

Bernanke is probably too generous in crediting large firms with the ability to substitute for small firm investment. Tugwell, a key Roosevelt advisor during the 1930s, felt that company managers were ill-suited to the transfer of investment from declining to rapidly expanding industries (Bernstein, 1987). In his study of English distillers in the interwar period, Weir (1989) found that both scientific and marketing considerations prevented the vast majority of firms from diversifying. To assume that large firms could replace the effort of small is to ignore the specialized nature of firms. There was, in fact, by modern standards, little diversification by large firms in the interwar period (Bernstein, 1987: 106). Still, Temin, as a test of Bernanke's theory, performed a cross-section analysis and found that industries with small firms had actually fared better rather than worse than industries with large firms in terms of industrial decline (1990: 53-4), though this may reflect a lesser ability to prevent prices from falling. Of course, firm size is not the only determinant of the cost of obtaining capital, and observation of such costs during a time of crisis is difficult. If large firms squeezed small firms out of capital markets, this could have slowed the growth of new industries. In this light, it is worth noting that the wave of bankruptcies in the early 1930s may have reduced the number of entrepreneurs capable of responding to new opportunities (if such existed).

Hayes (1951: 123) has made a similar argument to Bernanke's for the later 1930s. While gold inflows, Fed policy etc. should have made credit available, and for top-credit borrowers the situation was seldom better, the price of risk capital remained surprisingly high in 1936 and 1937. He attributes this to fluctuations in output, capital-gains taxes, and the increasing role of banks. He does not explain why banks do not wish to have a range of different types of investment in their portfolios.

When the same phenomenon is observed, and attributed to different supply-side forces, at different points in the 1930s, demand-side arguments seem increasingly plausible. In the face of critics who accused them of being overcautious, bankers in the later 1930s claimed that they kept high reserves due to a lack of safe investment opportunities (Bernstein, 1987: 111). By his own admission, the best evidence Bernanke has for his proposition is that it is plausible and that the empirical evidence is not inconsistent with it.[5] It is certainly true that there was a substantial drop in the volume of consumer credit (41 index points 1929-1933). However, this could be due to a decrease in the demand for credit, rather than the supply. Temin (1976: 136) finds no evidence of rationing and concludes that the decrease in short-term borrowing was due to a decrease in money demand. Barber (1978) has also argued that the decrease was on the demand side; both consumers and investors were less willing to borrow. He notes that the quantity of loans fell dramatically in Canada as well, although there were no bank failures there, and the banks could obtain reserves on demand under the provisions of the Finance Act. He also notes that while loans fell by $6.6 billion in the 12 months after June 30, 1929, in the United States, bank holdings of government and private securities increased $2 billion. Banks also borrowed less from the Fed. This demand-side argument is certainly plausible; we will argue that there were few new investment opportunities or consumer goods in the 1930s.

The failings of the international gold standard exacerbated the monetary situation. As Eichengreen (1992) argues, the international situation can help to remove some puzzles from the Depression picture. The sharp decline in American foreign investment from 1928 helps explain why other countries began their downturns before the United States (Eichengreen blames Federal Reserve tightening in 1928; we might instead point to the speculative profits being earned in the stock market). This in turn helps us understand why American export experience was so poor in the early 1930s. Moreover the failure of central banks to cooperate in the difficult political climate of the interwar period as they had before World War One left each central bank anxious to avoid or contain balance of payments deficits. This induced them to avoid expansionary monetary policy for the first part of the 1930s; only after successive countries abandoned fixed exchange rates would they do so (see also Temin, 1989: 7). Given the limited trade and financial links between the United States and the rest of the world in the interwar period (see Chapter 13), international factors had their predominant impact through domestic monetary authorities (Eichengreen, 1988, 1992): these, as we have seen, were far from the whole story.

In sum, there are good reasons for doubting that monetary phenomena can explain most if not all of the Great Depression. However, we can not dismiss the dominant role of the monetary approach unless we come up with a plausible alternative (Brunner, 1981). We will find in the next section that the Keynesian perspective faces difficulties as great as those of the monetary hypothesis. We should not, in any case, wish to totally dispose of the monetary argument. The Fed did behave suboptimally after 1930, and the credit-rationing arguments may well contain some elements of truth. At least for the period 1931-1933, monetary forces likely contributed in a significant way to the decline in output.[6] For other periods especially, we would do well to look for other explanations. Gordon and Willcox (1981: 94) argue that while there are hard-liners on both sides, most analysts are soft-line monetarists or non-monetarists who differ only in degree. St-Etienne believes that the literature recognizes that the Depression was not due to any one cause, either monetary or non-monetary (1984: 57). Lindert has written, "While debating other things, most of us would agree that the money stock and output are both endogenous variables, each partly affected by the other in a structural relationship including other variables as well". Both Schwartz and Temin concur with this appraisal (in Brunner, 1981). Calomiris (1993: 80) concludes that "one must combine financial influences with other factors to explain protracted underutilization of resources" (he also recognizes that non-aggregative analysis is the key to enhancing our understanding). In the last decade, research has tended not to be of the type which pushes either explanation as "the" monocausal explanation; rather researchers have seemed to recognize that we are far from the truth and have focussed on tackling smaller pieces of the Depression puzzle.

The Keynesian Approach

Though Keynesian theory dominated the macroeconomic literature in the 1960s and 1970s, it has only rarely been applied to the Great Depression. The Keynesian explanation begins with an autonomous fall in consumption and/or investment demand, which leads to a fall in output and prices. This pushes down money demand, and thus the interest rate, and, assuming some positive elasticity of supply, the equilibrium money supply. Observations that money and output move contemporaneously in some periods could be attributed therefore to output changes causing money changes. The banking panics of the 1930s were a result of declining money demand, and thus serve to prevent the build-up of excess balances (Temin, 1981). In an extreme Keynesian approach, monetary forces have no separate causal role to play. A more common position is that monetary forces may have been important at times, but can not explain the whole course of the Depression.

A falling average propensity to consume must mean a rising savings rate. This indeed had risen through the 1920s (Chandler, 1970: 16). Between 1928 and 1929, the growth rate in consumption tailed off, though output continued to rise. (We can not easily speak of desired savings in the 1930s, for savings by some are swamped by the dissaving of those experiencing catastrophic income declines.) Investment continued to fall to virtually nothing in 1933, and rose sluggishly thereafter. Theory tells us that if desired savings exceed desired investment, output will fall until these are equalized.

The first difficulty with the Keynesian approach is in understanding why consumption and/or investment expenditures would fall so dramatically in the late 1920s. "It is easier to show that non-monetary factors must have been at work in the first two years of the contraction than to determine what those factors were, much less to assign specific quantitative contributions to each of them" (Gordon and Willcox, 1981: 74). Unless we can provide a rationale for the original decline in consumption and/or investment, the Keynesian explanation must remain highly speculative.

Investment could be posited to decline due to a change in expectations. Even in the face of declining interest rates, entrepreneurs could have remained unwilling to invest if they were pessimistic about the future.[7] One problem with this line of argument is that it begs the question of what determines expectations. More centrally, expectations appear to have remained high until two or three years into the Depression.[8] The stability of prices and profits in the 1920s, following decades of steady growth, had convinced most people that they had entered a New Era where depressions were impossible (Ginzberg, 1939). Julius Klein in 1928 was confident that severe gyrations in the business cycle had been eliminated for all time (in Beard, 1928). R.W. Woodruff, President of Coca-Cola, in August, 1930, said "The general slump in business, in my opinion, has been greatly exaggerated." Julius Klein, then Assistant Secretary of Commerce, announced in June 1931 that, "The Depression has ended. The valley usually runs across six or seven months. If history repeats itself, this means that in July we go up" (in Angley, 1931: 52). Not knowing what we know now, and never having heard of a slump as severe as what they were to live through, the business community was sure through 1930 and 1931, and even 1932, that recovery must be just around the corner.

Keynes argued that it was also possible for investment to decline because previous investment had taken advantage of the most profitable opportunities.[9] Along the same lines, we could be in a situation where previous investment had resulted in excess capacity, and thus an eventual decrease in investment demand. Investment dynamics such as these may play a role in generating business cycles, but it is difficult to imagine (within Keynesian theory) how gross investment could be driven down to almost nothing for a period of years in the absence of changes in consumption demand. In an aggregate sense, a growing economy should continually provide outlets for more investment.

The output/capital ratio rose steadily from 1899 to 1953, with its fastest rate of increase from 1919 to 1929 (Bernstein 1987: 35). The accelerator mechanism — the amount of additional investment needed to achieve a given increase in output — was thus much weaker than it had been in previous decades. Since it is the level that is important rather than the rate of change, we should recognize that the accelerator was even less powerful at the start of the postwar boom. Also, we should note that unemployment becomes more of a puzzle in a low accelerator world, for the degree of capital creation required is reduced.

Investment also responds to changes in interest rates. Field (1984) has noted the important role of the volume of security transactions in increasing the demand for money, and causing high nominal and real rates of interest before the stock market crash. This could have depressed both investment and consumption expenditures. St. Etienne (1984: 58) also argues that the fall in investment could have been limited if the interest rate had been allowed to fall. While perhaps part of the reason for the initial downturn in mid-1929, this line of argument can scarcely deal with the period after late 1929 when interest rates fell steadily. Falling interest rates are, after all, the mechanism by which equilibrium in capital markets is supposed to be regained.[10]

At the aggregate level, then, it is exceedingly difficult to justify an autonomous decrease in investment of the extent and severity we see in the early Depression. Some attempts have been made to look at industry-level changes. The decline in (characteristically volatile) construction activity has been identified as a potentially significant contributor to overall decline. It began in 1926, but, once we accept that there is no single cause of the Depression, this does not prevent the construction industry from playing a key role in fomenting the Depression (Gordon and Willcox, 1981: 74).[11] No one sector of the economy could on its own generate a majority of the Depression experience, of course. Gordon (1961) has attributed much of the investment in the 1920s to (along with automobile and construction booms) various technological innovations; electrical products including radio, chemicals (especially rayon), new oil and rubber products, movies, and airplanes. Keynes himself, in his *Treatise on Money*, had accepted that the course of innovation would have a powerful influence on investment. He drew back from that position in the *General Theory*, and the Keynesian tradition has followed (Freeman, 1984). What is needed, then, is a number of studies of industry-level investment trajectories to determine to what extent the observed fall in investment was exogenous.

Much of the fall in investment might be attributable to an autonomous decrease in consumption expenditure.[12] In the crucial first year of the Depression the 4.6% decrease in consumer expenditure was the greatest part of the decline in GDP. However, this could be a result rather than a cause of GDP decline. The need to separate cause from effect in a world where all variables move in the same direction makes it impossible to establish conclusively whether the key is to be found in consumption or investment or both. Mayer (1978) has questioned Temin's conclusion that the fall in consumer expenditures was unusual in 1930, but the position has since been revived by Hall (Temin, 1989: 43). Romer (1993) notes that the decline in consumption was more significant in the United States than elsewhere. Moreover, the rate of growth in consumption had fallen sharply from 1928 to 1929. Gordon (1949) felt that it was consumption growth that supported economic growth until 1928, and that investment only played an equal role after 1928. A convincing explanation of an autonomous fall in consumer expenditure could thus provide a plausible description of the beginning of the Depression.

However, the fall in consumption is also difficult to explain. Expectational arguments were discussed above. Another possibility is that the stock market crash had a severely negative wealth effect. Consumption theory tells us that consumption decisions are dependent on wealth holdings as well as income. On its own, the market crash appears incapable of causing anywhere near the observed decrease in consumption. Mishkin (1978) has pointed out that in addition to the fall in real asset value associated with the crash, there was an increase in real indebtedness due to the fall in the price level. Individuals had not only been borrowing to finance stock speculation, but for mortgages and consumer durable purchases, in the late 1920s. Faced with a financial squeeze, these individuals shunned illiquid assets such as housing and consumer durables in the early 1930s (unlike today, one received no credit whatsoever if goods were repossessed). Mishkin felt that 45% of the decline in consumption could be attributed to this wealth effect, though Gordon and Willcox felt this to be overstated (1982: 79). Mishkin uses postwar data to estimate the effect of private debt on consumption; this causes him to overpredict spending in the early 1920s when consumer durables were new, and thus likely overestimates the consumption decline which can be attributed to debt in the 1930s. Mishkin does not recognize that much of the consumer debt incurred in the 1920s was short-term in nature, term or installment purchases rarely being in excess of a year. Much of the debt in the early 1930s was new debt, then. Wandersee (1981: 3) describes how families went into debt in the early Depression in order to support their usual standard of living (Mishkin does note that consumer credit becomes a larger proportion of total liabilities from 1930). Much of the observed relationship between debt and output is effect rather than cause, then. Indeed, the ability to borrow decreased the effect that falling incomes would have had on consumption, and thus more of this fall must be considered autonomous. Further, unless we conclude that consumers were forced to restrict all types of consumption to meet loan payments, Mishkin could in any case only explain the decline in durables expenditure but not in nondurables (Olney, 1989b). There is, again, much scope for industry-level studies to highlight and quantify the forces driving consumption decisions in the late 1920s and early 1930s.

The lack of an agreed-upon force pushing down consumption or investment is not the sole problem with the Keynesian approach. There is also no convincing explanation of why full employment is returned to so slowly. This is a problem faced by the monetary approach as well, but arguably to a lesser degree, due to a reliance on continued bungling by the Federal Reserve Board.[13] If we could identify some reason(s) for a sluggish recovery, this would reduce the scope of the problem in explaining the initial downturn. That is, if the forces which are supposed to take us back to equilibrium were somehow weakened at the time, then any shock to the economy would have greater than usual effects.

Bernanke and Parkinson (1989) have argued that the equilibrating mechanism worked well after 1933. Unemployment levels were so high in that year that it would naturally take some time for full employment to be restored. Keynes, of course, did not just speak of slow equilibrating mechanisms, but of the possibility of an underemployment equilibrium. He felt that we are generally below full employment (Sardoni, 1987: 136). Yet the experience of the Depression is better described by mechanisms taking us back toward full employment, though slowly and uncertainly (while the 1920s might be viewed as a period of underemployment equilibrium). It is, of course, a judgment call as to whether this recovery from depths we have never seen before or since was fast or slow. The general wisdom is that after a sharp downturn a speedier recovery would have been expected than that which occurred. Moreover, Bernanke and Parkinson likely overstate their case by focussing on manufacturing, which they recognize likely bounced back more quickly than the rest of the economy. The recovery in any case yielded to another sharp decline in 1937 long before full employment was reached. We can thus accept the dominant view that recovery was "tragically slow, halting, and incomplete" (Chandler, 1970: 1).

Fisher's Debt-Deflation Theory

We have argued above that neither indebtedness nor deflation could have been the predominant causes of the Depression. Irving Fisher concurred with this view in 1933. This point is worthy of note in its own right. As the creator of the MV=PQ equation, he is often thought of as the progenitor of monetarism. Yet by 1933 he had become convinced that a radically different theory was necessary to understand (and, more importantly, combat) the calamity which surrounded him. It is also noteworthy that he casually accepted that the deflation which he had just lived through had been unanticipated.

Fisher argued that while neither debt nor deflation were too harmful in their own right, combined they reinforced each other (thus the severe deflation of 1920-1921 had had little effect because debt loads were much smaller). Attempts at debt liquidation decreased both the level of outstanding loans and the velocity of circulation, and thus lowered the price level. Deflation in turn decreased the real value of debt. There was thus a disequilibrating mechanism whereby the very attempts by debtors to pay off their debts caused their real debt load to increase. There was no reason for this process to stop until everyone was bankrupt; Fisher was forced by events to abandon his faith in the economy's equilibrating mechanism, and thus became one of the first economists to urge Roosevelt to reflate.

Fisher's theory has few echoes in the modern literature, perhaps because he never formalized it. Mishkin's and Bernanke's approaches are each similar in some respects, and the flaws detected in their approaches apply also to Fisher. Fisher argued forcefully that the negative impact of deflation on debtors would outweigh the Pigouvian real balance effect by which deflation is supposed to increase real aggregate demand through increasing the real wealth of those with money holdings. Debtors were forced to react to new circumstances, while creditors were not. Still, given that assets exceeded debts by a wide margin, and that outstanding debts were a small proportion of national product, one could easily doubt that the debt-deflation model can carry the whole weight of the Depression.

Fisher, perhaps, would not have been surprised by this judgment. Laudably, he declared his openness to alternatives, simply finding his own theory the most compelling of those which existed. Moreover, he pointed the way to a more comprehensive technological explanation. He recognized that the debt-financed investment of the 1920s resulted from new product innovation in the early 1920s (though he failed to see that continued innovation would have sustained that investment). High profits from new technology in the early 1920s made investors susceptible to con men in the real estate and stock markets later in that decade. Fisher, unlike later monetarists and Keynesians, thus realized that the time path of innovation could be the key to economic fluctuations. We would be well advised to build upon this insight, and explore other than monetary mechanisms by which the level and type of innovation could affect economic performance.

Labor Markets

Whether the initial causes of the Depression were real or monetary they would have exhibited themselves in the labor market as a decreased demand for labor. This should have put downward pressure on both employment and wages. Empirical research generally finds that both employment and wages will respond to external shocks. Hamermesh (1989) has argued that due to the adjustment costs incurred by firms in reacting to shocks, they will not adjust employment levels in response to small shocks, but will make large sudden changes in employment in response to large shocks. Empirical studies of labor markets have, he suggests, come to misleading conclusions by lumping together these two distinct cases. The large adjustments in employment we observe in the early years of the Depression will be more readily understood, then, if we can first establish what the exogenous shocks were.

Eventually, though, we would expect the wage rate to fall to restore full employment.[14] Faced with a sea of available labor, employers should lower their wage offers until a new equilibrium is reached where (almost) everyone who wants to work at the going wage does. There may be some frictional unemployment — people between jobs will not find a new one instantaneously. There will also be some voluntary unemployment — people who are not willing to work at the existing wage. That should be all (we discuss the possibility of structural unemployment in Chapter 5).

Blinder (1988) has noted that wages represent not just income to a worker but prestige as well. Laid-off steelworkers will be very hesitant to accept lower-status work elsewhere. Then, as now, workers producing consumer durables had a relatively high status, and thus this psychological reaction had a role to play in the early days of the downturn. When unemployment rates of 25% or more are sustained for years, however, we must look elsewhere.

Contrary to our expectations, real wages actually rose through the early 1930s. Nominal hourly wages fell 17.6% in manufacturing and 20.8% overall between 1929 and 1933, but the consumer price index fell 28.2% and the wholesale price index 31.2%.[15] After 1933 real wages and nominal wages both rose. Between 1933 and 1937 real wages rose 25.6% in manufacturing and 10.8% overall. From 1937 to 1941 real wages rose a further 12.8% in manufacturing and 12.2% overall.

How could this be? Lucas and Rapping (1972) argued that the high level of unemployment was due to unanticipated deflation. Employers made wage offers with respect to the prices they could receive for their output. Potential employees looking for a job continually rejected job offers as unreasonably low, because they did not recognize that prices and nominal wages were falling. That is, the supply of labor is conjectured to have reacted more to the falling nominal wage than to the rising real wage. We thus observe chronic unemployment. Real wages are observed to rise because workers rejected those jobs in which wages had fallen as much as prices. Attributing the trends in employment and real wages to the process by which workers search for jobs may work at the very start of the downturn. However, we would expect that workers would eventually realize that nominal wages were falling. As well, we might expect workers to become increasingly desperate as the Depression drags on. Once this happens, workers will no longer be willing to trade off present leisure for future work (if they ever were). Thus, certainly after 1932, the Lucas and Rapping model is a poor description of reality (Eichengreen, 1987).[16] Even Lucas recognizes its inappropriateness after 1933 (Bernanke, 1986: 83). As well, the quit rate falls from 1929, indicating that workers were well aware of the fact that nominal wages were falling (Baily, 1981: 48).

Another way of viewing the real wage increase is as a secular rather than cyclical phenomenon. Bernanke and Parkinson (1989) note that, while unanticipated deflation caused a spike in real wage growth in 1931-32, real wages rose throughout the decade. This could reflect continued increases in productivity or changes in labor relations. Following on the work of Jacoby (1985) it is suggested that employers had begun to pay efficiency wages by that time. That is, higher than necessary wages were paid to workers to encourage a greater work effort (they might also attract a superior crop of applicants and discourage those whom the company has trained from quitting). Wright (1987) feels that business support of the National Industrial Recovery Act, and of tight immigration policies, was symptomatic of a shift from a low-wage low-productivity labor force to a high-wage high-productivity labor force. Jensen feels that decreased immigration, along with complex technology, encouraged efficiency wage practice (1989: 561). Slichter, without using the phrase of course, felt that efficiency wage considerations meant that firms were less likely to lower wages when prices fell (1929: 433). If productivity is felt to be dependent on relative wages firms may be hesitant to be the first to cut, but others may follow quickly. Efficiency wage theory does not necessarily imply wage stability; Ford, whose $5 day has been credited by Raff and Summers as an efficiency wage, had still cut wages during the recession of 1920-1922 (O'Brien, 1989: 732). Thus, efficiency wage theory itself provides no explanation of mass unemployment. (And Kimball (1994) finds using calibration techniques that efficiency wages have a significant but very small effect on unemployment; see also Margo (1993: 47) for a discussion of problems with applying efficiency wage theory to the 1930s.) Bernanke and Parkinson wonder why this real wage growth did not impede economic growth from 1933 to 1941. A more appropriate question would be why the high rates of unemployment did not interfere with real wage growth. That is, even if we recognize that secular forces are at work, wages could still fall to equilibrate the labor market.

Alternatively, we could posit that there were some sort of rigidities preventing wages from falling.[17] Garraty (1986: 84) has suggested that we observe solidarity among farmers because agricultural prices were flexible so that all suffered equally, but discord among workers because wages were not flexible so that some workers lost nothing and others everything. The most sophisticated approach is implicit contract theory. Employers being less risk-averse than workers implicitly agree to maintain employment and wages through times of adversity in return for a lower overall wage. Implicit contract theory is unable to explain the simultaneous movements in wages and employment, however, and does not answer the question of why wages were not slashed once the Depression was well under way (Baily, 1881: 44-50). Implicit contracts were regularly broken in the

1930s; a number of pension plans were simply cancelled. Moreover, we must recognize that the labor market operated quite differently in the interwar period than in the modern era for which implicit contract theory has been designed. The long-term attachment of blue-collar workers to a particular firm was much less. Jacoby (1985) describes how the labor market of the late nineteenth century was characterized by casual employment relations, and by foremen rather than firm bureaucracies making the hiring, firing, and wage decisions. World War One caused a shift toward more modern employment relations, but the unemployment of the 1920s caused an (incomplete) reversion to older methods. Thus, when the Depression began, only a very small progressive minority of firms had the simple policies which we tend to take for granted today. In the 1927 recession, only 40% of firms had any sort of seniority-based layoff rules, and these generally operated only at a departmental rather than a plant level (Jacoby, 1985: 196). Very few firms had a policy of purposely hiring back previously laid off workers. Firm stabilization plans affected less than 1% of the labor force. Only 32,000 workers were members of firm unemployment insurance programs in 1933. Jacoby concludes that "the quasi-permanent employment relation with which we are familiar was largely a post-1929 development" (Jacoby, 1985: 205). By 1933, job security was a major concern of both government and the revitalized labor movement. To fend them off, business increasingly turned to seniority based layoff and promotion rules, pension plans, paid vacations, health plans, and various other policies which diminished the dictatorial power of foremen.

Still, it is possible that employers to some extent laid off their least productive lowest-wage workers first. Jensen (1989) argues that bureaucratized hiring and firing allowed firms to significantly improve the average quality of their workforce in the early 1930s. Thus data on average wages paid will exaggerate wage trends (see Margo, 1993: 344-5), for we will have seen an increase in the average productivity of employed workers.

Weinstein (1980) has argued that much of the Depression can be attributed to the National Industrial Recovery Act, which did not allow wages to adjust after 1933, and to earlier efforts by the government to keep wages up through moral suasion (we discussed in Chapter 1 the effects of government relief efforts). Margo (1993: 46) argues that the empirical evidence for this contention is weak. In the heady days of the 1920s, it was widely believed that business should stabilize employment by maintaining high wages in downturns (Ginzberg, 1939). Hoover called on business to maintain nominal wages in late 1929, and many firms, especially in non-competitive industries, attempted to do so,[18] though they generally gave up by late 1930 or 1931 (Jacoby, 1985: 216-17). In May 1931, Secretary of the Treasury Mellon was still exhorting firms to maintain wages: "the standard of living must be maintained at all costs" (in Angley, 1931). A National

Industrial Conference Board survey of firms in 20 industries taken in 1937 found that nominal wages had only fallen 2% by January 1931, fell another 1.5% by August, and 25% between 1931 and 1933 (O'Brien, 1989: 720). Neither this nor the increase in union membership (from low levels in the 1920s) seems sustained or large enough to explain the lack of real wage adjustment (Baily, 1981: 55-6).

Bernanke (1986) has suggested that it may have been necessary to increase hourly wages somewhat in order to achieve desired decreases in the work week. It is estimated that hours worked fell 24% 1929-1933. We have seen in Chapter 1 that labor was often hostile to this idea, favoring layoffs to hours reductions. Even before the Depression, shorter hours were often opposed by workers, especially immigrants who wanted to earn money fast and return home (Wright, 1987: 336). Presumably, workers would be willing to make some leisure-income tradeoff. It would make no sense, however, to reduce hours unless weekly costs of labor could be reduced (it is noteworthy in this regard that postwar experience indicates that hourly productivity tends to rise almost enough to compensate for at least small decreases in the workweek; Sinclair, 1987: 275).

Insider-outsider theory argues that unemployment is a result of the exercise of bargaining power by workers with jobs. Because hiring (including screening and training) and firing impose costs on the firm, existing workers can bargain for a wage significantly above the going market wage. While this theory too may have less applicability to the 1930s than today, it provides a further partial explanation of the experience of both wages and hours in the interwar period. Even with a large mass of unemployed, firms might not find it advantageous to replace workers earning high wages. Beyond the direct turnover costs, there is the possibility that insiders may refuse to co-operate with outsiders, and thus severely lower the potential productivity of the latter. Or, high rates of turnover themselves may encourage a lesser work effort (Lindbeck, 1988: 81). Lindbeck has suggested a number of policies which might reduce insider power and thus alleviate the unemployment problem, but concludes that perhaps encouraging new firm formation may be the best remedy (1988: 265). New firms entering extant industries may not be profitable, especially in the short run, even if they can pay lower wages. Thus, insider-outsider theory can help to explain why wages in particular industries do not fall. Lindbeck suggests, though, that there is always a competitive sector where wages are determined by market forces.[19] We would also suspect that any new industries would be in an ideal position to take advantage of the low wage demands of the mass of unemployed. We still lack an explanation of economy-wide high unemployment in the 1930s.[20]

If Jensen (1988) is right that firms had jettisoned their least productive workers, then we have a further potential explanation. Firms would not hire the unemployed, even at much lower wages, because they correctly viewed them as being less productive. While this might apply to the long-term unemployed, we saw in Chapter 1 that a substantial proportion of the population rotated in and out of employment. Blaming the unemployed for their own plight can at best be only part of the story.

All of these theories provide only partial explanations. They may describe why the still-employed saw their wages rise. They also may explain why the unemployed were unable to steal jobs from the employed by working for less. At the aggregate level, though, they fall far short of being able to explain why the mass of unemployed do not obtain work at substantially lower wages and thus lower the observed average wage in the economy. Empirical analysis across various countries shows that unemployment had only a slight negative impact on real wage trends, especially in the case of long-term unemployment (Eichengreen and Hatton, 1987: 24-6).[21] If the fault lies not with workers, then we are left with the question of why employers for so long did not take advantage of this vast pool of labor willing to work for low wages. "The Great Depression was a time when... men would work for 10 cents an hour, and employers could not profit from their labor" (Hughes, 1987: 446). If we could understand this, we would be able to understand the movements in wages and employment throughout the 1930s.

Other Market Imperfections

The delayed return to full employment need not result from circumstances within the labor market. Demand for labor depends on the demand for output which can be artificially depressed by price rigidities. As demand for output depends on income, and thus employment, an underemployment equilibrium is possible (see Patinkin, 1956). This requires, of course, that prices never adjust to their appropriate levels, despite the pressures that would bear upon them.[22] Kalecki noted that if there are imperfections in product markets, a declining real wage may serve only to decrease aggregate demand without restoring full employment (Bernstein, 1987: 11n). The latest thrust of Keynesian macroeconomics is once again to focus on sticky prices causing the goods market not to clear, due to both theoretical and empirical problems in attributing business cycles to sticky wages; often nominal price rigidity is coupled with real wage rigidity (Mankiw, 1990).

It was common in the 1930s to attribute the severity of the Depression to the growth of monopoly power resulting from the merger movement of the 1920s.[23] Arthur Burns, the Federal Trade Commission, and the President all at least partially accepted this explanation (Cox, 1981). Especially after the collapse of the New Deal recovery plan in 1937, the administration became convinced that the Depression stemmed from the abuse of power by big business; one result was a tremendous surge in antitrust activity (Hounshell, 1988: 346). The Temporary National Economic Committee concluded that some industries had restrained production in order to increase prices (1941: 23-4).

According to Krooss (1972: 149) the two decades after 1890 witnessed the emergence of big business, with firms coming to dominate national markets. In 1885 there were only five companies worth more than $10 million, outside of railroads. By 1897 there were eight companies worth more than $50 million. This number rose to 40 in 1907 and 100 in 1919. Once we recognize the predominance of oligopolistic structures, we see that firms might have hesitated to lower their prices for fear that they would lose market share to competitors who did not (Baily, 1981: 46-7).[24] Blockades, bombs, and sabotage were occasionally used by competitors to prevent price cuts in the 1930s (Krooss, 1972). With government blessing, and with wartime experience in cooperation to guide them, the 1920s had seen the number of industry trade associations increase from 700 to 2000; these may have served to further blunt competitive instincts. Olson (1982: 227) attributes much of the early Depression experience to these, noting that they were powerful enough to convince the government to introduce tariffs. "Is business rapidly developing, in the course of the evolution of large-scale enterprise, corporate price policies of a character that make the system of free enterprise unworkable?", wondered Hansen in 1938.

Empirically, however, this case has not stood the test of time very well. The degree of concentration was no greater in 1939 than 1900; 25% of national income was monopolized (Cox, 1981). Thorp maintained that mergers in the 1920s were not aimed at monopolization as in the nineteenth century, but rather were aimed at increasing borrowing power and gaining national advertising potential (1929: 217). While the nineteenth century mergers had involved the biggest firms, the 1920s mergers were medium-sized firms pursuing vertical integration and diversification; little oligopoly power resulted (Krooss, 251). Leuchtenberg (1958: 193) noted that some mergers actually increased the degree of competitiveness.

The modern macroeconomic literature focusses on monopolistic competition and the costs firms face in changing prices (changing catalogs, advising customers, etc.) in its explanation of sticky prices (Mankiw, 1990). It could easily be that in the interwar period, when, as we shall see, methods of communication, distribution, and a host of other factors were changing, concentration ratios are a poor indicator of how flexible prices were. However, the 1930s evidenced greater price (and wage) flexibility than the postwar world (Sachs, 1980). According to Balke (1989), the postwar period has witnessed price volatility similar to the pre-World War One era. Thus, price rigidity would seem incapable of explaining why the Depression was so much more severe than earlier or later twentieth century contractions. Still, measures of average price flexibility could be misleading. Gardiner Means in 1935 characterized the Depression as "a general dropping of prices at the flexible end of the price scale and a dropping of production at the rigid end" (in Garraty, 54); some prices were stable 1929-1933 while others fell 60%. Fabricant found that large firms tended to make money in the early 1930s, while smaller firms generally lost money (Chandler, 1970: 28).[25] A government study in 1939, however, found that industries with high and low levels of concentration experienced quite similar trends in both price and output during 1929-1933 and 1933-1937 (Cox, 1981: 180).

Comparison with other countries is also useful. The United States does appear to have had sluggish price adjustment relative to Europe before 1933 (Gordon and Willcox, 1981: 98). In the postwar period as well, American prices have been sticky relative to Britain and Japan, and were sticky relative to Japan in the interwar period (Gordon, 1983). Thus, while price stickiness can not be portrayed as the main villain, it is undoubtedly true that the Depression in the United States was somewhat worse than it would have been if prices had been perfectly flexible.[26] We could concur with Ginzberg: "The economic fluctuations of the last decades can largely be explained by the interaction of technological, psychological, and monetary factors, though the fluctuations were intensified by the imperfections of the pricing mechanism" (1939: 243).

The Role of the Stock Market Crash

In the popular mind, the onset of the Great Depression is often seen as closely tied to the stock market crash of 1929. The scholarly literature has attributed considerably less importance to it. In Mishkin's model it contributes to decreased consumption in concert with three years of severe deflation. Bernstein blames it for disrupting financial markets and wiping out business and consumer confidence (1987: 30). However, commercial loans did not start to decline until well after the crash (Gordon, 1961: 427). The first wave of bank panics was in late 1930, and is attributed to a wide range of causes. As for expectations, we have seen that these seem to have held up well into the 1930s (and even if business confidence had fallen at the time of the crash it would be difficult to separate cause and effect). Those of us who have lived through the crash of 1987 can recognize that such an event need have no major impact on the macro-economy. The crash of 1929 only returned stock prices to the levels they had been at in 1927-28 (though further price declines would occur in the 1930s). Undoubtedly, the ability of some investors and consumers to spend was inhibited.

Yet it seems unlikely that the crash itself played a major role in fomenting the Depression. It is clear that the downturn into the Depression began in the summer of 1929. The crash may have added some impetus; the downturn does accelerate, though the biggest decline comes two years later. Or, as Aldcroft asserts, since stock market activity trails rather than precedes economic activity, it should be attributed no causal role. White (1990) has argued that the onset of the recession is the most plausible trigger mechanism to have brought about the stock market crash.

Romer (1992) has championed a novel argument as to how the crash might have affected economic activity. She finds an inverse correlation between market fluctuations and consumer durables expenditure for both the 1920s and the postwar period, and concludes that the former causes the latter (this echoes Bernanke's claim that uncertainty caused a drop in investment). Even if this causal link exists, as consumers because of increased uncertainty postpone durables purchases, it still takes a leap of faith to believe that the correlation coefficient which holds during minor fluctuations would hold during the major perturbations of 1929-1930. To explain away the 1987 crash, Romer argues that it is not so much the crash itself but continued stock price oscillations in 1930 which make consumers uncertain. Given the small percentage of the population which participated in the market, and the evidence discussed above that expectations were unchanged, we must remain sceptical that increased uncertainty as a result of the stock market can explain much of the observed consumption decline in 1930. Since Romer is inclined to blame the Fed for both the preceding and following phases of the 1929-1933 decline, we can be even more confident that we do not have the complete explanation.

It might seem remarkable that two rare events such as the crash and the Depression would occur at roughly the same time without some link between them. It is worthwhile to look at what the causes of the crash itself might have been. Eichengreen (1987: 65), in a survey article, concludes that the origins of the stock market boom remain shrouded in mystery. There is, though, one line of analysis which runs through much of the literature. Waters (1977) and Soule (1947: 332) attribute the boom to a combination of high profits with few investment opportunities. Douglas (1930: 590) blamed speculation on high profits which were in turn due to stagnant wages. Zevin feels that the stock market boom, like the earlier land speculation boom, reflected excess investment (1982). The idea, then, is that for some reason(s) there were not enough real investment opportunities in the economy to satisfy those with money to invest (there was also much American foreign investment in the 1920s).[27] The idea is generally not rigorously formulated. Excess investment should depress interest rates to the equilibrium level (there is some positive level below which the interest rate cannot fall, of course.) Once a significant overflow into stock markets occurs it has a self-perpetuating effect by creating high initial profits there. There is strong evidence that the stock market boom was a bubble — a self-fulfilling speculative price path in which rising prices rather than fundamentals encouraged further price increases: prices rose much slower than dividends, the premium on brokers' loans rose, and firm managers even announced that their stocks were overvalued (White, 1991). Closed-end mutual funds were overpriced by 30-60% in the late summer of 1929 (De Long, 1991). Gordon (1961: 417) describes how the major part of the new capital issues went for erecting a financial superstructure of holding companies and investment trusts. The speculative boom arguably kept a sick economy functioning for a couple of extra years.[28] How could a paper transfer — for every buyer of stock there is a seller —have such an effect? There are two ways, both outlined by Gordon (1961). Promoters made huge profits and may have had a high propensity to consume. More centrally, the stock market boom acted as a mechanism to transfer money from buyers who wished to forego present consumption to sellers who wished to consume. "The rise in security prices created capital gains, particularly in the upper income groups, and thus put an artificial support under the demand for luxuries and durable goods that would collapse with the eventual and inevitable break in security prices" (Gordon, 1961: 418). That such an extreme price rise was necessary to achieve this is perhaps indicative of the state of consumption opportunities.

While the crash did not directly cause the Depression, then, the preceding boom may have had some effect on its timing. If we grant that possibility, it could help us to comprehend the severity of the 1929 downturn. If speculation was artificially supporting economic activity, deflationary forces hitting the economy in the late 1920s may have been submerged under apparent vitality. While their aggregate effect may have been the same in any case (depending on how we think shocks interact in the macroeconomy), the boom served to hide their depressing effects until 1929. Much of the trouble encountered within the Keynesian approach in defining the initial cause of decreased consumption and investment expenditures may be due to the fact that various forces of importance were spread over the previous years. This point has been, at least implicitly, recognized by those who have looked at the downturn in construction after 1926.

Real Business Cycles

Both Monetary and Keynesian explanations of the business cycle have tended to view the forces causing cyclical disturbances as being distinct from the forces causing economic growth.[29] As Hicks long ago noted, however, there is no theoretical reason to expect that these phenomena have different causes and good reason to expect that the forces driving growth also determine fluctuations. There should, after all, be no business cycles in a stationary economy (Schumpeter, 1939: 138). The starting point for Real Business Cycle theorizing is the idea that fluctuations and growth are joint outcomes of the same forces. While various forces could be analyzed, almost all work has treated labor-augmenting technological change as the key force at work. As Mankiw (1989) points out, technological disturbances are the only type likely to cause consumption and leisure to both move in the same direction, as is observed.

This presents a great difficulty, for technological change is, by its very nature, difficult to measure. RBC theorists posit that technological change occurs discontinuously, with periods of radical advance interspersed randomly with periods of little, no, or even negative, change. Since technology itself is unmeasured, reliance has been placed on Solow's estimates of American labor productivity for the postwar years. It has been found that these estimates were strongly correlated with aggregate variables such as output and investment, which was taken to indicate that fluctuations were due to technological shocks. There are at least two problems with this. One is that Solow's figures contain a couple of years in which productivity falls by 3%. This is unlikely to be due to technological backsliding (Mankiw, 1990: 1653). This leads to the second point. Productivity is expected to move pro-cyclically. However, if firms hoard labor in a downturn, output per worker naturally falls. Thus, much of the observed relationship is effect rather than cause (Mankiw, 1989). Even Plosser (1989) recognizes that RBC models will have to pass much stricter empirical tests than they have so far.[30]

There are theoretical problems as well. The mechanism through which technological disturbances are expected to affect employment is intertemporal substitution. That is, workers voluntarily reduce their employment in one period in anticipation of larger wages in the future. Empirical studies give virtually no support to the idea that intertemporal substitution is a significant determinant of labor supply. This is perhaps one reason why RBC models do a much worse job of explaining movements in wages and hours worked than they do of output. Ignoring the possibility of widespread involuntary unemployment, this approach trivializes the social cost of business cycles (Mankiw, 1989). Inappropriate even to the postwar world in which it has generally been applied, RBC theory has little to say about the Great Depression.

Some real business cycle models have broken the economy up into sectors. Since the models rely on infrequent large shocks, these models can only deal with a few sectors subject to large disturbances. Otherwise, independent disturbances should tend to balance out across sectors, and no cycle would emerge. Extensions along these lines, then, are very similar to the one-sector analyses, and suffer from the same weaknesses (Mankiw, 1989). Hosios (1994) questions the validity of the empirical evidence used to support these models.

A second set of extensions, though, focus on the difficulties of moving capital and labor from one sector to another. Human and physical capital are recognized as being highly specialized. Shocks to any sector (positive or negative) will induce unemployment as labor moves from one sector to another.[31] We need no longer rely on intertemporal substitution to generate cycles. Periods of widespread technological change will be periods of higher unemployment because it takes time for workers to move from industry to industry. More plausible than previous RBC models, these extensions still appear a poor description of reality. They predict that labor mobility will be highest in recession, and that job vacancies should increase in times of high unemployment; neither accords with observed reality. It could be argued in response that sectoral adjustment, by creating unemployment, reduces aggregate demand and thus vacancies; it becomes difficult to test this empirically for its prediction is similar to Keynesian theory (Mankiw, 1990).

At present, therefore, even the best of Real Business Cycle models appear unrealistic. Some of the building blocks on which they are based are not, however. Presuming an identity between the causes of growth and of fluctuations seems reasonable.[32] Recognizing that shocks are likely to spread from one or more sectors to the rest of the economy is a step forward. Further adjustments are required to make these models accord with reality. One is to recognize that technological change, while beneficial in the long run, could have a negative impact in the short run. Workers may be replaced by labor-saving technology and have difficulty finding work elsewhere. Thus, we could generate increased employment while seeing decreased vacancies. A second would be to expand our scope of analysis to include non-technological factors. These could have the same effect of releasing workers involuntarily from one sector and forcing them to search for jobs elsewhere.

Either of these developments would involve major breaks with the model outlined by Lilien (1982). He assumes that the sum of employment shocks across sectors must be zero. Thus an increase in layoffs in one sector will be matched by an increase in vacancies in another. While this may be true for a shock like the oil crisis which helped some sectors while hurting others, it need not always be the case. In the long run, workers losing jobs in one sector will decrease wage demands in the rest of the economy (and if prices fall in the first sector, and demand is elastic, demand for other products should increase), and full employment will be restored. However, the mechanism which equates the sum of changes in labor demand to zero (in a static economy) is far from instantaneous.

Underconsumptionist and Marxian Theories

Policymakers in the 1930s did not make the distinction that we make today between microeconomics and macroeconomics. They attributed the Depression to structural problems as much or more than to problems of economic aggregates (Temin, 1990: 107).

Some once-popular theories concerning the Great Depression, which viewed it as a natural outcome of capitalist development, are rarely heard mention of these days. This can scarcely be attributed to the success with which more mainstream theoretical structures have dealt with that event. It is thus worth looking at these ideas to see if they truly deserved to be so casually discarded. We will find that while the overall theories may have been flawed, they contain insights which we can build upon in succeeding chapters.

Underconsumptionism had been around for a long time before the Depression. Malthus had long since suggested that the mass of humanity would become indolent once their basic needs were satisfied (though he doubted per capita incomes could rise in the long run). In the late nineteenth century, Hobson became the champion of the cause. As the ability of capitalist economies to produce continues to grow in per-capita terms, a point is eventually reached where consumers no longer wish to consume the bulk of what is produced in the economy. Until that point, all income not consumed is saved, and all savings are invested — hoarding is ruled out. We thus see steady growth until sudden disaster. When production exceeds consumption by some extreme margin producers respond by cutting back, and the economy spirals downward. The recovery from the downturn was never fully explained, but it seems that the glut on the market would be removed through a period of dissaving, and the process could begin again. While the crisis might be inevitable, it could be hastened by an unequal income distribution which would encourage savings while limiting effective demand (Hobson further suggested that people were more likely to consume from income they earned than from income generated by exploiting market power). The action of investors in failing to anticipate the coming crisis could be attributed to competitive forces, which were alleged by Sismondi and others to cause them to overestimate the potential demand for their product.

Underconsumptionism blossomed in the 1930s. Terbogh (1945) remarked on the sea change in attitudes between Hoover's promise of boundless prosperity in 1929 and the pessimistic view of the Temporary National Economic Committee in the late 1930s that the American economy had "matured" and was suffering from secular stagnation. Hope of a market-driven recovery to full employment had been abandoned. Terbogh, though hired by the Machinery and Allied Products Institute to oppose underconsumptionism, had to recognize that mainstream economists had provided little opposition, since they had no compelling explanation to offer. In the 1930s, as macroeconomics was born, the profession found itself listening to pundits it had previously considered crackpots; Hobson himself was published in *Economica* (Dimand, 1991). Alvin Hansen, who at first thought Keynes was crazy and then popularized his ideas in America, infused these with the idea of stagnation. Hansen attributed the 1920s to a fortunate combination of expansion in construction, automobiles, consumer durables, and exports; such good times might never return. Terbogh was sure that Hansen and Hobson were both wrong, but had no alternative to offer: "It is a fundamental error to assume that the gravity of a Depression is necessarily a reliable measure of the scope and importance of the 'maladjustments' which initiated the movement. A downswing tends to become a self-aggravating phenomenon, and given a few unlucky 'breaks' in the evaluation of events, including inexpert handling by government, it can expand to proportions out of all relation to the magnitude of its original causes" (1945: 174).

It is not surprising that underconsumptionist (or overproductionist)[33] ideas were popular during the Depression. They soon disappeared in the postwar era.[34] Partly, this was due to the fact that the main ideas were never rigorously formulated and therefore it was often difficult to be sure of what particular authors were saying. Moreover, as economics as a discipline moved away from attempts at intuitive understanding toward formal model construction, underconsumptionism was left behind. Another likely reason is that as economic growth took off in the postwar era predictions of inevitable stagnation appeared less plausible. Bleaney (1976) refers to an ideological bias against theories which predict inevitable stagnation.

Thus, the question is still open as to how much merit these ideas possessed. Bleaney (1976: 151-8) notes that underconsumptionists must assume that investment is tied to consumer demand in order to generate their crisis. He maintains that investment, dependent only on profits, can expand forever. Even if consumption stagnated, therefore, the steady increase in income would simply be devoted to investment. It is difficult to imagine as stable a world where output outstrips consumption by ever-widening margins. Rather, it seems that investment must depend on expectations of future demand, and must eventually crumble if consumer demand falls far short of available output.[35]

The best answer to underconsumptionism is perhaps to be found in Keynes. Once we recognize the difference between actual and intended investment — that savings is not naturally transmitted into investment — the world changes markedly. Rather than having a lengthy boom followed by a sharp crisis, entrepreneurs are likely to react at any time that they perceive output to be outstripping consumption. Unless investors loose touch with reality for an extended period, we should not see an underconsumption crisis. A tendency toward underconsumption could nevertheless be a chronic weakness of the economy.

If so, then underconsumption theory might well point to potential sources of structural slumps. Even Terbogh recognized that innovation and population growth are important determinants of investment and employment, and that major innovations are distributed unevenly through time (1945: 96-8).[36] While the theory appears extreme in its predictions, it does highlight potential problems in unequal income distribution, sluggish population growth, and the time path of innovation.

One further argument against underconsumption is that while we do see declining average propensity to consume in cross-section we do not find it in time series (Bleaney, 1976: 221). Why is this so? It is likely due to the fact that increases in output have been associated in modern times with an expansion in the range of consumer goods. Over time, people are exposed to new goods and devote a significant portion of income to goods that used not to exist (or advertizing may alert consumers to the existence of previously unknown goods). This may be an important means by which the macro-economy avoids the problems of chronic underconsumptionism.

While bearing some similarities to the work of underconsumptionists, Marxian interpretations differ in important respects. Marx, indeed, emphasized the dynamic nature of capitalism rather than the tendency toward stagnation. Crises are due to fluctuations in investment. Again, capitalists are expected to expand investment without regard to purchasing power. Wages cannot rise fast enough without pushing the rate of profit below some critical level. The crisis occurs when expected price falls below average variable cost. As Sardoni (1987) has noted, if we allow for differences across firms, the crisis will not hit all at the same time, and it is possible that some firms will fold while the rest of the world carries on. As with underconsumption, Marx felt that after the crisis and a period of stagnation, the upswing would start again, unless the crisis had triggered a revolution. As with real business cycle theories, he felt that periodic fluctuations were due to the same forces that caused growth.

The Marxian assertion of investment at maximum levels between crises was based on the assumption of a constant marginal efficiency of capital. That is, entrepreneurs would receive the same rate of return for each additional unit of investment. Marx was following the classical economists in this respect. With the assumptions of perfect competition and constant returns to sale, this will be the result; firms will face the same price and average cost for each additional unit produced (they will always choose to operate at capacity) and thus the same profit. Kalecki has argued that this is a poor characterization of reality. In an oligopolistic world firms face a downward-sloping demand curve for their output, and would thus not naturally invest as much as possible. Mandel (1980: 178), provides a partial counter-argument; while stupid in the aggregate, oligopolistic firms may still invest as much as possible as they are fighting for market share. This is a severely exaggerated view of oligopolistic behavior. Morever, capital market imperfections might prevent firms from borrowing as much as they would like to (Sardoni, 1987: 6). Once again, we might wonder how expectations are formed, and why some capitalists (or bankers) do not react far in advance. No matter what shape it takes, the marginal efficiency of capital schedule will fall as expectations worsen. There are powerful reasons, then, to suspect that investment need not absorb all of savings between crises.

It is worth noting the prediction of Marx with respect to prices. When the original crisis hit, capitalists would have to lower their prices in order to sell their entire output. The process of hoarding rather than investment would be aided by expectations of further price decreases (Sardoni, 1987: 39). While Marx was sure deflation would be anticipated in a crisis, though, it does not appear that this was the case in the 1930s.

A second stream of Marxian theory deviates more from underconsumptionism, and has a more explicit role for technology. Mandel (1980), for example, feels that during the boom profits fall due to labor-saving technology which decreases the surplus value, decreasing unemployment (despite the labor-saving technology) which pushes up wages, and rising natural resource prices due to inelastic supply. After the crisis,[37] which devalues the capital stock and depresses wages, profits are once again high and the cycle begins again.

The analysis of wage movements through the cycle is little different from that to be found within Keynesian theories, though it is expressed by Mandel in harsher language. The 1920s, however, were characterized by chronic unemployment and wage stagnation. Rising relative resource prices, while perhaps important at other times, is the exact opposite of what was observed in the 1920s. While there were some differences across countries and commodities, in general the terms of trade turned against primary goods after 1920, and especially after 1929, as demand fell for inelastically supplied goods (Rostow, 1978: 205-7). Rostow notes that primary producers, except for those who were able to turn to high-value production such as citrus, had little income with which to buy manufactured goods. Hansen and Timoshenko are among the authors who have attributed much of the Depression to problems within the primary sector (Aldcroft, 1977: 271). We must explore this possibility in later chapters.

New technology need not depress profits. Baran and Sweezy have, indeed, attributed to epoch-making inventions such as the steam engine or railroad the role of extending the life of capitalism (Bleaney, 1976: 226).

New technology may save on capital as well as labor. As we will see, much of the new technology of the 1920s was capital-saving. New technology might also improve the quality of a good and thus improve its market potential. Indeed, one must wonder why entrepreneurs would choose technology that would lower their profit rate. A more plausible mechanism through which new methods of production might affect aggregate variables is through their effect on the labor force. Workers who are displaced from their jobs, and have difficulty finding work, will suffer a decrease in income which will show up as a decrease in aggregate demand.

Long Waves

Mandel has noted that the literature on long waves is itself subject to long waves. During the postwar secular boom few writers paid much attention to the idea that modern economies might be governed by long-wave cycles of fifty to sixty years. During troubled times in the 1970s, long waves became a more popular subject of research, just as they had been in the 1930s. Mensch has suggested that long waves would have wider acceptance but for the fact that most leading economists are too young to remember the last downturn (1979: 6). This problem may plague all approaches to the Depression; as the event becomes more distant, scholars are likely to increasingly forget that it happened, unless awakened by a similar crisis in the future. Long waves also suffer, of course, from their non-compatibility with the static analysis which characterizes most economic theory (Goldstein, 1989).

Despite a surge in research in the last decade or so, it is still not clear if long-wave or Kondratieff cycles actually exist. There do appear to be alternating periods of price increases and decreases. Saul (1983: 11) has noted, though, that the price series are dominated by peaks in 1812, 1873, and World War One, which could just be isolated events. Long cycles in output or other real variables are even less clear. Different researchers have come to different conclusions. The analysis of long waves is far from straightforward. Cycles, if they exist, should be a worldwide phenomenon, and should be better represented by aggregated data than by the experience of particular countries. Even worse, it is only possible to discern long cycles if the effects of shorter-term cycles are removed from the data. While Berry (1990: 4) finds Kondratieff cycles in the rates of change of many variables, Solomou (1987) has maintained that instead it is the much shorter two to three decade Kuznets cycle that exists. Most of the profession could be characterized as sceptical of both results. There is one overriding problem which makes it impossible to accept the existence of regular cycles with certainty. If they are a natural characterization of the modern (i.e. post-Industrial Revolution) world economy, the most we can hope to observe in the data is three or four cycles. Thus, even if we were to accept the results of those who argue in favor of long cycles, we would have to face the possibility that we are looking at a co-incidence, as Kleinknecht (1987) has noted. We must wait a couple of centuries before being sure of the existence of long waves.

We will be in good company, then, if we remain sceptical of the existence of regular cycles. Nevertheless, it is obvious that in the history of the developed world there have been some decades-long periods which have experienced rapid growth with only minor recessions, and other decades-long periods which have seen stagnation or even economic decline. We can, then, speak of upswings and downswings much longer than the normal usage with respect to the short three-to-five-year business cycle. Characterizing the Depression as a particularly bad cycle and a half is not likely to yield much insight. Looking for forces which can explain more than a decade of poor economic performance will be more fruitful.

In that light, it is still worthwhile to look at elements of the long wave literature. Gordon (1983) has suggested that business cycles behave differently in a long-term upswing than in a long-term downswing. In particular, unemployment rates rise much faster in a downswing recession. Thus, we should perhaps see the initial downturn as a combination of short and longer term forces. Forrester (following Schumpeter) has gone further and suggested that the sharp decline is due to the unfortunate co-incidences of Kondratieff, Kuznets, Juglar (eight to ten years) and Kitchin (40 month) cycles. Such an analysis is only helpful if we understand the forces which have generated the longer-term cycles.

Various explanations of Kondratieff or Kuznets cycles have been proffered over the years, including demographic cycles, investment cycles, credit cycles, raw material price cycles, and infrastructure cycles. Generally, there is a feedback mechanism such that these variables both affect and are affected by aggregate variables. Often, two or more of these variables are assumed to work together. We have discussed the possibilities of credit rationing, raw material shortages, and over-investment during the interwar years above. We will discuss the potential effect of demographic variables in Chapter 3, when we look at consumer demand. Berry (1990: ch. 4) notes that the working of the capital market guarantees that infrastructure investment will not occur evenly through time; the possible effects of infrastructure overbuilding in the 1920s will be discussed in the industry studies to which it is relevant.

Forrester has suggested that long cycles could be generated by a natural cycle in the capital goods sector, which must expand rapidly to satisfy new requirements in the consumer goods sector, but contracts thereafter.[38] It is important to recognize that when traced back through all the inputs the vast bulk of the cost of any good is labor; capital itself is mostly embodied labor. Thus, in the long run, reduction in the cost of a good almost always implies a reduction in the labor input required. In the short run, this need not be the case; labor may be expended today to produce machines which will produce consumer goods in the future. The link between output today and employment today is not independent of the historical record of investment. It will be worthwhile, therefore, to pay special attention to the capital goods sector.

The most common explanation of long waves involves the clustering of technological innovation. A burst of new technology supports a decades-long economic boom. When the technological potential is exhausted, the economy slumps. Mensch (1979) and others have argued for feedback mechanisms from economic activity, but the historical record does not justify their assertion that periods of Depression or early stages of recovery are most conducive to innovation. More recently, writers such as Perez (in Dosi, 1988) and de Bresson (1991) have argued that interlocking technological systems emerge in succeeding generations, and these generate waves of economic activity, though not necessarily of equal duration. Among the difficulties with this approach (see Chapter 4) is the fact that the Great Depression does not seem to have occurred as it should at a transition point between economic systems. As with the variables discussed above, we need not accept the existence of regular cycles in order to recognize that technological clustering could be behind lengthy periods of growth and decline. We will see in what follows that the most plausible view of the Depression relies on technological behavior.

The Resurgence of Structural Approaches

Few modern economists style themselves underconsumptionists, Marxians, or believers in long waves. Yet structural approaches to the Depression have increased in popularity in recent years. The limits of standard macroeconomics can be discerned in this definition of macroeconomics by James Tobin (1971, in Bernstein, 1987: 1n):

> Macroeconomics concerns the determinants of entire economies... The theoretical concepts and statistical measures involved are generally economy-wide aggregates or averages such as national income, total employment, or a national cost-of-living index. The objective is to explain ups and downs of these magnitudes and their interrelationships. The basic assumption is that this can be done without much attention to the constituents of the aggregate, that is, to the behavior and fortunes of particular households, business firms, or regions

While modern macroeconomic thought is characterized by a desire to supply microeconomic foundations for macroeconomic relations, macroeconomic approaches to the Depression are still limited by their focus on aggregate variables.

We have already seen in Chapter 1 the attempts by Margo and others to investigate the behavior of microeconomic variables during the Depression. We will discuss later the efforts of authors such as Bresnahan and Raff to investigate the behavior of key sectors. We are far from alone, then, in suggesting that mere macroeconomics will not give us a satisfactory explanation.

There is one work in this vein which might seem at first to have much in common with our own. Bernstein (1987) also argues that the time path of innovation played a key role in exacerbating the Depression experience. Beyond this, however, our approaches have little in common. He focusses only on the sluggish recovery, accepting the view that the stock market crash induced the initial downturn. As we have seen, much of the initial downturn remains to be explained. Even in terms of the sluggish recovery, Bernstein argues that the time path of innovation is only important in conjunction with the financial crisis. He argues that American industries were either old and in decline or young and reliant on loans. We have seen above, though, that there is good reason to believe that the decline in loans was due to decreased demand rather than supply. The expansion of refrigerators and rolling mill construction are the exceptions which prove the rule: when the technology to drive expansion was there the money was found. This is not surprising; Kuznets noted that throughout history profits have been high and capital requirements low in the early days of a new industry (1966: 51). Hicks has argued that credit constraints have only prevented entrepreneurs from exploiting new technological opportunities in developing countries (Katsoulacas, 1986: 146). More generally, Sharpe (1994) has found that while financial contraction has caused smaller firms to lay off more workers in postwar contractions, there is no relationship between financial expansion and small firm growth in upswings. Bernstein provides no examples of new technology which is put on the shelf for years due to a lack of financing. Moreover, he provides no explanation of why American industrial structure is as he says: why were there no middle-aged industries possessing both growth potential and financial reserves? A convincing technological explanation must answer such a question. Otherwise, like Keynesian theories, it can easily be dismissed as too ad hoc. In terms of evidence too, Bernstein surveys only a small sample of manufacturing. If we are to explain the Depression — even just the sluggish recovery — we must understand why investment and workers did not flow to other sectors, both inside and outside manufacturing. A technological explanation which depends also on a financial crisis suffers inevitably from all the criticisms which can be lodged against purely financial explanations. A focus on recovery (and manufacturing) alone leaves too many questions unanswered. And a convincing technological approach must not ignore the question of why the time path of innovation was as it was.

More recently Phelps (1993) has also suggested a structuralist approach to understanding medium-term economic fluctuations. Like Fisher in the 1930s, his book is worthy of note as the admission by one who has long worked within the monetary tradition that there is much more going on than can be captured by that approach. Phelps distinguishes between the short business cycle which has been the major focus of postwar macroeconomics, and may be amenable to analysis using modern macroeconomic theory, and booms and slumps of more than a decade in length which had been the focus of classical economists and which macroeconomic theory does little to explain. At the very least, then, Phelps agrees with our suggestion that "structural" forces could operate in a manner that was largely distinct from the behavior of macro variables such as the money supply.

It is thus no surprise that Phelps recognizes that, "...the experience of the Depression needs to be restudied with alternative interpretations, structuralist theory being a strong candidate" (1993: 162). Much of his attention is focussed on the long-term course of capital accumulation and government debt growth. He also discusses at some length the effects of changes in the growth rate of the labor force on economic performance, a matter we will both discuss and reach similar conclusions with respect to in the next chapter. While technology is absent from his survey of postwar structural behavior, he does derive theoretically the result that technological "shocks" can affect equilibrium employment through changes to desired investment. This will be one of our key arguments in the next chapter (though our results will not depend on restrictive homogeneity assumptions as do Phelps'). Further, Phelps notes that previous business cycle theories had ignored the potential effect of technology on unemployment simply because those theories could not separate short from longer term influences (1993: ix-x). While Phelps does not make the distinction between product and process innovation which we must, he does recognize that Ricardo's conclusion that [process] technology biased toward conserving labor relative to capital would increase unemployment in fact holds for [process] innovation with no factor bias (1993: 357-8). However, the mechanism by which this happens in his model, via expectations and the interest rate, is different from that which we will posit in Chapter 5. Still, Phelps provides a valuable service in emphasizing that standard macroeconomics is poorly equipped to deal with medium-term slumps or booms, and pointing to technology and demography as potential key structural forces.

Questions Unanswered

In surveying this vast literature, we can do no better than to borrow Phelps' own advice from an only slightly different context: "The truth is that every one of the extant theories has something to tell us about some of the effects of some shocks over some time frame." Our purpose here must be to focus on the questions which they collectively leave unanswered, while borrowing that which is best from them all. The unanswered questions prove to be interconnected, and thus point the way toward a comprehensive answer.

If we can arrive at an explanation for falling consumption demand from the late 1920s, we will be able to understand why investment fell as well. This provides insight into the crash of 1929 and therefore allows us to consider a variety of forces hitting the economy over a period of years. The lack of investment opportunities will show up in credit markets as a decrease in transactions. It will show up in the labor market as long-lasting high unemployment and will underpin rising real wages, as the willingness of the unemployed to work for low wages will be severely underrepresented in wage data.

Why might consumption autonomously fall? We might look first at how the available product mix was expanding with income. We are directed to suspect that the causes of fluctuations are the same as the causes of growth. Pressure on primary producers could negatively affect investment and employment. Technological change could cause employment levels to fall as well. If so, we must understand why technology evolved in the way it did. If new industries do not arise to counteract the negative influences on existing sectors of the economy, a protracted depression results.

If we accept that stock market speculation "artificially" supported economic activity in the late 1920s, then we can temporally widen the scope of our search for causes. Still, the precipitous decline would seem to indicate that some of these forces must act in 1929 (and perhaps cause other longer-term hidden weaknesses to become apparent).

The Depression happened. Yet decades of application of economic theory and the latest statistical techniques has failed to tell us why. The lesson could be drawn that greater flexibility in terms of both theory and evidence is required.

Notes

1. Friedman and Schwartz have been accused by the Romers of focussing only on historical episodes in which a decrease in output followed a decrease in the money supply, and ignoring episodes in which output did not fall. One problem with much of the empirical work is that models are constructed to explain interwar experience which would fit poorly the postwar economy. While admittedly there are major structural differences between the two worlds, the reasons why a model fits one period but not another should be made explicit. As well, the limited data for the interwar period makes it especially difficult to distinguish between different hypotheses. Eichengreen accepts favorably the position of Cagan that time-series econometrics is fundamentally incapable of answering questions of historical causality; compelling answers are only possible through case studies of historical episodes.

2. Temin asserts that one of the difficulties of testing the monetary hypothesis is that the transmission mechanism is not specified (1981: 122).

3. Gordon and Willcox (1981) have also suggested that the finding of a contemporaneous correlation between money and nominal income is curious, as monetarists generally argue that there should be long lags between policy changes and real effects.

4. The line of argument which we will pursue in this book suggests that deflationary pressures, bankruptcies, and a decreased demand for money could all be the result of technological forces. At the very least, this placed the Fed in a difficult position.

5. He quotes some contemporaries who felt credit was tight (there will always be some), and suggests that the credit crunch is more severe in the United States than in other countries experiencing a similar downturn. Bernanke (1989) has developed a model which shows that business cycles can be greatly exacerbated if lenders look at the balance sheets of borrowers when making lending decisions, for these will naturally look worse in a downturn. This draws on the work of Mishkin which we will discuss below.

6. Cross-country comparisons do indicate that countries in which the money supply contracted less also suffered less severely in terms of output and employment. Causation in both directions is possible.

7. As Weil (1989: 889) has recently shown, in a world of increasing external economies, there are an infinity of equilibrium growth rates associated with different levels of expectations. In this model, then, expectations become self-fulfilling.

8. Nor should they have done otherwise. Neither the econometric models then or now would have forecast anything like the downturn of the early 1930s until well after the stock market crash (Dominquez, 1988). Jaeger (1972: 141) is at first willing to attribute the fall in investment to a fall in expectations, but then recognizes that expectations lagged behind economic activity. Temin, after surveying writings of the time and bond ratings, concluded that expectations only fell well into the Depression (1976: 72-7). In more recent writings, Temin has attributed the entire downturn to inflationary expectations which flow not from individual gov-

ernment policies but the underlying international policy regime (1989: 91). This approach is hard to prove or disprove, but hard to believe.

9. Sardoni (1987) objects to the Keynesian concept of declining marginal efficiency of capital. He notes that within a perfectly competitive framework all firms would always be at capacity and investing as much as possible. As Kalecki had noted, it is necessary to abandon the assumptions of perfect competition to obtain a finite equilibrium level of investment.

10. Strict Keynesians (including Keynes himself) would object to this. They would feel that both savings and investment decisions are too loosely dependent on the interest rate for this to equilibrate the capital market. Only changes in output would bring desired investment and savings back in line (Laidler, 1989). Businessmen of the time affirmed that prospective profits were a more important determinant of investment than interest rates (Hansen, 1938: 311).

11. In the downturn of 1929-1930, the fall in construction investment is a greater proportion of the total fall in investment than in other interwar downturns (Temin, 1976: 61-5). The fall in inventory investment is smaller than would be expected. This lends some support to the idea that the fall in investment is at least in part a reaction to decreased consumer demand.

12. Investment has often been found to be the key variable in postwar recessions. This will naturally be the case if investment is treated as an exogenous variable, since all relevant variables move together; many empirical studies have assumed the correctness of central hypotheses rather than testing them (Temin, 1976: 41, 34). One incentive to treat investment as exogenous is that it is the variable which econometric models predict the worst. This may reflect the fact that investment can not be understood at just the aggregate level.

13. Lindert (1981), while recognizing that Fed policies after 1931 were misguided, doubts that this was the key to the slow and painful recovery. Gordon and Willcox (1981) note that neither the monetary nor non-monetary approaches do a good job of explaining the sluggishness of the equilibrating mechanism.

14. Keynes had recognized that wage cuts might not restore full employment because the resulting change in distribution would lower the marginal propensity to consume. Casson (1983) dismisses this difficulty, arguing that profits will rise more than wages fall, unless output also falls. Nell (1988: 94), in a model where prices are sticky and are based on a markup over long-term variable costs, generates the result that cutting wages reduces sales, has no effect on profits and investment, and thus actually reduces employment. We will see that for those still employed, real wages actually rose in the 1930s, as Nell would recommend. We could still ask why falling wage offers to the unemployed did not return the economy to full employment. In postwar recessions, unemployed workers are often seen moving from high wage to low wage sectors; high wages and no jobs in some sectors need not generate unemployment if there are jobs elsewhere.

15. Some versions of Keynesian theory, suggested but not formulated in the 1930s, could deal with this situation. Keynes had noted that prices might fall faster than nominal wages. Even if prices and wages both fell, full employment could be restored by the increased demand which would result from the increased value

of money (Temin, 1989: 140). There is much theoretical dispute about how powerful this mechanism could be. In practice, the money supply fell as fast as the price level. Unemployment has had little effect on real wages in the postwar period, and nominal wages fell less than prices in the 1890s as well.

16. Lucas would disagree. "If you look back at the 1929 to 1933 episode, there were a lot of decisions made that, after the fact, people wished they had not made; there were a lot of jobs people quit that they wished they had hung on to; there were job offers that people turned down because they thought the wage offer was crappy" (in Klamer, 1988: 41). Even allowing for quoting out of context, this indicates a lack of familiarity with the empirical reality of the period.

17. Sundstrom (1989) has argued that wages were more rigid early in this century than they are today. Relying on Ohio labor survey data, he finds that only 22% of workers received nominal wage cuts in the 1893 downturn. Given the likelihood of under-reporting in the survey, the fact that the downturn only affected particular industries, and that wages were generally rising over the period so that employers could react without actually cutting wages, the 22% could rather be interpreted as evidence of wage flexibility. Gordon (1983) finds an increase in wage stickiness in the 1920s, which he attributes to longer contract length, itself a response to changing labor relations.

18. O'Brien (1989) has put forward a behavioral explanation. Wage cuts in the recession of 1920-1922 had seemed to make matters worse by reducing aggregate demand. Thus, in 1929 firms were ready to experiment with a different way of doing things. Mitchell (1929: 866) felt that executives had come to recognize that mass production required high wages. Given the small role played by the employees of any one firm in aggregate demand, we can be sceptical that employers would have supported wages for this reason for very long.

19. Four sectors, mining, agriculture, construction, and domestic service, do see real wage declines in the early 1930s. These are among those in most serious decline, and tend also to be competitive.

20. Temin cites insider-outsider theory as an explanation of high wages, and suggests that one of the keys to German recovery was the low wages which resulted from destroying the unions (1989: 118).

21. Benjamin and Kochin (1979) argued that high unemployment in interwar Britain was largely due to the liberal provisions for unemployment insurance there. Their results do not stand up to international comparison (Eichengreen and Hatton, 1987: 26). Hatton (1986) notes that the number of jobless far exceeded vacancies; even if insurance shortened the queue it did not affect the rate at which jobs were taken up. Only 10% of workers received UI benefits in excess of 70% of their previous income (Thomas, 1987: 131). Postwar studies find that while higher benefits do increase unemployment the effect is not large (Sinclair, 1987: 271-4).

22. Later developments argued that uncertainty about the future could induce disequilibrium in any market — especially the financial market. Rational expectationists would argue that a permanent disequilibrium could not result.

23. While the merger wave of the 1920s was less important than that of 1898-1902, it re-established the concentration ratios achieved in that earlier wave. While

this second wave reflected the emergence of new industries, and while industrial and technological changes are possible long run causes of mergers, it is also possible that capital market behaviour is the key proximate cause. Though an increase in merger activity is visible from 1916, the second wave is concentrated in the years 1926-30 (Nelson, 1959).

24. This could be why manufactures prices vary less than agricultural and raw material prices (Olson, 1987: 205). Domowitz (1988) has found with postwar data that price markups are actually more procyclical in concentrated industries. While nondurable markups are procyclical, durable markups are slightly countercyclical.

25. Steindl has argued that increasing industrial concentration would act to decrease rates of investment in a depression, as firms would be less likely to go out of business and thus create opportunities for entrants (see Bleaney, 1976: 243-5). This is difficult to test, for some concentrated industries are relatively vibrant in the 1930s while some competitive industries are moribund (Bernstein, 1987: 19).

26. Mayer, commenting on Cox in Brumner (1981) notes that while no serious modern economist would agree that monopolization caused the Great Depression, Cox was too severe in attributing no role to it. Gill (1939) summarized the view of the WPA likewise: "Price movements and price disparities cannot be singled out as the causes of the ups and downs of business. They are, however, important contributing factors."

27. White (1990) downplays the importance of the availability of funds, at least in terms of the easy credit offered by brokers at the time. He argues that the speculative surge drew in funds, rather than easy credit causing the surge. He argues further that speculative bubbles are most likely to occur when it is difficult to predict future returns because of the rise of new firms with new products, and when new investors are entering the stock market. This bears a similarity to Irving Fisher's view of the stock market boom.

28. "If it had not been for the wave of speculation the prosperity of the 1920s might have ended much earlier than it did. Coolidge's deflationary policies had withdrawn government funds from the economy, consumers had cut spending for durable goods in 1927, and the market for housing had been glutted as early as 1926. But with the economy sparked by fresh funds poured into speculation, a depression was avoided and the boom continued" (Leuchtenberg, 1958: 242-3).

29. Tests of business cycle theories assume that there is a trend growth rate about which fluctuations occur. They are thus joint tests of the growth model (trend) as well as of the cycle model supposedly being tested alone (Plosser, 1989). If output growth is a random walk rather then a trend, all of this empirical work is mistaken.

30. Mankiw (1989) notes that there is often observational equivalence between the predictions of RBC and Keynesian models. However, RBC models cannot account for the observed Phillips curve relationship between unemployment and inflation. Rather, the models generate countercyclical price movements. Cooper (1990) has shown that RBC models relying on factor movements predict an inverse correlation of employment and output movements across sectors. Other

models which rely on demand-side linkages predict the observed comovement of output and employment across sectors.

31. Since labor is constantly moving in and out of the labor force, the shocks should have to be of greater magnitude than could be handled by normal attrition in order to generate unemployment. Leonard has estimated that 11% of all jobs are lost and 13.8% of all jobs are created every year; there should therefore be considerable flexibility with which to respond to technological change (1988: 44, 51). Most of this employment flux is within rather than between industries.

32. The vast bulk of research in the economic history of the eighteenth and nineteenth century focusses on industry-level factors. When attention turns to the Depression, the emphasis becomes macroeconomic, and the focus shifts to an entirely different set of factors.

33. These two concepts yielded the same result. They differed in whether they sought the cause of the crisis in too little consumption or too much production (Bleaney, 1976: 14-5). Note that for Keynes an increase in investment would be a way out of Depression, but here it just makes matters worse.

34. Bleaney places Baran and Sweezy in the under-consumptionist tradition. They also speak of a surplus obtained through monopoly power like Hobson, and assume this grows over time. They feel this surplus has to be wasted in advertizing, military spending, and government. It is not clear theoretically or empirically whether the economy is characterized by increasing surplus.

35. Hobson further argued that investment opportunities were limited by barriers to entry. The problem of excess investment had to be solved through imperialism, then. That part, at least, of his argument was borrowed by Lenin.

36. Terbogh notes that secular stagnation would have described the 1870s and 1880s reasonably well, but would have failed during the next three decades of steady growth (1945: 188). He does not explore the possibility the theory could be altered to explain intermittent crises.

37. Mandel feels that the inherent instability of the system is exposed by some exogenous shock. This shock does not cause but, given that other conditions are in place, triggers the crisis. He ascribes such a role to the stock market crash of the 1920s and the oil price hikes of the early 1970s (1980: 169).

38. Haberler felt that overcapacity could easily develop in both capital and consumer goods sectors, and that the resulting dislocation of the production process could trigger a cumulative process resulting in Depression (1957: 111-2).

3

Building a New Theoretical Approach: Consumption and Investment

As with any complex theoretical formulation, analysis of the theoretical approach to the Depression taken in this work could begin with any of several causal links. We will, in this chapter, argue that a lack of new product innovation could have caused the declines in both the propensity to consume and desired investment which in turn were the proximate causes of the Great Depression. We will defer to Chapter 4 a detailed discussion of the time path of technological innovation during the interwar period. While we discuss consumption and investment in turn, we should note that factors which serve to depress consumption in a particular sector will naturally tend also to depress investment. We avoid repetition by discussing particular arguments under the most appropriate heading. In particular, while understanding of both consumption and investment requires a sectoral approach, we will perform a brief sectoral survey only under the investment heading.

We will also discuss some non-technological influences on consumption and investment behavior. We will find that demographic trends and movements in the income distribution both served to exacerbate the negative impact of the time path of innovation. On the other hand, we will suggest that changes in advertizing, instalment credit availability, fiscal policy, and export demand were much less important in their impact on aggregate demand.

It should be noted that this chapter can be read as providing the rationale for falling consumption and/or investment needed by Keynesian models. This is not precisely the author's intention, as the discussion of technology and labor markets in the next two chapters will indicate. Dosi has suggested that Keynesian macroeconomics could be generated by evolutionary microdynamics, but this has not been attempted (1991: 1). While the author does not think the Keynesian framework a perfect fit for the

Depression experience, he does recognize the compatibility of this chapter with that approach.

Consumption Demand

We all recognize that in order for production to increase over time in a market economy, aggregate demand must also increase. While in the short run government spending and investment have a role to play, in the long run it is per capita consumption that must rise in order for increases in per capita output to be sustained. Given the insatiability of humanity this would not seem problematic. However, the reason that we consume many times more than our great-grandparents is not to be found for the most part in our consumption of greater quantities of the same items which they purchased. To be sure, we may have a few more clothes in our closets, and indeed, a somewhat bigger closet, and may also consume marginally more and better food. The bulk of the increase in consumption expenditures, however, has gone toward goods and services those not-too-distant forebears had never heard of, or could not dream of affording: cars, radios, TVs, VCRs, stereos, vacations in distant lands, movies, PCs, drugs, plastics, etc., etc. This is clear from statistics on consumer expenditure.[1] Would we as a society of consumers/workers have striven as hard to achieve our present incomes if our consumption bundle had only deepened rather than widened? Hardly. It should be clear to all that the tremendous increase in per capita consumption in the past century would not have been possible if not for the introduction of a wide range of different products. Consumers do not consume a composite good X. Rather, they consume a variety of goods, and at some point run into a steeply declining marginal utility from each. As writers as diverse as Galbraith and Marshall have noted, if declining marginal utility exists with respect to each good it holds over the whole basket of goods as well. It should seem surprising then that the discussion of particular episodes in the behavior of aggregate consumption demand is never related to the particular bundle of goods available at that time and place. Even in the case of the Great Depression, where deficient consumption demand is often painted as villain, the standard analyses have uniformly been highly aggregative. The simple fact is that, in the absence of the creation of new products, aggregate demand can be highly inelastic, and thus falling prices will have little effect on output.[2]

Moreover, as we enter a world of consumer durables, it not only becomes potentially difficult to extricate an economy from a situation of deficient demand, but quite straightforward to enter such a world.[3] We commonly measure not consumption but consumption expenditure. A person buying a durable in one year is unlikely to do so again for several years. If for whatever reason there is a bunching of durable purchases, most likely as new such products are introduced, the following years may witness a substantial drop in consumer expenditure unless still newer products are introduced. Competitive markets provide no mechanism by which market saturation can be prevented.

Old Goods; New Goods

A cursory look at the fate of different industries in the 1930s bears out the previous analysis. Food and nondurable consumer goods showed a relatively good performance throughout. Indeed, as Romer (1992) notes, expenditures on nondurables actually rose in late 1929 as consumers shifted away from durables; while she claims that her theory alone explains this, it flows naturally from ours as well. The slight drop in production in 1930 can be easily interpreted as reflecting the decreased purchasing power of the unemployed and a continuation of previous consumption levels despite falling prices by the employed. Consumer durable production dropped to 25% of the 1929 level by 1933.[4] Major ripple effects were felt in industries that supplied inputs to the durable sector. In the 1920s the automobile industry utilized 15% of the steel in the nation, and was the major consumer of rubber, lead, nickel, and plate glass. The transport sector, especially railroads, was hurt by the decrease in production. A dramatic fall in railroad investment rocked the steel industry. Employment by construction contractors fell to 50% of what it had been in 1929 by 1933 (the 1920s boom in housing and road construction being largely due to the automobile-induced move to suburbia). The only bright points were the chemical, electronic, and aircraft industries. Here, there was still some technological potential to be realized and scope for expansion and prosperity. This was not enough to spark economic recovery. (We will discuss the rise and fall of industries in greater detail when we look at investment possibilities below.)

A falling average propensity to consume will first show up as a decrease in the rate of growth in consumption, even though income continues to rise.[5] This is observed to happen between 1928 and 1929 in the United States (Aldcroft, 1977: 280).[6] Unless new products are introduced, we would expect the situation to worsen. Jaeger describes the Depression as entailing "shrinking consumption of goods even at rapidly falling prices" (1972: 134). This will be exacerbated by the fall in investment and employment which will result. A further factor of importance in the particular case of the late 1920s is that retailers, reacting to the growing importance of fashion in consumer decision making, and to improvements in the speed of transportation, began to hold smaller inventories (Copeland, 1929: 340; Cunningham, 1929). Sheldon feels that wholesalers in return cut back their inventory levels and moved toward more but smaller shipments from manufacturers (1932: 26). This provided a one-time shock (per retailer) to demand facing producers,[7] and ensured that changes in consumer purchases were more quickly transferred to producers than ever before.

Income Distribution

The fact that consumer demand was falling in the late 1920s does not, of course, mean that people possessed everything which they could ever have dreamed of. As at all places and times, consumption depended on income. We have merely been describing why the average propensity to consume of an average consumer might have fallen. To go from this level to aggregate consumption requires that we discuss distributional issues. Further, when we progress to industry-level studies, we cannot speak of market saturation in a vacuum, but must relate it to the effective demand of the population, which obviously depends on how total income is divided.

We do not have data on income distribution during the interwar period in the United States. Income taxes were only paid by those with very high incomes. No comprehensive records exist on the income of the rest of the population. In a complicated world, we could be easily misled by individual indicators, such as that income taxes fell, that some mergers created very high profits, or that purchases of luxuries expanded rapidly. Batra (1987), for example, focusses on the fact that the share of wealth owned by the top 1% peaks in 1929 at 36.3%; though he may be right that worsening distribution encouraged speculative investment by the rich, we would wish to have firmer knowledge of distributional matters. While it is impossible to accurately calculate a Gini co-efficient for a particular year, there are two methods by which changes in distribution can be estimated. The first is pursued by Holt (1977). From the income tax data, and estimates of GNP, Kuznets had been able to calculate that the top 1% of the American population earned 13% of total income in 1923, 19% in 1929, and only 7% at the end of World War Two. It is also possible from agricultural statistics to estimate the income gains of farmers during this period. Subtracting the gain to the top few percent calculated from tax sources, and the gain to farmers from total income growth during the period, yields the result that the bottom 93% of the non-farm population received no income gain — and may even have lost — even though real economic income increased 10% and real disposable income increased 14% during the 1920s (the 4% difference being due to the treatment of capital gains). Holt also finds that this group was relatively more subject to the business fluctuations of the 1920s. As with any residual, Holt's estimates will bear the effects of any errors in calculating total income or the income of other groups. Smiley (1983) has suggested that the income tax records underestimate the incomes of high income earners in the early 1920s, since falling marginal tax rates induced increased compliance over the decade, and thus Holt overestimates their income gain and underestimates the gain to the lower 93%. Smiley does not, however, attempt to estimate this bias, and we are left to suspect that the bulk of the labor force shared little in the prosperity of the 1920s. While Smiley also speaks of micro-level studies which show real wages to have been rising, the fact is that these appear to have been almost stagnant 1923-1929.

The second method is to calculate the functional income distribution; that is the share of income which goes to capital versus that which goes to labor. Keller (1973) performs this exercise for mining, manufacturing, transportation, and public utilities. In other sectors of the economy, unfortunately, entrepreneurial and rental income, which are difficult to attribute appropriately to either capital or labor, loom too large. He finds that the 1920s is the only decade in the twentieth century in which capital's share rose. Looking at peak to peak comparisons (since factor shares vary with the business cycle), we see that whereas capital's share fell 18.1% 1899-1907 and 11.6% 1948-1957, it rose 25.5% 1923-1929. Noting that interest rates fell more in the 1920s than real wages, Keller goes on to estimate that the predominant reason for the rise in capital's share was that energy costs fell dramatically. This in turn was due both to falling oil and coal prices and increased efficiency in use. A labor-saving bias in technological change may also have played a role. Woolf (1984) has re-affirmed the Keller result that decreased energy cost led to an increase in capital's share for almost all manufacturing sectors; he also notes the role of labor-saving technology in the worsening of the income distribution. These results then provide corroboration for the Holt finding, and provide some rationale.

There are other rationales. During periods of economic growth, wages tend naturally to lag behind profits. As well, the chronic unemployment of the 1920s might have inhibited wage increases. Industrial concentration and union-bashing may have affected the relative bargaining power of firms versus workers. Mellon, as Secretary of the Treasury through three Republican presidencies, reduced the taxes on high income earners that had been installed under Wilson (Leuchtenberg, 1958: 98). The government in the 1920s may have been the most business-oriented the United States has ever seen. There were hopes that social welfare provisions by firms and widespread stock ownership would help solve the distribution problem (Leuchtenberg, 1958: 199-200).

It is clear, then, that income distribution did worsen in the late 1920s. The results of Keller, though, warn us that some of what we see in the 1920s may simply be a reversal of earlier trends. It is important to recognize that the income stagnation of the 1920s was built on to the extraordinary wartime peak (Wright, 1987). As well, we must look beyond the aggregate estimates. Wright feels, along the lines of efficiency wage theory (see Chapter 2), that some firms were opting for a high wage, high productivity, low turnover strategy. While most incomes stagnated or fell, other workers recorded large gains. Stricker (1983) surveyed numerous micro-level studies and concluded that white collar and skilled workers often fared well, while unskilled workers generally suffered during the 1920s. Thus, those who suffered were those who already had the lowest incomes.

Williamson and Lindert (1980: 5) have used General Equilibrium models to estimate and explain trends in inequality over the long sweep of American economic history. While their results must be viewed with caution, they do provide reinforcement for what we have seen above, and insight into what went before and after. They estimate that the degree of inequality was basically unchanged through the last four decades of the nineteenth century. The period 1900-World War One saw increasing inequality. The War administered a brief dose of equality but this was eaten away in the 1920s (1929 level roughly equal to 1910). Their strongest result, which holds no matter what technical adjustments they make to their model, is that equality steadily improved from 1929 to 1950. Thus, while the 1920s may not be unique, they did, along with the immediate prewar period, experience the worst inequality in the last 130 years.

Williamson and Lindert attribute much of the observed trends to unbalanced technological change which aided skilled workers disproportionately before 1929 and unskilled after. In the earlier period, total factor productivity growth was most rapid in skill-dominated industries like automobiles, appliances, tires, and electric utilities. While they do not mention it, it would appear likely that market saturation in some of these sectors would have contributed to the post-1929 trend toward equality. Slower population growth also contributed to greater equality. As capital accumulation tends to benefit skilled workers more, decreased investment aided the move toward equality as well (1980: 247-9, 289).

The increasing inequality in the 1920s likely affected the economy adversely. Concentrating income increases in a relatively few hands increased the difficulty those individuals faced in finding outlets in consumption for it. The rest of the population, which might have had many uses to which it could put additional income, was prevented from adding to effective demand. Thus, while some industries only faced demand problems after they had supplied almost the entire population, others faced difficulties at a much earlier point. Leven (1934), while recognizing that conclusive answers about the effects of distribution on growth were impossible, calculated that "reasonable" improvements in distribution, such as a minimum family income of $2500, would have increased consumption beyond existing capacity. He noted that families with incomes below $3000 were not spending the recommended minimum amount on food, that nearly half the expenditures on non-necessities were made by the top 10% of income earners, and in particular that amounts spent on personal transport, education, medical care, and recreation and travel rose rapidly with income. A 1929 *Business Week* survey estimated the average propensity to consume was 94% for those with incomes below $1000, 83% for those with incomes between $5000 and $10,000, 50% for those with incomes between $100,000 and $150,000, and 6% for those with incomes over $1 million.

Many in the 1930s felt that the income distribution had worsened in the 1920s, and that this had important deleterious effects (while Mitchell (1929) and Copeland (1929), perhaps fooled by the widening market for consumer durables, felt that workers had at least held their own in the 1920s). These included Paul Douglas, Major Douglas of Social Credit fame, and Foster and Catchings (Fano, 1991: 284). The much maligned Temporary National Economic Committee argued this case forcefully. Scoville and Sargent, in their attack on TNEC, refused to believe either that this could happen in a competitive environment or that it would hurt consumption (1942: 333-6). TNEC, interestingly, concluded that increased competition was the best way of ensuring improved distribution in the future (1941: 22). Taylor (1933) concurred that productivity increases had not been passed on to labor; this had caused savings rates to rise to a level which investment could not match in the long run in the absence of population growth or product innovation. Bell (1940) felt that both capital and certain segments of labor had received the gains from increased productivity, rather than the entire population as would happen if prices fell as they should. This, Bell argued, naturally depressed both consumption and employment (in part due to the inefficiency associated with price rigidity).

In the long run, distributional issues might appear only to affect the timing of the downturn. If, however, the problem of the interwar period truly was that new products were introduced too slowly, then an improving income distribution (both level and trend are important) could have brought the growth in consumption demand back in line with income growth and averted disaster.[8] Some writers (e.g. Waters, 1976) have suggested that the Depression was largely due to the worsening distribution.[9] Many (e.g. Hughes, 1987: 432-3) suggest that it is at least partly at fault. Worsening distribution is likely not the sole cause of the Depression. In concert with other factors it probably played a role. It certainly did not help matters.

Allen (1952: 286) has suggested that "the great American discovery" was that the economy functioned much better in the postwar era precisely because the income distribution had become much more equal. While the trend was regressive in the 1920s, there was a dramatic improvement overall from 1900 to 1950. In dress, hiring of servants, purchase of consumer durables, and various other ways, differences between rich and poor had been narrowed (Allen, 1952: 220). Economists who proffer a demand-driven explanation of the postwar boom (pointing to veteran's benefits etc.) neglect the fact that income growth can collapse, as it had after previous wars, if consumption does not keep up. Along with the creation of a range of new products, the improvement in the income distribution (and the demographic changes discussed below) acted to ensure that consumption expenditure rose with income.

Finally, we should note that by modern standards the bulk of the interwar population was exceedingly poor. Works such as Wandersee (1981) or Cowan (1983) provide much insight into the widespread poverty of the era. Nor were these people satisfied with their plight. A Yale survey in 1928 found the not-surprising result that people at all income levels found their salaries inadequate to support the standard of living which they aspired to (Wandersee, 1981: 22). Wright (1987) argues that we must compare incomes in the 1920s with incomes before rather than after. As poor as they were, people were better off in the 1920s than they had been before the War.[10] Cowan, who argues that half of American families were below the poverty line in the 1920s, recognizes that sheer subsistence was not the problem it previously had been (1983: 181-2). Widespread house and car ownership, electrification, and running water were among the gains made in the 1920s.

We will appear in retrospect to have been abysmally poor to our descendants, I suspect. Life before VCRs, color TV, and pocket calculators already appears primitive. This does not mean that our economy can not suffer from deficient aggregate demand from time to time. Likewise, the fact that people in the 1920s were poorer than us does not mean that at times they had little desire to consume out of incremental incomes, given the prices and range of goods with which they were faced.

Once the Depression began, consumption demand would receive a further blow. The bulk of the unemployed would have been perfectly happy to agree to spend every penny they could earn on the existing basket of goods, rather than receive no income at all. An omniscient central planner might have been able to get them back to work on such a basis. The market, of course, separates choices about spending and working. Once a population is unwilling to spend all of its income on the existing range of goods, some unemployment will result. Then, the multiplier effect kicks in, and still more are let go. Through the early 1930s, if those who still worked and thus had purchasing power provided no market for increased production, those who were unemployed could not create their own effective demand.

A Hierarchy of Goods?

Stanback (1981) has proposed a tripartite division of human needs. First, basic needs such as food, clothing, and shelter are satisfied. This is followed by convenience needs: goods and services which make life easier are sought. The third stage involves lifestyle/identity needs; it is at this stage that humanity pursues the ultimate goal of realizing human potential. As this requires time, it helps explain the importance of labor-saving consumer durables in stage two. The interwar period might be viewed as witnessing an important shift upward in this hierarchy. Such a transformation need not happen smoothly. In general, "People tend to have poorly developed preference functions for things they have not tried. Accordingly, the dynamics of changes in preference functions will involve adaptive and learning processes" (Stanback, 1981: 33). Keynes (1930) realized that consumers would have to learn "how to use his freedom from pressing economic cares, how to occupy the leisure..., to live wisely and agreeably and well." Producers and consumers both would have to feel their way. "A new world has been created within ten years. It must be explored, charted, and business adapted to it" (Calkins, 1932: 14).

Linder has questioned the degree to which people in the modern era have used their newfound free time to pursue self-fulfilment (Clewett notes the expansion in amusement parks and country clubs in the 1920s). Idleness actually decreases with income; despite the ideal of freedom from activity those who can most afford it rarely pursue it. As productivity has risen in both paid work and housework, it is natural that people have sought to increase the value of their leisure time as well (one should on the margin equalize the value of time spent on all activities). We can pass over the varied implications of his analysis — less time devoted to courtship, foreplay, reading books etc. The important point is that people will tend to concentrate leisure time in activities in which the value can be increased. Reading books provides little scope for this. Watching television can be enhanced with better equipment, as can listening to music. Linder does feel that there is a certain degree of irrationality involved in the choices people have made, and feels society must strive to regain its appreciation for self-actualization.

While the goals of the human spirit appear less laudable here, we still get a hierarchy of goods. There will be a threshold level of societal productivity necessary for widespread adoption of both new labor-saving consumer durables and consumer goods associated with leisure. Producers will be faced with continued difficulty in predicting how consumers will optimize as the value of their time increases. As the productivity gains of the 1920s were not passed on to the bulk of the population, it would not seem on the face of it to be a period in which this problem would have been unusually severe. In a longer term perspective, again, the early twentieth century does appear as a period in which non-necessities were greatly expanding their market. The rise and fall of durables could leave producers quite uncertain of where the market was moving. Moreover, the sluggish decline in the work week could have limited the market for time-intensive products such as sporting goods or travel.

Market Saturation

Some economists may object to the idea that markets could become saturated as argued above. Surely, they would claim, demand curves slope downward and therefore producers need only lower their price in order to expand their market. Yet the foregoing argument, it must be stressed, does not require perfectly inelastic demand curves. It merely requires that demand be inelastic, and thus that price cuts yield only slight increases in output. In industries where firms exercise considerable market power, theory tells us that they will not lower prices for then total revenue will fall. Even in more competitive environments — which would include oligopolies in new product lines fighting to establish market share — there will be a limit to how far prices can fall. While firms may be willing and able to incur small losses in the short run, they can not lower their price much below their costs of production especially in the longer run (and remember that real wages actually rose through the 1930s). Thus, there is very little scope for expansion in production across industries. Before prices could fall very far, firms would be forced to shut down.

If in such a world consumer durables are a large part of production (automobiles alone absorbed one eighth of industrial output in the 1920s) and their demand curves shift to the left, other sectors will simply be unable to expand to make up the difference. As output and employment fall (and we will discuss below further downward pressures on these), all demand curves shift further to the left.

It might be thought shortsighted of industrial leaders to have saturated their markets. However, in a competitive setting where firms were striving for market share this was not possible. General Motors executives began to worry about market saturation early in the 1920s, but without the cooperation of Ford and others could do little about it. Melvin Taylor, President of the First National Bank of Chicago, recognized this point (1930, in Angley, 1931):

> Business leadership, had it read the barometer properly, should have noted the storm that was gathering and trimmed sail accordingly, but ambition for place, power, and profit blinded leadership to the obvious dangers ahead and prevented the preparation of a safe harbor against the storm.

Some may still doubt that this argument can be transferred from the single industry to a general equilibrium context. This is far from the case. When we buy expensive items, we often think not in money terms, but in terms of how many hours it took us to earn enough to pay for the item. So let's think for a bit in terms of hours. The system works as long as people are willing to "spend" most of the hours they work, and thus desired saving equals desired investment. As productivity increases, this balance is interrupted. If we become five times more productive, we are not going to want five times as much food and clothing (even allowing for a shift toward better quality). We would no longer wish to spend our hours. We would eventually reach a new equilibrium with a shorter work week, but this would take time. New products, though, increase our willingness to spend our hours (and also increase desired investment), and thus can overcome the problem of excess saving. In the absence of new products, what at the aggregate level appears as excess saving appears at the industry level as market saturation. Once you have a car, how many hours are you willing to spend on a second car? Assuming away labor quality differences to make things easier (they could easily be incorporated, but this is just a heuristic exercise), the auto manufacturer can not lower the price below the number of hours of work embodied in the car (in the short run, the hours embodied in fixed capital will be ignored, but not in the long run). Thus, it is quite easy for a durable industry to be unable to sell as many units as the previous year at any price above cost. For non-durables, the case is less dramatic, of course. Still, the important point is that it is theoretically quite possible for full employment supply to exceed demand at the aggregate level, even if all firms lower prices to the level of costs.

The theoretical possibility of market saturation has been argued by many previous authors. Establishing its empirical validity has proven much more difficult. Mercer and Morgan (1972) and Bolch (1970) attempted to estimate the degree of market saturation in the 1920s and 1930s for automobiles and housing respectively. The lack of attention which their work has since received in the literature indicates that the profession feels that econometrics is incapable of satisfactorily providing such an estimate. We are thus left with the options of ignoring the strong theoretical possibility of market saturation, relying on a couple of very imprecise estimates, or expanding the scope of inquiry to include non-quantitative analysis. In Part Two of this book I pursue the latter course. In a series of industry studies, I will provide many indicators of market saturation (and also of limited investment opportunities due to the time path of innovation, and the employment prospects which resulted from all of these forces). In doing so I will be able to draw upon a vast literature of industry studies by scholars with industry-specific expertise. Given that time series econometrics has proven incapable of telling us why the Depression happened, this seems the best course to pursue.

Finally, what of exports? Saturated domestic markets need not be a problem if foreign outlets can be found. The simple historical fact is that exports did not surge. In a world of fixed exchange rates and much American foreign investment (though this falls off dramatically in the late 1920s as money pours into the stock market), exports might have exceeded imports by a wide margin for a while. Tariffs, on top of the natural barrier of transport costs, served to severely limit this. In the important case of automobiles, tariffs in European countries started near 30%. Ford had thus opened a plant in England before the First World War. Chronic unemployment was characteristic of Europe of the 1920s, and thus protectionist sentiment ran high. Different tastes (smaller cars) and standards (electrical goods) also placed barriers to exports. Moreover, the downturn in foreign economic activity, which in some countries began even before that in the United States (we will discuss why other countries suffered in Chapter 13), further limited the scope for exports.

A surge in exports to counterbalance falling domestic consumption would require not only the cooperation of foreign governments but a fall in prices. We have discussed above the limits on price cuts by American producers. Moreover, we will see in Chapter 13 that demand conditions across a range of goods were likely similarly inelastic in other developed countries.

Ours is not the only explanation of the Depression which needs to deal with the question of why increased exports did not pick up some of the slack. Ours, though, has the advantage of pointing to an important force which should have been acting to depress exports for years before 1929. Product cycle theory is now almost universally accepted; it predicts that new products will at first be produced near where they are invented, but as production technology is standardized and simplified it will move to lower wage countries. The United States in the 1920s was the clear technological leader and had the highest real wages in the world. Its position as an exporter depended on continued product innovation. American producers were quite ready to establish production overseas. If we are correct that there was a shortage of new product innovation in the late 1920s and early 1930s, then it is not surprising that export markets were softening just as consumption expenditure at home was collapsing.

We will accept through this book the general consensus of the profession that the American Depression triggered a worldwide Depression, though we will see in Chapter 13 that many of the "depressing" forces at work in the American economy in the 1920s were also at work in other developed countries. Since most other countries began their downturns shortly after the United States (and this soon evoked further protectionist

Other Interwar Characteristics of Consumption

Before leaving this section, we should consider some other potential determinants of interwar consumption behavior. The First World War might be blamed for causing a clumping of consumer purchases in the 1920s. Even in the United States, whose entry into the war only occurred in 1917, there was a period of a couple of years of economic dislocation and concentration on war production, which must have induced the delay of some consumption decisions. This can only be at best a partial explanation. The Second World War causes an even greater dislocation, but consumption expenditure expands steadily for decades after that conflict.

Contemporaries often commented on the popularity of both instalment buying and advertizing during the 1920s. These were not new phenomena and would become even more prominent later. More importantly, their popularity should be viewed as a result rather than a cause of market saturation.

The ability of consumers to borrow against the future in the interwar period was much less than today. In 1926-1927 only 12-3% of credit purchases involved a term longer than one year, though longer terms were more common earlier in the 1920s. Only 5% of instalment purchases in 1927 called for less than a one-third down payment (Copeland, 1929: 391). While instalment buying was only 15% of retail sales, it was involved in over half of all car sales, 75% of radios, and 90% of vacuum cleaners, and was important for furniture and other consumer durables. Of the total of $5 billion in instalment borrowing, $3 billion was for automobiles, $7-800 million for furniture, and about $100 million for washing machines. Instalment buying had existed for decades for farm machinery, furniture, and sewing machines. Its effect was to reduce sales resistance somewhat, though advertizing likely had a bigger effect in fostering expansion in most of the industries affected (Copeland, 1929: 401).

The total amount of retail instalment credit fell by two-thirds between 1929 and 1932-1933. It then increased by 50% to 1937. Credit granted fell even more rapidly than total credit outstanding due to delays in repayment of existing debt. The decline was due both to a decreased demand for goods typically purchased on instalment, especially automobiles, and the increased difficulty faced by many consumers in qualifying for credit. The vast bulk of instalment credit was extended to consumers with annual incomes below $5,000.00. Thus, while total credit amounted to 6% of the income of those who used it in 1933, and 11% in 1937, Holthausen

(1940) is able to conclude that instalment credit only played a very small role in exacerbating the cycle.

There was a considerable expansion in advertising expenditure during the interwar period. Some of this may be due to technology; the invention of radio and superior methods of reproducing photographs in newspapers. However, many writers, both in the 1920s and since, have treated increased advertising as a response to problems of market saturation. "Until comparatively recent times, the problem of industry was to produce in sufficient quantity to supply the demand. Today the problem of industry is largely that of disposing of its products" (Kimball, 1929: 81).[11] Krooss (1972) has attributed both the advertising boom and increased use of instalment credit to excess capacity resulting from slow market growth, and the rise of consumer durables which quickly saturated markets.

This is also indicated by the growing importance of style changes. Most readers will be familiar with the story of the Model T, and how in the late 1920s even Ford was forced to realize that yearly style changes were necessary to compete in the auto industry. The necessity for stylistic diversity was felt across many industries at that time, even though this interfered with the pursuit of economies of scale in production. Many companies set up special offices to match markets with products; "Without belittling the need for economy in production, the experience of recent years has especially emphasized the fact that the first requisite of success for an individual company or for an industry is to be able to cater effectively to the demands of the consumers" (Copeland, 1929: 329). Chase bemoaned the inability of firms to decrease costs of production; "But the hands of management are not free. The technician is constantly undone by the sales department, which, floundering in a pecuniary economy, sees no other way —and indeed there is no other way — to maintain capacity than by style changes, annual models, advertising misrepresentations, and high pressure merchandising." (1932: 29). He recommended government planning as a solution.

Frequent style changes and large advertising expenditures have characterized the postwar boom years as well. Writers such as Galbraith have claimed that the continuance of economic growth depends on the artificial creation of consumer desires through advertising. Without needing to take such a position, we can still recognize that increased advertising does not necessarily mean that a depression is imminent. It does indicate that difficulties in keeping consumer demand in line with income growth is a characteristic of the entire modern epoch. If we accept that the 1920s mark a turning point where the role of non-necessities came to play a significant role in total output, it is perhaps not surprising that entrepreneurs of the time were not able to fully cope (collectively) with the new world. Several decades later, the world has adjusted — perhaps in part along lines suggested by Galbraith. The key to success, though, has been that the range of consumer goods has widened steadily since the interwar period. This has served to keep consumer demand increasing. It is not, then, that the mere existence of advertizing or style changes, or instalment buying, is evidence of a sick economy, but that their simultaneous emergence in the late 1920s is symptomatic of an important, and dangerous, structural transformation.

We should recognize that the relationship between advertizing and economic growth is still a matter of great dispute. One school of thought holds that by providing relatively inexpensive information to consumers, it encourages new firms to enter industries. An opposing school contends that advertizing aids market consolidation and discourages entry. The empirical evidence is unclear (Albion, 1981: 69). It is also not clear whether advertizing has a major impact on the allocation of demand between sectors. Galbraith has suggested a major impact, but many empirical studies show it to be relatively unimportant. Advertizing may slow a decline in demand or increase the rate of growth, but these will largely be due to other factors. Advertizing may have played some role in the structural experience of the interwar period, but not a dominant role.

The Decline in Investment

> Theorists tend to analyze what can be understood with the methods available and, unfortunately, this may lead to concentrating on less important but more tractable questions at the expense of what really counts... The problem here is that what has been left out is neither inessential nor a detail. It is an absolutely fundamental aspect of the capitalist industrial system, namely, that to work properly the system must grow, and to grow it must continually transform itself through the introduction of new products and new processes, creating new life-styles, redistributing incomes and generating new markets.
>
> Edward Nell (1988: 159)

We have seen in the previous chapter that we lack an acceptable explanation for why aggregate investment should have fallen from the late 1920s. Bernstein shows how the proportion of total investment varies across industries over the interwar period (1987: 41). By looking at proportions, the inter-industry effects are isolated from macroeconomic impacts on total investment, though industry-level effects can themselves determine the aggregate level. These numbers indicate that much of what was happening occured at the industry level. It would seem that the causes of these wide and longstanding inter-industry differentials in investment are to be found in differences in opportunities rather than in access to credit. Some industries, such as oil, were, after all, able to borrow more in the

1930s than they had in the 1920s. Both Rostow and Hansen have referred to the lack of investment opportunities (Aldcroft, 1977: 272-8). Gordon (1961: 449) attributes stagnation in the 1930s to the fact that investment opportunities were so thoroughly exploited in the 1920s.

Growth theory often focusses on a sort of growth in which every element in the economy grows in the same proportion. Yet, one would think that such growth would tend to tail off, even if one ignored the problem of consumption, because existing plants would take a long time to wear out and thus investment would be sluggish (Nell, 1988: 162-4). "Only if allowance is made for the rise and subsequent decline of certain goods and industries can one get to the heart of the matter of growth" (Cornwall, 1972: 19).[12] Freeman (1982: 133-4) identifies depression as a period in which the industrial structure is fundamentally unbalanced by an increasing number of declining industries and a decreasing number of expanding industries. Klein (1977) has developed a model in which economic growth is not a natural equilibrium but the result of a coincidence of booming sectors. It should come as no surprise that in the real world investment rates differ markedly.

As R.A. Gordon (1949) suggested we need to look at investment opportunities across industries in the late 1920s. He recognized that this would in turn depend on consumer outlays by sector. "We need more study of the timing and extent of the apparent drying up of investment opportunities at the end of the 1920s. For each major industry what was the nature of the market situation and the prospects for further investment in 1928-29, and how did the changes which occurred during the 1920s affect developments from 1920 on?" He recognized the importance of comprehending the role of secular forces, and pondered whether World War One may have upset these. In Part Two of this book, we will survey in detail the investment opportunities which existed in 1929 industry by industry. It is useful here to establish in general why these would have been almost universally low.

Necessities; Demographic Change

The traditional necessities certainly provided little if any scope for entrepreneurial activity. As we would expect, expenditures on necessities were a shrinking proportion of national income. A *Business Week* study in 1932 found that whereas only 36.6% of national income had been devoted to items other than food, clothing, housing, and home operating expenses in 1919, this proportion had jumped to 48.4% by 1929 (Wandersee, 1981: 23). Investment in necessity production could not be expected to keep pace with the growth of the economy for long. Basic needs were increasingly satisfied, and consumers desired other outlets for their income. Food output grew by only 2% 1922-1928 and textile production by 9% (Sylos-Labini, 1984). "If manufacturing industry should devote all of its energies to the production of necessities alone, it would be difficult to dispose of the output intelligently. The problem of industrial production has been temporarily solved and as a consequence we have passed from a seller's market to a buyer's market" (Kimball, 1929: 81).

Not only were individual consumers unlikely to provide a rapidly growing market for these goods, but the number of people was growing much more slowly than previously. This was largely due to a tightening of immigration laws,[13] though partly also to the secular decline in the natural rate of increase.[14] Hansen (1938: 312-3) felt that the tailing off of population growth was a major force behind the fall in investment. Barber (1978) has analyzed population growth rates from 1870 to 1930. There was a sharp downward spike during World War One, and a downward trend after 1920. This is observed world-wide, but is especially marked in North America and Australia. In the United States, most of the decline occurs outside of the 15-64 age group, that which adds much more to consumption than production. Fair (1991) has found that an increase in the proportion of the population within this prime age group is associated with poor economic performance. Those with children can readily attest to the effect that the dependency ratio has on the propensity to consume. Just as some of the postwar economic boom might be attributable to the baby boom (some reverse causation is also possible), some of the interwar experience could be blamed on the negative effect on consumption of a falling dependency ratio.

The relationship between population growth and per capita income is ambiguous, but it would be expected to be positively correlated with investment. Svennilson found that in interwar Europe the countries with the fastest population growth also had the most buoyant investment and output performance (1954: 43). He notes that if population grows quickly all industries will grow. If population is not growing fast, different income elasticities for different products will mean that there will be a much greater variance in industrial growth rates, and some stagnating industries will be hit severely. Phelps (1993: 341) has argued that slow labor force growth also limits the demand for capital goods, and decreases the importance of young dissavers in the economy; he concludes that this was a major cause of high European unemployment rates in the 1980s. Krooss has suggested that producers were facing problems of excess capacity throughout the early 20th century due to the absence of the market-widening forces of population growth and railroad construction which had been so powerful in the nineteenth century. While we cannot be sure of the existence of chronic excess capacity, we could expect that investment levels must fall at some point.

Structural change is easier to handle in terms of employment as well if the population is increasing. The speed at which declining industries decline is lessened, and young workers can be steered toward growing industries (Pasinetti, 1981: 95). Henry Wallace, the Secretary of Agriculture, lamented in 1937 that it was more difficult to respond to technological change as, "The population is growing ever more slowly, so that some of the readjustments must be made by forcing adults out of the industries in which they had been trained." He noted approvingly that the Rust Brothers, the inventors of the cotton picker, had suggested that some of the profits from new machines be put toward alleviating some of the economic problems they create (1937: 18, 20). Immigration should, of course, affect both the demand and supply sides of the labor market. The supply-side effect was, however, outweighed by the surge of labor off farms. "Despite the restrictions on immigration, therefore, the supply of labor has been large in relation to the demand." (Slichter, 1929: 426).

We will look at food, clothing, and housing in greater detail in later chapters. Some new niches in processed foods were to be found through the interwar period. These would stimulate some investment. The agricultural sector itself suffered from oversupply at both national and international levels. Some new textiles would be introduced, but most textile manufacturers would have to rely on fashion-management for market extension. In the housing market, the secular trend was exacerbated by a boom in housing construction in the mid-1920s which saturated that market. Further, Barber (1978) has argued that the housing sector was especially hard hit by the slowing of population growth.[15] Thus, while the number of opportunities was not zero, they were severely limited. The economy would be sluggish at best if the only source of growth was expanded production of these goods. Stagnating industries can generate only limited investment to compensate for depreciation and firm exit. Even there, improvements in organization and distribution increased the efficiency of existing capital, so that replacement investment was reduced (Waters, 1976).

As the *Business Week* figures show, necessities had dropped from almost two thirds to barely one half of total expenditures in just one decade. This important transformation had been occurring at an accelerating rate for decades. It is important to recognize that the Great Depression occurred in a world essentially different from the eighteenth and nineteenth centuries. "In the nineteenth century, consumer perishables and producer durables dominated manufacturing. In the twentieth century consumer durables achieved an equal prominence. New products followed each other in rapid succession, and production was so large that in a surprisingly short time what had been a novelty became a commonplace" (Krooss 1972: 282). Improvements in the textile, iron, and pottery industries in eighteenth century England can be seen as increasing the range and quality and lowering the cost of traditional items of consumption. The mid-nineteenth century revolution in railroads and steelmaking served largely to enhance the ability of the economy to produce much the same mix of goods (often of superior quality). At the tail end of the nineteenth century, though, we enter a new era where the production of entirely new products plays a significant role in industrial growth. These include bicycles, automobiles, and various electrical and chemical products. The Great Depression occurs early on in this new era in which necessities no longer totally dominate the process of growth. While this opens up great new opportunities, it carries with it the risk of greater economic calamity. In the eyes of at least one contemporary, deficient demand could only become a problem in "an economy in which the output of goods is considerably more than enough to provide for the primary needs of the population" (Lough, 1935: 1).

Consumer Durables; Excess Capacity

After the interruption of War, the early 1920s saw a startling diffusion of new products based on new technology. "While the old established industries founded on the inventions of a generation ago have gone into a decline, the recent wave of prosperity was mainly due to the sudden flourishing of industries that grew out of the development of a general technical advance" (Sheldon, 1932: 95). In the words of Wolman there had been an "observed revolution in the things people buy and use in the country" since 1922 (1929: 13). Olney (1991) concurs that there was a consumer durables revolution. Mitchell (1929: 874) listed as basically new products in the 1920s cars (and accessories), trucks and tractors, radio, rayon, oil furnaces, gas stoves, various electrical appliances, watches, airplanes, and, of course, propeller pencils. Most of these "new" industries boomed spectacularly in the 1920s. "The rapid increase in consumer expenditure occurred, in the main, in the markets for goods which are more or less durable, and it was the swelling production of these goods which gave the period its characteristic tone" (Temporary, 1941: 60). A rarity in 1910, there was almost one car per American family by 1929. The same can be said of radios and others of the new consumer durables. It is noteworthy that electric power generation doubled in the United States between 1923 and 1929; virtually all urban households gained access to electricity in the 1920s. Consumer investment as a proportion of Gross National Product had been 4.4% 1899-1908 and 5.2% 1909-1918, but jumped to 9.3% 1919-1928, similar to the 9.8% recorded 1946-1962 (Cornwall, 1972: 25).

Booming industries based on the production of durables could, and did, saturate their markets. "The sales growth that occurred in the group of industries making consumers' goods of the highly durable type points significantly to the relatively high state of "consumers' inventories" that prevailed at the time of the 1929-30 collapse" (Epstein, 1934: 48). The 1930s were destined to be a period of contraction for the glory industries of the 1920s. Both Burns and Kuznets suggested that new industries were experiencing more rapid growth and decline than in the past (Lough, 1935: 223-4). They would be victims of their own success. Selling automobiles to a world where everybody already had one would be a more difficult task than cracking a fresh market. "American businessmen have been slow to grasp the fact that virtually no industry which starts out developing rapidly can maintain for long the same rate of growth. In consequence many American industries tend to be overbuilt" (Douglas, 1931). Investment would start to tumble before output, if there was either a decline in the rate of growth in output, or previously excessive investment. Krooss (1972) has described the rising importance of consumer durables, the quick shift from a virgin to a replacement market, and the resulting increased emphasis on non-price competition. Allen (1952: 138) speaks of the canonization of the salesman in the 1920s due to the requirements of the new industries. We have noted before that increased advertizing can be taken as symptomatic of deeper problems. No matter how hard they tried, manufacturers would face declining markets once the period of initial penetration had passed.

Even while they were booming, these industries had not driven the American economy to full employment, although they did come to employ millions. The "necessity" sectors we have discussed before were releasing labor (at least in the sense of not expanding employment as fast as the labor force grew) faster than these new industries could absorb it. Some of the new goods, notably rayon, accelerated the decline of traditional products through direct competition. The problems of the old industries were largely independent of the rise of the new, though. Rostow has noted that if not for the handful of booming industries, the United States (and France) would have likely experienced the high rates of unemployment of most of Europe in the 1920s (1978: 204). Similarly, levels of investment would have been pitiful throughout the 1920s if not for the booming new industries.

As investment and employment tumbled in the new industries, then, it reinforced the trends in the old, which had been partly, but not totally, obscured in the previous boom. The two sectors together yielded a substantial drop in both aggregate investment and employment in the late 1920s. If this were not compensated for by another generation of new industries with new investment needs, we would be well on our way into the Depression. In the words of Ginzberg (1939: 217-8):

Just as rapid advances in the construction and automobile industries swelled the national income, the retardation in their rate of growth was certain to depress the economy. The automobile had been especially stimulating, for it conditioned the growth of suburbs, the radical alteration in distribution, and the diversification of recreation.... Only the emergence of new industries, or the more rapid growth of old industries, could have prevented the decline in general economic activity...

Cornwall (1972: 22, 228) has argued that most recessions are mild because not all sectors decline at once. During the preceding boom, available investment funds will have been rationed in some manner. As some industries cut back, there will be others in which continued expansion is desirable. The 1920s appear then as an unusual period of growth. The speculative fevers of the period are easily attributed to a shortage of viable investment opportunities. The handful of new products achieved overcapacity without at all straining the financial resources of the economy. An economic boom that relies almost exclusively on a handful of sectors sows the seeds of future decline. There simply were not any sectors which had had their growth constrained in the 1920s, and thus the existing range of industries provided little scope for investment in the 1930s.

Process technology as well as product technology can be an important cause of investment. New methods of production can require not only new equipment but entirely new structures (e.g. developments in petroleum cracking). We can too easily forget that a good deal of investment in our economy simply renders obsolete existing capital which is far from fully depreciated. Its effect on aggregate demand in the short run is thus much greater than its effect on the long-term productive capability of the economy. New process technology need not, however, always require extensive investment. While that which occurred in the 1920s often did, we will find that process improvements of the 1930s generally did not. Moreover most firms adopted the new generation of machine tools early in the 1920s; these were much better than previous machines and improvements over the next years would not require machine replacement (see Chapter 8). Thus, there was little scope for process technology to stimulate investment in the late 1920s.

Chandler's (1963: 385) description of the development of the American corporation is instructive:

> Thus, four phases or chapters can be discerned in the history of American industrial enterprise: the initial expansion and accumulation of resources; the rationalization of the use of resources; the expansion into new markets and lines to help assure the continuing full use of resources; and finally the development of a new structure to make possible continuing effective mo-

bilization of resources to meet both changing short-term market demands and long-term market trends

For most corporations, the initial accumulation occurred between 1880 and 1914, and the initial administrative structures were put in place in the first couple of decades of the twentieth century. The rush to conquer the new markets in the first stage often led to the collection of excessive facilities and personnel (1963: 388). Expansion through diversification began for some in the 1920s, but for most was delayed until after the Depression. Diversification generally involved either new products from existing lines, or entirely new products from the research labs which possessed some complementarity in production with the existing product line.

While Chandler's focus was quite different from ours, we can nevertheless observe much of importance in his typology. Many firms had entered and consolidated their hold on new industries before the First World War. They were devoting their efforts toward rationalization in the early twentieth century. This could well have played a significant role in fostering the increases in productivity during the 1920s which we will be discussing below. For our present purposes the key point is that most of these firms had accumulated serious excess capacity. This boded ill for both investment and employment. The long-term solution was diversification, but most of these firms did not do so until after the Depression. If we accept Chandler's time frame, we must suspect that this was due to lack of opportunity rather than desire. This leads us naturally to the next question of why there were not new sectors with growth potential in the early 1930s.

The Shortage of New Industries

> New technology based industries did not spring from the labs to take up the economic slack left by the drop in the production of autos, steel, houses, and the other established products of an industrial society (Wise, 1985: 282).

Why would there not be other industries with growth potential which could pick up the slack? Kuznets (1966: 4-5) argues that the history of modern economies is full of industries rising and falling away. The obvious importance of the rise and decline of industries was widely recognized in the 1930s (the idea of growth retardation inspired some of the case studies we will encounter later), but was largely forgotten by the mainstream literature postwar with the rise of Keynesianism and as prosperity slowed industrial decline (Norton, 1986: 1). Mitchell (1929: 909) recognized that there were problems in a number of sectors. He wondered whether the prosperous industries would continue to prosper, and whether technological change would hold up in the future. He viewed national growth as the sum of industrial growth rates, and noted that the economy would expand after the discovery of new technology (1929: 845).[16]

Mitchell was correct in recognizing the connection between the growth of new industries and the discovery of new technology. Pasinetti (1981) feels that an alternative way out of the Depression was for producers and consumers to discover new wants which could be fulfilled with existing technology. He notes, though, that investment might lag considerably due to difficulties in predicting in which direction consumers are willing to move. Indeed, Lough argued that the real problem of the 1930s was that entrepreneurs were hesitant to invest because they could not predict consumer desires in this new non-necessity world; "Consumers, lacking the individuality and discrimination required in order to make diversified choices, tend more and more to join in a massed clamor for a small number of highly standardized commodities and services, and to swing violently from one favorite to another" (1935: 150).[17] Clewett (1951) speaks of the 1920s as an era of fads and novelties.

Calkins (1932) called for the development of a new discipline of consumer engineering. He defined the economic problem as "how to persuade a people to consume more goods." He noted with approval that some entrepreneurs in the 1920s had begun to cater to consumer tastes, rather than simply choosing what goods to produce. Simple changes in color and design had created new markets for bathtubs, kitchen ware, and fountain pens. "Obsoletism is another device for stimulating consumption...Wearing things out does not produce prosperity but buying things does...Any plan which increases the consumption of goods is justifiable if we believe that prosperity is a desirable thing" (six pages later, he asks if artificial obsolescence can be created). He urges entrepreneurs to get out and find new product niches; he feels they are everywhere. Sheldon and Arens build on these ideas. Obsolescence can not be a dirty word; the replacement must be superior or it would not be purchased. Junking a perfectly good car just because it does not suit modern tastes is not wasteful (1932: 56). They bemoan the lack of market research relative to industrial research. Consumer engineering must "create new needs and stimulate consumption" (1932: 55). They believe that examination of the evolution of consumer tastes, combined with the redesign of existing products with artistic sensibility, would engender an increase in production. Consumers would be induced to consume rather than save.

While stylistic changes could encourage more rapid replacement, and these authors are correct in predicting the growing importance of market research in modern society, they appear to horribly exaggerate the growth potential of taste manipulation alone. In particular, they ignore the problem of substitutes; marginally different goods will largely replace production elsewhere. They also neglect the role of technology; enamelling of kitchen ware and typewriters could only spur consumption after it became possible.[18] At times they believe, like Schumpeter (see Chapter 4), that there is a lot of unused technology ready to be harnessed.

Freeman concludes that taste manipulation will not work, and the only way out is through new technology (1982: 141); not just product but process technology, which allows existing products to reach a wider market, might work. We could hypothesize that if there were a lack of new product technology in the economy, the potential for new industry growth would be severely curtailed. Calkins' entrepreneurs could be expected to take a very long time to find enough truly new product niches to lift the economy out of depression. On the other hand, if there had been a number of dynamic sectors, they could have allowed the economy to avoid the Depression and continue to grow. Kalecki has argued that the natural rate of growth of a capitalist economy would be zero if not for some developmental factors of which the main one is technological change (see Bleaney, 1976: 246). Economic historians generally recognize that the dominant engine of growth in the world economy since the Industrial Revolution has been technological change. It is at least plausible, then, that the absence of new technology could underpin economic decline.

This point was emphasized by Hansen (1938: 313-4):

> The Industrial Revolution for a time gave rise to a great burst of investment in plant and equipment. There followed a period of readjustment and relative stagnation. Then came the railway age and another prolonged period of vast capital expenditures. The railway age ran its course and gradually reached a saturation point. Then came the electric and automobile age, and with it new outlets for investment, involving a host of accessory industries, including roads. Extremely important for the history of economic evolution is the emergence, growth, maturity, and decline of dominant industries. Thus the investment outlets created by technological progress have come not in a steady continuous stream, but in long waves of expansion and stagnation.

Gordon (1949) suggested that one of the foci of a study of the Depression should be those industries which led the recovery after 1933. We could then presumably try to understand why these industries had not grown faster earlier. With the advantage of four additional decades of hindsight, we could extend our analysis to those industries which would drive the postwar boom. "The industries that fared relatively well during the thirties were in most cases the same ones that secured larger and larger shares of national income in the postwar era" (Bernstein, 1987: 89). Why did the industries which did so well after 1945 not do well enough in the 1930s to prevent the Depression? If they truly did underpin the tremendous aggregate performance of the 1950s and 1960s they should have been able to do so in the 1930s. Of course, there is the ever-present danger of confusing cause and effect. Some sectors such as durables or construction may have been in secular decline until rescued by the war and the stimulus of the baby boom (Bernstein, 1987: 89). Yet if we look at the handful of industries that dominated the postwar period the role of technology virtually leaps out at us. Aircraft, computers, television, pharmaceuticals, plastics, and synthetic materials are all getting the vast bulk (sometimes all) of their sales from products which were not technologically ready in 1929. (Even the Corning Glass Company had 50% of its sales in 1950 from products which did not exist in 1940; Allen, 1952: 198). It is not at all surprising that they, and the variety of industries which supplied these with inputs, could not grow as fast in the 1930s as in the 1950s (independently of the decreased aggregate demand as other industries declined in the 1930s). By adding to the bundle of goods available to the American consumer, they were responsible for over two decades of nearly full employment. We will discuss in the next chapter the reasons for the shortage of technology in 1929.

It should be emphasized that some industries did succeed in the mid-1930s. Success was possible, even once the Depression was underway, if technological conditions were right; clearly the success could only have been greater if the technology had come on stream earlier when market conditions were more propitious.[19] In Britain, where new industries such as automobiles grew slowly in the 1920s, these proved capable of expanding rapidly in the 1930s, and thus explain much of Britain's relatively good performance in that decade. The replacement of old industries by new has received more attention in Britain than America. The British literature has tended to emphasize the beneficial effects on productivity which the new industries were supposed to have, through higher internal productivity growth rates as well as by attracting workers from low-productivity sectors. Von Tunzelman (1982) has found that these alleged productivity effects were insignificant, and suggested that the major impact of the new industries was to absorb resources which might otherwise have remained idle. The approach taken here is in line with that result.

An alternative to the technological argument would be to maintain that new industries grew sluggishly due to credit constraints. We discussed in Chapter 2 the major shortcomings of Bernstein's (1987) argument that this was the case. His evidence is far from conclusive. He points to the fact that financing from retained earnings became more important in the interwar period, rising from 30% in the 1920s to 78% 1934-1937. However, part of this was due to a secular trend, and the rest could be attributed to decreased investment opportunity. Again, Bernstein focusses only on recovery, and thus has little to say about financial restriction in 1929. He notes that the Depression itself depressed both savings and retained earnings. He also speaks of tougher securities regulations after 1933 which made it harder to float equities (though the cost of bond issues fell), and corporate taxes later in the decade. Implicitly recognizing that bank finance was still an option,[20] he argues that the new industries were those which had relied most on stock issues in the 1920s (Bernstein, 1987: 108-9). Bernstein does not explore the forces behind the creation of new industries (as we must). As well, Bernstein himself is forced to admit that in some industries investment does recover quickly in the 1930s. Among these are aircraft and chemicals (1987: 30-2). They must have financed the investment in some way. Industries, we might expect, should have a strong ability to obtain finance at the beginning, and early high profits should encourage further supplies of capital.[21] We discussed in Chapter 2 the fact that the evidence on credit markets is compatible with either restricted supply or restricted demand. Highly aggregated analysis can not determine which was the case. Again, the answer can only be found at the industry level. In Part Two, we will show for individual industries that it was not primarily a credit constraint but a technology constraint that held back investment.[22]

This is not to deny that the downturn itself did not depress investment once it set in. The imbalance between declining and rising sectors in the early 1930s created an environment which would slow new industry growth as technology became available. Beyond the obvious aggregate demand impact, we should not neglect the fact that some firms which might have been sources of investment demand were forced into bankruptcy. Falling price levels induced bankruptcy in two ways: the real value of debt increased, and the real wage increased (Casson, 1983: 76-8). Bernstein also discusses how the Depression may have skewed political decisionmaking (1987: ch. 7). Since new industries are naturally politically powerless relative to established industries, the NRA codes may have impeded growth by the former.

Inventory Investment

We would expect that producers in many industries would have found their inventory holdings increasing to undesirable levels in the late 1920s as sales fell below expectations. They would then be expected to reduce inventory levels in the early 1930s (we saw earlier that changes in distribution methods may have provided further downward pressure on inventories). Indeed, the period 1929-1935 is characterized by inventory liquidation (Hughes, 1987: 461).

Much recent research has been devoted to the puzzling fact that production is more volatile than sales for virtually all industries (Blinder, 1991: 76). Firms tend always to accumulate inventories in upswings and unload these in downturns. Yet if they face increasing costs of production, they should have an incentive to smooth production and accumulate inventories in bad times. Arguments based on the costs of holding inventories, or on firm decision rules, are possible but have not yet taken the profession by storm. Another line of argument suggests that firms may not in fact face increasing marginal costs, and thus have no incentive to smooth production. Ramey (1991) has suggested that if firms have excess capacity, hoarded labor, implicit contracts in which a certain amount of overtime is expected, and/or can increase the return on capital with double shiftwork, they can find that cost per unit falls as output increases over the relevant range. They should then purposely bunch production. While the last two criteria may be unimportant in our period, we have seen that the first was a major characteristic of the 1920s, and will see in Chapter 5 that labor hoarding was likely also widespread in that decade. Thus, producers may have faced a greater than normal incentive to overproduce in the 1920s and then cut back in the 1930s.

A Note on Fiscal Policy and Export Demand

Aggregate demand has four components: consumption, investment, government spending, and net exports. The total of government spending plus exports is virtually unchanged 1929-1931 and only drops by 10% from 1931-1933. Thus, we seem correct in focussing on consumption and investment. The early Depression was characterized by a lack of active fiscal policy (Temin, 1976: 143). Tax reductions and veterans benefits which had a mild stimulative effect in 1930-1931 were not introduced to fight the Depression (Peppers, 1973: 202). In 1932 taxes were increased in order to balance the budget (and restore banking confidence). From a Keynesian perspective the government might be faulted for not stimulating an economy in decline. Yet the government merely acted as previous governments would have, and in line with the wisdom of generations of economists (Bernstein, 1987: 9). While arguing that fiscal policy in the 1930s was contractionary, Temin (1989: 28) succeeds only in showing that it was not expansionary. Hoover, though blamed for doing less than Roosevelt, actually did more than ever before; Roosevelt criticized him for not balancing the budget (Higgs, 1987: 163). While it perhaps could have done more, the government cannot be blamed for causing the Depression through fiscal policy.

The government eventually did become more active. Hoover's committee on recent economic changes had in 1929 advocated public works as a tool to deal with future recessions. Congress was willing to appropriate billions for this purpose in the early 1930s but Hoover felt this should be done by state and local governments (he urged these in vain to at least maintain previously planned public works). Most cities did launch some work relief efforts, but their funds were limited. In rural areas, local government was either nonexistent or incapable. New York State introduced a program in 1931, which was copied by others. The belief in balanced budgets, and local responsibility, faded as the natural equilibrating mechanism failed to work, and local resources were run dry. The Reconstruction Finance Corporation was set up in 1932 over a Hoover veto to aid state relief efforts. Roosevelt's New Deal finally established direct federal expenditures on relief.

After 1933, policy may have come to play some role. Roosevelt believed that the Depression was due to structural problems and doubted that stimulative spending would solve the problem; he viewed relief expenditures not as pump-priming but rather as direct support of those most in need (Peppers, 1973: 204). One fifth of government expenditures 1933-1941 were on relief. Social security taxes, along with a levelling off of expenditure, caused government policy to become less expansionary in the late 1930s. Total government deficits in the 1930s amounted to only $25 billion, a figure exceeded almost yearly during World War Two.

Galbraith feels the New Deal had a positive impact through improving the income distribution (Allen, 1952: 154). Temin (1989: 96-108) has recently suggested that the New Deal marked an important regime change. While it, like Hitler's policies in Germany, were not very expansionary in a Keynesian sense, they induced people to have expansionary expectations. Beyond difficulties with any expectations-based explanation, it is worth noting the different perception of the New Deal from that of Roosevelt himself. The major impact of the New Deal may have been exactly what Roosevelt envisioned; by 1938 government relief efforts employed 10% of the labor force. Although the Works Progress Administration worked through private firms, other relief efforts employed workers directly. Others (e.g. Bernanke in Chapter 2) have disputed that the New Deal played a positive role in recovery. It is possible that government interference slowed recovery somewhat. Saint-Etienne even feels the government deficits of the period crowded out private spending, and applauds the restrictive fiscal policy pursued by the British (1984: 37-8). On balance, given the limited scope of the New Deal,[23] it would seem the entire American Depression can be viewed as largely the result of market (and monetary) forces. However, many, including Bernstein and Hughes, attribute the final end of the Depression to war-related government spending.[24]

We discussed above why exports did not expand to replace the decline in domestic consumption. Could the American Depression be blamed on falling export demand? There is widespread agreement in the literature that the American Depression was homegrown. The international situation was in many ways not propitious. The United States had grown faster than Europe in the 1920s, and this had hindered her performance (Mitchell, 1929: 857). Particular industries, such as rubber and textiles, were having trouble dealing with foreign competition. Once we recognize the structural problems of the American economy, the successful efforts of dying industries to achieve tariff increases become more comprehensible (in the 1930s, all countries would pursue tariff policies and/or devaluation to artificially increase demand). Whereas foreign trade accounts for 10% of American GNP today, it accounted for a mere 5% in the 1920s. Then, as now, the major trading partner was Canada, which was an even more dynamic economy in the 1920s. A slight decline in exports due to decreased European demand in the 1920s may have helped depress incomes, but the effect was dominated by movements in domestic consumption and investment. As noted above, even if international conditions were observed to turn against the United States, this likely was not an exogenous shock. The United States, as the technological leader, was exporting high-tech goods, and thus a fall in the rate of product innovation would hamper export behavior.

Notes

1. Expenditures on transport, household operation, recreation, and medical care now comprise half of consumer expenditure; they comprised less than 40% in the interwar period, and barely a third at the start of the century. The bulk of this increased expenditure is devoted to twentieth century products such as cars, radios, drugs, and appliances. Conversely, the percentage of income devoted to housing fell in the 1940s and only returned to interwar levels in the 1960s (about 14.5%), that for clothing declined from 12-4% interwar to 10% in 1970, and that for food, after rising in the 1940s, fell to 23% in 1970 versus 26-32% 1929-1939 (*Historical Statistics of the United States*, 1976: 316-20).

2. If consumers find limited utility from increased consumption, they might be expected to substitute toward leisure. We discuss elsewhere the reasons for a decrease in the workweek during the 1930s. Despite much talk of shortening it, the workweek had remained virtually unchanged in the 1920s. Institutional constraints might have prevented some workers from making the optimal hours-worked decision. Greater flexibility in hours decisions might allow smooth reaction to structural change in the economy in the future.

3. According to Pasinetti, as income rises, each increment of consumer demand tends to concentrate on a particular group of goods and services (1981: 79). If this was consumer durables in the early 1920s, we introduce a large dose of instability to the economy. Stigler (1956: 20-4) speaks of the increased homogeneity of the American market in the 1920s, due to decreased immigration and mass communications. A homogeneous national market carries the risk of an economy-wide downturn in average propensity to consume.

4. Consumer durables tend naturally to be hurt worst in any recession as consumers postpone purchases. Harberger (1960) argued that durable holdings depend on expected rather than present income, and that purchases in a year should be seen as replacing stock plus one-third to one-half the difference between actual and desired stock; this dampens the cyclical responsiveness of durables purchases. Bernstein asserts that there were secular dynamics at work as well and attempts to disentangle secular and cyclical factors statistically (1987: 30, 100-2). Romer (1992) argues that the decline in durables purchases is unusually large in 1929-30.

5. Economists and laypeople both tend to oversimplify the tradeoff between consumption and savings. Thus, they feel that consumption can squeeze out investment. Except at the height of expansion, though, there is considerable slack in the economy. Thus, if consumption of new goods stimulates investment, total output expands. Since the marginal propensity to consume is below the average (in the short run), this increase in total income will naturally act to bring saving and investment back in balance. Because of this, annual statistics must give us a restricted impression of the volatility of the propensity to consume as new products are introduced.

6. Goldsmith has studied saving behavior in the United States. He finds that, while saving is much more volatile than other economic aggregates, there is a discernible secular increase of about 2% per year in per capita saving from the late nineteenth century to 1929 (1955: 5). He feels that the volatility reflects the fact that saving tends to be a residual in household expenditure. He suggests the existence of twenty year waves in saving, with one peaking in the 1920s. The 1920s are characterized as an irregular high plateau for both national and personal savings. Olney (1989) has re-estimated savings rates, and finds these to have been lower in the 1920s (4.43%) than 1900-1916 (7.01%) or 1948-1974 (6.66%). Batra (1990) asserts that savings rates rose in the 1920s.

7. Ramey (1989) has recently shown that inventories of inputs and goods in process are highly volatile; changes in inventory holdings are not just a propagation mechanism for fluctuations but an important source of fluctuations themselves. The same could be argued for changes in finished goods inventories. Mitchell (1929: 865) felt that some of the decrease in inventories might be attributed to business caution. The observation by Temin that inventory investment falls less in the Depression than might have been expected may be due to this previous decrease.

8. It is possible that a more equal income distribution might have caused some industries, such as automobiles, to boom and decline even more spectacularly. A process of gradual distributional improvement could alternatively serve to stabilize sales. Sylos-Labini (1984: 236) thinks that both the housing and automobile booms resulted from the skewed income distribution.

9. Sylos-Labini (1984: 236) develops a model in which the deviation of real wages too far from the level of labor productivity causes a crisis. The two major impacts of wages on investment are through aggregate demand and costs of production. Wages falling far below productivity will result in deficient demand and a drop in investment. He also attributes speculation in stocks to the combination of excess profits and limited investment opportunities, "not the unexplained ex-

haustion of a wave of optimism, followed by an equally unexplained wave of pessimism."

10. Indeed, a Bureau of Labor Statistics study in 1936 suggested that standards in 1935-1936 were higher than in 1917-1919, as people could travel more, due to better and less expensive automobiles, ate better due to refrigeration and better knowledge of nutrition, and had better lighting (Wandersee, 1981: 37). See Allen (1954) for a description of how life was in 1900.

11. "There was little need for advertising in an economy of scarcity where manufacturers could sell all that they produced...Manufacturers no longer sought to sell a supply of goods, but rather to create a demand for their products" (Bryant, 1990: 190, 187). He dates the transformation to the turn of the century. Already by 1910 advertizing absorbed 4% of national income. Clewett (1951) notes that marketing research was ignored before the 1920s in large part due to the "comparative ease with which goods had been sold prior to the early twenties." From the 1920s, special researchers were employed to predict sales volumes. Major advances in market research occurred in the 1930s (see Albion, 1981).

12. Svennilson, in his classic work on the Depression in Europe (1954), also recognized that investment was not a simple function of aggregate output, but varied with the degree of structural transformation.

13. The War had heightened racist sentiment toward immigration, which had for decades been coming from non-Protestant southern and eastern Europe. Unions had long favored restrictions. With unemployment rising in 1921, business opposition was overcome and a bill was passed limiting immigration to 3% of the total of each nationality already resident in the United States. In 1924 this was lowered to 2% and migration from Asia banned. Further tightening occurred in 1927 (Fearon, 1987: 22-3).

14. The Temporary National Economic Committee's research monograph no. 22 recognized that this phenomenon had a technological root in contraception (1941: 175). Rowthorn (1990: 66) finds that since 1973 changes in population growth rates explain one third of cross country differences in unemployment in the OECD.

15. Terbogh (1942: 180) notes that 5 million new families were created during the 1930s. By the time saturation would have been overcome by this, depressed incomes themselves served to limit demand. Rostow has noted that the boom in automobile production accelerated the move to suburbia, and thus can largely explain the mid-1920s surge in housing and road construction as well (1971: 84-5).

16. G.C. Allen, introducing the series in Sturmey (1958) notes that "Their starting point was a proposition of great practical importance, namely, that continued economic progress depends upon the capacity of industry to introduce a rapid succession of new products and processes." The purpose was to understand the factors conducive to technological change.

17. Mass production technology may have exacerbated the problem. "The bulk of the output in many industries at the present time comes from large-scale specialized plants which cannot be changed quickly, and which have not heretofore maintained organizations which were prepared to cope without embarrassment

with swift changes in demand (Copeland, 1929: 328). He also speaks of changes in taste which resulted from the spread of cars, radios, and movies.

18. Despite a fall in demand for toilet preparations of almost one quarter in the 1930s, ongoing chemical research yielded a range of new perfumes, emulsifiers, preservatives, hair waves, depilatories, sunscreens, and cosmetics (Haynes, 1948: ch.19). This proved not to be the path out of the Depression.

19. We must be careful of the fallacy of composition. Simply because a couple of industries are able to grow during the Great Depression does not of itself mean that several could have if the appropriate technology were available. As always in economics, the action occurs on the margin. Each additional industry with growth potential is able to find a market for its goods among those who are still employed (even if this were to involve some shift in expenditures from other products, we have seen that industries losing their markets are slow to disgorge workers). This causes a net increase in employment, which, with multiplier effects, increases the market available for the next industry. Can anyone doubt that the path out of Depression would have been appreciably quicker if the jet engine, television, pharmaceuticals, diesel engine, etc. had been ready in 1930? If they were ready in 1929, the experience might have been avoided entirely.

20. Holthausen (1940) found that industrial banking companies expanded their loans until 1931. Kimmel surveyed 1755 corporations and examined banking and financial statistics for the N.I.C.B. in 1939 and found little evidence of credit restriction between 1933 and 1938. There were perhaps some firms whose ability to borrow was weakened by their experience during 1929-1933. In the forward, Conference Board President V. Jordan noted that, "Increasing the supply of credit funds and the sources through which they are available is meaningless unless there is an active or latent demand for them, based upon prospects of sound and profitable business expansion".

21. Seltzer (1928: 267-73) notes the difficulty the auto industry had in attracting external capital in the early days. Re-invested profits were the key to success. By relying on suppliers who were already equipped to supply other industries, the capital requirements of the auto industry were greatly reduced (parts suppliers at first simply used existing equipment to capacity).

22. Bernstein himself attributes the concentration of research efforts on nondurable goods to the existence of greater potential there rather than to credit restrictions; he also recognizes that the supply of credit is meaningless unless there is an active or latent demand for it (1987: 112, 120).

23. As an expansionary device. Reforms in such areas as labor law, financial regulation, and unemployment insurance (which Wallace in 1937 noted would have been rejected out of hand six years earlier) were of longstanding importance. Higgs feels that the Depression was the most significant modern episode for changing public attitudes toward the role of government; "Nothing had a greater effect on the twentieth century economy, polity, and society of the United States" (1987: 159, 196).

24. "We must not be misled by the propaganda that a war will give jobs to those now unemployed. War might produce prosperity and full employment for a time, but, as sad experience has shown, it is a very unhealthy and shortlived type of prosperity and leads only to great depression in the future" (Wallace, 1937: 25).

4

The New Theory: Technological Change

I have argued elsewhere (1991) that the English Industrial Revolution was a turning point in history primarily because from that point on technology has been the dominant engine of economic growth. This view, that technological change is the primary cause of rising incomes in the modern world, is widely held. Gomulka (1990: 19) has summarized the results of empirical studies of the sources of economic growth. While there is some dissent, he asserts that the primacy of technology is clear. It emerges as the main determinant of differences in growth across countries and over time. Its importance is even greater if it is recognized that technological change often induces investment. It is not a huge leap of faith, then, to suspect that the course of innovation may somehow be connected with the worst period of economic decline in the modern era:

> Although economists have seen fit to ignore or minimize its importance, technology is the true master of the modern world .... Not only is technology largely in control of the long-term trend, but it plays an important role in conditioning economic fluctuations of shorter duration. In the first instance, new industries depend upon technological progress, and new industries have always left their impression upon cycles of economic expansion (Ginzberg, 1939: 214-5).

We have seen in Chapter 3 that an understanding of the course of technological change is essential to our comprehension of the lack of growth industries in the 1930s. In this chapter we begin by differentiating between new product technology on the one hand and process technology which allows us to produce existing goods at lower cost on the other. While the precise mechanism by which labor markets were unable to clear in the 1930s is the subject of Chapter 5, we note here that there is good reason to expect that an abundance of labor-saving process innovation coupled with a lack of new product innovation would cause economy-wide unemployment. We show that in fact the interwar period was characterized by rapid process innovation.

Why should the interwar period have been characterized by so much of one sort of innovation and so little of the other? Schumpeterian arguments relying on entrepreneurial zeal are not compelling, nor are a number of other hypotheses which can be found in the literature on technological change. I argue that the main reason for this unfortunate combination of events is to be found by examining the path by which the technological potential of the Second Industrial Revolution was realized. Within this broader perspective it is no surprise that there was a temporal gap between the radio and television, the car and the airplane, or rayon and nylon, nor that automobile production and developments in the electrical products and chemicals sectors triggered major advances in process technology. Still, we must recognize that technical considerations alone do not determine the rate or direction of technological advance. Since many of the key innovations of this period emanated from industrial research laboratories, and these had only emerged in the twentieth century in the United States, it is natural to focus attention there. I argue that these laboratories tended to emphasize process over product technology much more in their first decades than at present, and that this served to exacerbate the Depression experience.

Product Versus Process Technology

One of the major shortcomings of previous technological arguments, including Schumpeter (1939) and Bernstein (1987), has been the failure to adequately distinguish between product and process technology. The former comprises innovations whose main effect is to create qualitatively new products. The latter comprises those innovations whose main effect is to decrease the cost of existing products. As suggested above, these two different types of innovation can reasonably be expected to have opposing effects on employment. Thus any effort to describe some general technological impact on employment, or on other aggregate variables such as consumption or investment, is destined to fail. This simple distinction will be central to the analysis of this work.

We should note, though, that process innovation need not always have a negative employment effect, even within the industry directly affected. If lowering the price of a good raises the quantity demanded by a greater proportion — demand is elastic — then total expenditure on the good increases, and employment may increase as well. The elasticity of demand for individual goods usually falls as price falls and quantity expands (we have seen that demand for most goods was inelastic in the interwar period). In the early days of the automobile, small percentage decreases in price yielded larger percentage increases in cars sold. As the market became saturated, this was no longer the case. What we perceive as product innovation can in fact be process change: the rendering of a product commercially feasible. In the early days of a product, then, process improvements can easily increase employment. As these continue, however, they have a steadily more negative impact on employment.

Alternatively, product innovation need not always have a beneficial effect on employment. While the definition of a good as "new" must imply that it is significantly different from that which has gone before, it may nevertheless substitute to some extent for existing products. Its production and distribution could well employ fewer than those it displaces elsewhere. The degree to which a new product increases employment depends then on its novelty, the degree to which it creates a new want among consumers.

Only in Part Two will we be able to establish that there were exceedingly few new products in the late 1920s. Also at that time we will be able to determine the full impact of process innovation on employment. Historically, high rates of productivity growth have been associated with high rates of output growth (Fabricant, 1945); in the 1920s productivity growth swept even industries with stagnant or declining output and this almost guaranteed employment losses. Since process innovation is reflected in aggregate productivity figures, we can discuss the broad trends in productivity here, and point to their technological roots.

Process Innovation in the 1920s and 1930s

There is no doubt that widespread process innovation occurred in the 1920s. While the number of workers in manufacturing was relatively constant through the 1920s, output increased 64%. For the period analyzed by Kendrick (1899-1953) total factor productivity growth was fastest in the 1920s. Woolf (1984) has revised Kendrick's estimate downward slightly, but the 1920s still see the most rapid productivity growth. While labor productivity would grow slightly faster in both the 1950s and 1960s, the rates of growth in capital productivity have not been duplicated.[1] Various estimates find that the rate of labor productivity growth doubled from 1% to 2% per year in 1920 (Fano, 1987: 252-5).

We are aware of the gains in productivity achieved by Ford's assembly line. Mitchell (1929) hailed conveyor belts as one of the basically new products of the 1920s. The counterpart to the conveyor belt for homogenous output was continuous processing; this technology spread through industries such as food processing and oil in the 1920s. Electrification, which allowed for the first time the development of self-powered machinery, was also of great importance.[2] Jerome viewed electrification as a major cause of mechanization; he noted that less than half of industry was electrified in 1919 but almost three-quarters in 1929 (1934: 20). The price of electricity fell 50% 1920-1929 while that of coal tripled (Woolf, 1984: 178). By 1929, almost all of urban America had access to inexpensive electrical power. Schumpeter (1939: 413-4) despaired of his ability to give an adequate picture of the ramifications of electrification; new plants were constructed in many industries to take advantage of the new power source. Plant layout could now more easily encourage rapid throughput. Motors could now be designed with speed and power to meet specific uses. The value added by the electrical machinery industry doubled between 1919 and 1929 (Rostow, 1978: 336). Largely as a result of this, horsepower per worker rose 50% over the period (Aldcroft, 1977: 196). Woolf (1984) notes that electrification greatly encouraged a bias in American innovation toward labor-saving (at least before 1929). Electrification, in conjunction with improvements in alloy steels for cutting and grinding, unleashed a new generation of machine tools in the 1910s; most industrial establishments retooled in the early 1920s (see Chapter 8). Among other innovations which served to increase labor productivity in the 1920s were tractors and excavating machines, and paint sprayers (Mitchell, 1929: 874). Despite the feeling at the time that the age was already mechanized, Jerome (1934: 45-8) discovered that most improvements involved the replacement of hand processes by machines;[3] most of the examples he cites involve electrical machinery.

While technological advance was central, there were also improvements in management techniques which further increased productivity. We could hope that this was aided by the growth of business schools; the first undergraduate program had begun in 1881 (Wharton) and the first graduate programs in 1900 (Dartmouth) and 1908 (Harvard); associated with this was, not surprisingly, a proliferation of magazines and journals on management. The revolution wrought by Frederick Taylor and scientific management from 1895, in which workers were approached like machines and it was ascertained whether various tasks could be performed faster or eliminated, had given way in the early twentieth century to psychological approaches which recognized that workers had non-economic motives, and in the 1920s to industrial sociology which stressed group dynamics.[4] Chase (1932) speaks of the efforts of trade associations to share information on cost-cutting measures, of new employment systems which encouraged quicker learning and lower turnover, and continuation of the long run trend toward standardization and interchangeability; others spoke of benefits from Prohibition (in Jerome, 1934: 38). Experiments with company unions might also have increased productivity (Bernstein, 1960: 172-3). On a more somber note, Jacoby (1985: 177) argues that some of the productivity increase during the interwar years was simply due to workers putting forth more effort out of a fear of losing their jobs in a world where foremen decided layoffs. Thus, some unemployment served to induce even more.[5]

Despite the availability of a massive amount of inexpensive labor, process innovation would continue in the 1930s. Output per man-hour in manufacturing rose by 25% in the 1930s, dipping only in 1932 and 1937-8; for the entire non-agricultural sector productivity fell to 1933 and then rose from then on (Baily, 1982). For manufacturing, Mills (1938) estimated that half the decrease in hours worked was due to rising productivity rather than falling output. Overall, national output was higher in 1939 than 1929, while employment was over two million less (Gill, 1940: 4). Per capita output was higher in 1940 than 1929, but unemployment four times greater. Some of this might be due to stagnation itself: a tendency to lay off the least productive workers and shut down the most antiquated plants first. Much of it, though, was due to a continued process of process innovation. A writer in the "Revue d'Economie Politique" in 1932 argued that "Mechanization is an essential element in the worsening of the Depression" (in Garraty, 1986: 5). Some elements of the NIRA were designed to restrict mechanization, including a tax on machinery (Jerome, 1934: 21). Bernstein has described how the food products, petroleum, chemical, and glass industries especially saw process innovation during the 1930s (1987: 140).[6] The Works Progress Administration undertook a series of studies of industrial productivity and re-employment opportunities in the late 1930s, for they recognized that new technology was exacerbating the unemployment situation (see Gill 1940: 5-6). Gill notes both the continuation of earlier developments such as the assembly line and continuous processing, and basically new process innovations. While revolutionary changes were relatively rare, there was a continuing process of detailed improvements to established techniques. "Their cumulative effect on labor productivity and their threat to the security of workers has been far greater than the occasional spectacular, revolutionary changes" (1940: 9). Still, tungsten carbide steel allowed machine tools to work up to 30 times faster. Increased knowledge of chemical processes cut the time needed to cure rubber tires in half.

Of special interest is the fact that much of this technology required little or no capital expense for introduction.[7] Thus, limited stimulus to investment came from these process innovations (Gill, 1940: 6-10). This point was emphasized by the Temporary National Economic Committee (1941: 179-80), which noted in particular the introduction of new alloys, instruments, and chemical processes. In some cases, the Depression itself spurred entrepreneurs toward greater efforts at cost cutting. Thus, Chase (1932: 26) relates that "A manufacturer of bricks told me recently of two smashing technical improvements which he would never have thought of when he was more prosperous" (alternatively, Cohen (1987) argues that the paper industry came to emphasize product innovation as the 1930s wore on). While paving the way for unheard of prosperity after the war, such innovation could only slow the return to full employment in the 1930s.[8]

The nineteenth century view that technology was the answer to humanity's problems had worn off in the early twentieth century as it was recognized that it could cause social, economic, and political problems (Williams, 1982: 1). The perception of technology as the cause of widespread unemployment in the 1930s encouraged many writers to point to the negative effects of modern technology (Pursell, 1979). Pursell speaks of an ambivalent attitude toward technology; both good and bad effects were recognized: "It was a deeply held belief that if such research could turn up just one device such as the radio or the automobile the resulting new industry would solve the problem of Depression." (only the overriding concern with immediate results prevented New Deal notions of publicly supporting research from becoming law). Though they (and Pursell) did not use the words, contemporaries were clearly aware of the differential impact of process versus product innovation.

Why This Time Path of Innovation?

Why should the interwar era have been hit by the unfortunate combination of substantial process innovation but limited product innovation? Isn't technological change a fairly random process such that both product and process technology should be spread fairly evenly through time? Can we really attribute the major Depression of all time to technological forces? To do so, we must be able to provide a reasonable explanation for innovation not occurring smoothly through time. We will, as in Chapter 2, survey the existing literature, and then from it develop our own approach.

There is widespread and increasing recognition that technological innovations do tend to cluster. Mokyr (1990) has compared technological clustering to the observation in biology that some periods see more rapid evolution than others. However, there is widespread dispute about both the reasons for clustering and the timing of clusters. Mensch (1979) has suggested that radical technological innovation clusters in depression as it requires significant social adjustments which are only possible then. Freeman (1982) has found that clusters of innovation occur during long-term upswings as well. The determination of the timing of clusters is rendered difficult by the fact that we must distinguish basic innovations, which either introduce a new product or a radically different method of production, from incremental innovations. We would expect theoretically, and observe generally, that a basic innovation will be followed by numerous minor improvements. Patent data, then, may show a burst of inventive activity when all that is happening is a number of (important but) slight improvements.[9] While we should recognize the overwhelming cumulative importance of incremental innovations, it is still obvious that not all innovations are equal. Establishing which innovations qualify as basic is a judgement call. Freeman relies on Baker's list of significant innovations to generate the finding above, but recognizes the difficulties in statistically establishing the timing of clusters. Only by detailed industry-level research can the relative importance of different innovations be accurately discerned.

In particular product lines there is of course a natural progress from product to process innovation. Once a product has been made commercially feasible, the profit-maximizing strategy (especially if there is competition) is to reduce production costs. This may be delayed at first as minor product changes designed to more clearly match market demand dominate innovative effort. At all times, there is a tradeoff between cost reduction and product flexibility (modern computer-controlled production technology has considerably reduced this tradeoff). Despite these caveats, in general we would still expect a period of productivity advance to follow a new product innovation. However, as long as the economy as a whole sees a relatively steady stream of new products, there is no reason why that which characterizes individual product lines should also characterize the aggregate economy.

Why might innovations cluster? Perhaps, the solution is to be found in business organization, which is more conducive in some periods than others to technological advance. The early twentieth century can be characterized as a period of industrial concentration, after all. It is not entirely clear, though, what form of organization is most conducive to new technology (see Lodewijks, 1990). Klein (1977) has argued that small, competitive firms are best. Entrepreneurs in a competitive marketplace will constantly strive for ways to beat the competition. Open organizations where diverse people interact freely and competitively will be the most innovative. The dynamic gains from competition will be much greater than the static gains. Unfortunately, as an industry matures, and its future become less certain, there is a natural tendency toward bureaucratization, which stifles innovative impulses. Dearden (1990) has developed a model which shows that larger firms are relatively poor at innovating because of the moral hazard and adverse selection problems besetting hierarchical organizations. The aggregate economy can remain innovative if new industries emerge, but individual industries will become dominated by routine. For Klein, "there is a good deal of evidence that imaginative scientists and engineers are being replaced by business school graduates and lawyers; that is, by people who perform the same function in modern societies as did genetic inbreeding in feudalistic societies"; he attributes the postwar boom to the existence of a large number of small firms competing with each other, plus a technological backlog (1977: 182-3).

Galbraith, on the other hand, has forcefully argued that in the modern world the requirements of technological innovation can only be met by large firms secure in their markets with considerable capital at their disposal. Competition reduces the potential gain from innovation. Recent studies have found that large firms do spend relatively more on research and development (Freeman, 1982: 42).[10] While Mowery has noted that the threshold firm size for undertaking research is not unduly large, it is still above the median firm size, and thus small firms are largely excluded. Some had hoped that contract research firms could supply research services to small firms, but the uncertainty surrounding research makes contracting difficult, and successful research efforts generally involve interaction between researchers and both the manufacturing and marketing arms of the firm.[11] Nor is this trend toward corporate research just a postwar phenomenon. Kimball (1929: 95) noted that, "A few years ago new ideas came into existence largely as the result of independent inventors experimenting empirically, or as the result of work in university laboratories. Today the idea of organized industrial research is thoroughly established both in university laboratories and in many industrial enterprises." Gill (1940: 10) argued that new technology in the 1930s had been primarily suited to large firms, as they had funded the research. Following Edison, General Electric opened an R&D laboratory in 1901, Du Pont and Parke-Davis in 1902, Bell in 1911, and Eastman Kodak in 1913. By 1921, 526 U.S. firms had done so. There is some evidence that oligopoly is the most conducive to R&D (Sawers, 1978: 38), perhaps combining the necessary financial resources with some competitive spur. On the other hand, it could be that research helps firms increase market share, and in industries where innovation is costly, is thus more a cause of concentration than a result (Sawers, 1970: 39-42, Reich, 1985: 3-4).[12]

It is possible that both positions might be partially correct. Large corporations may devote their energies mostly to incremental innovation. Development is a much greater proportion of Research and Development expenditures than is Research. There may still be an important role for small firms in basic innovation. A study of the seventy most important inventions of the first half of the twentieth century found that over half came from independent innovators (Basalla, 1988: 128). Sawers still feels, though, that new firms were a much less important source of innovation in the twentieth century than they had been in the nineteenth. Have innovative firms been absorbed into larger entities with the resources to pursue innovation, or have these made it harder for new firms to get their feet on the ground? Gort (1962: 4-6), studying diversification between 1929 and 1954, felt that companies had diversified largely into industries characterized by rapid technological change (this was an even more powerful determinant than the rate of growth of the target industry); he argued that this plausibly aided the transfer of much-needed capital to these industries. He also noted that firms with a large proportion of technical personnel were most likely to diversify (1962: 135). Thus, we might expect a generally positive impact on technological change.[13]

Nelson and Winter (1982) have argued that it is impossible to draw a simple relationship between industrial structure and innovation, as firms will act differently depending on their circumstances. Clustering might thus be due to the state of the economy. There may be a tendency to postpone innovation in the downswing. Schmookler has argued that innovation is largely demand-driven (though he recognized that this might not apply to basic innovation). Most of the controversy surrounds the idea of Mensch noted above, that a severe depression will encourage radical innovation. He argues that stagnation causes people to lose their identity, and thus become more open to radical social and technological innovation (1979: 90-101).[14] Firms in desperate straits may try anything. Van Duijn speaks of various barriers to innovation; these may be lessened in depression (1983: 131). Nelson and Winter have developed a model of evolutionary change which recognizes that one of the forces driving firm behavior is the urge to establish a routine. Only when profits are low will firms be likely, then, to pursue radical innovations. Likewise, Amendola (1988) argues that in times of uncertainty firms will increase their research to widen their options.

Against this could be marshalled the idea that high unemployment creates a bunker mentality, restricts creativity, and slows down the rate of structural change in the economy (Blinder, 1988). Certainly, at General Electric, some areas of new product research did not seem worth pursuing in the poor economic conditions of the 1930s (Wise, 1985: 294). While incremental innovation will likely be the greatest victim here (Kleinknecht, 1987), radical innovation may not be immune. Thus, it is not at all clear theoretically whether we should expect basic innovations to increase or decrease in a depression.

There was certainly a tendency for overall expenditures on research, and patenting behavior, to be curtailed in the early 1930s. The best estimate, based on examination of a large sample by Weart, is that total research employment fell 20% 1929-1933 (Wise, 1985: 301). Perazick (1940) suggests that there were large economies of scale in laboratories due to division of labor; thus this should have reduced discoveries by more than 20%. Over the entire Depression period, though, the research effort did increase significantly, though not as rapidly as it had in the 1920s. Laboratories rose from fewer than 300 in 1920 to over 1600 in 1931 and more than 2200 in 1938; personnel employed rose from 6000 to over 30,000 to over

40,000 at the same dates (Fano, 1987: 262). Research expenditures increased from $100-160 million in 1930 to $225-250 million in 1940 (Reich, 1985: 2). Perazick (1940) estimated that industrial research thus qualified as one of the 50 top industries by employment. The thirties thus appear to continue a secular trend toward increased research effort. Patents applied for, though, fell from 89,000 each in 1929 and 1930 to 56,000 in 1933 and then rose only to 66,000 in 1938.[15]

The fact that the rate of increase in R&D is less than in the 1920s or postwar period indicates that the Depression itself likely depressed overall research effort.[16] Mowery feels that the Depression did cause some retardation, but mostly among the larger firms (1983: 978n). In terms of laboratory foundation, rather than expenditure, the 1930s, while still inferior to the 1920s, actually look better than the postwar period (Mowery, 1989: 61). It is unclear whether the relative importance of radical innovation within the overall research effort increased during this period. Weart was unable from his sample to tell. Some companies, such as General Electric, may have expanded fundamental research, but the G.E. experience might be more due to personnel change than the Depression. However, since we know that the degree of cost overrun increases with the degree of technical advance sought (Sawers, 1978: 31), there is some reason to expect a reduction in radical research in hard times.

Some authors have tried to posit explanations for why the interwar years in particular would see so much process innovation and so little product innovation. Jerome suggested that a rise in the ratio of unskilled to skilled wages would encourage mechanization (machine-making tending to be a more skilled occupation), but himself noted that while this ratio rose to a 1920 height of 76.8, it fell to 70.7 in 1925 and 72.8 in 1929 (1934: 348). Others suggested that declining firms might intensify their efforts at improving productivity.[17] War is another possibility. Military research may have spillover effects, while civilian innovations may be delayed while the war is on.[18] Given the limited time period of American involvement in World War One, this effect may not have been large. These can at best be only partial answers. The same would appear to be the case for the more general explanations of clustering discussed above. If our approach to the Great Depression is to be convincing we need a more powerful explanation of what can be termed "The Technological Riddle."

Solving the Technological Riddle

We have touched on a possible solution above. A particular innovation may trigger a series of innovations along the same line. This, in a simple case, would occur entirely within a particular industry. However, it could also occur across a range of industries. Indeed, de Bresson (1991) develops a model of how revolutionary changes will work themselves through different sectors; technological systems thus rise and fall. Assembly line technology was diffused across a number of industries and sparked a series of industry-specific improvements in each. We might think of waves of mechanization, electrification, and automation. There is an element of technological determinism possible here. Whatever the causes of the original facilitating innovations were, the others naturally followed. These in turn fostered further innovative effort. If these particular "webs of technology" (Freeman 1982: 6-8) are found to be of major importance, we will have no difficulty in explaining a considerable degree of clustering. Yet clustering seems, at first glance, to involve more than just the bunching of interrelated innovations. Mensch has certainly emphasized the simultaneous emergence of diverse forms of technology. This requires a deeper form of explanation. Freeman has pointed to exogenous breakthroughs in science as a possibility, but a cursory look at the course of technological progress shows that it has generally occurred far from the frontiers of science, even in the twentieth century.[19]

Since our concern here is with the Depression, we need not try as other authors have to build a theory which explains long waves. We can treat the Depression as a one-off event. In this we can be guided by those such as de Bresson or Freeman (1989) who speak of technological paradigms which intermittently sweep across economies. De Bresson, in particular, notes that if there are waves there need be no regularity whatsoever in their occurrence. While we will speak briefly of the 1870s "depression" below, and of the 1970s in our conclusion, we will do best if we focus here on the technological forces which created the Great Depression.

The answer, once again, is to be found by stepping back somewhat in history. Many scholars refer to the last couple of decades of the nineteenth century as the Second Industrial Revolution, largely because of the emergence of radically new technology. Three "new" areas of technology are usually pointed to: chemicals, electricity, and the internal-combustion engine (see Cross and Szostak, 1995). All had antecedents of course. Production of chemicals had a very long history; still so many new chemicals are developed from the late nineteenth century that the use of the word revolution can not be deemed totally inappropriate. The previous centuries had witnessed a series of important discoveries concerning electricity, but only in the late nineteenth century, with the development of low-cost generation (and then transmission), as well as the electric lamp, did electricity become an economically advantageous power source for purposes other than communication.

We will take as a premise in this work that much (though not all) of the technological history of the twentieth century must be understood as the development of these three rich veins of technological potential. Indeed, much of the entire economic history of our century must be understood in this light. We will see that changes in industries such as food processing or oil can be traced to developments in our three groups. It appears that the most important innovation of the twentieth century which owes nothing to electricity, chemistry, or the internal combustion engine is the zipper (or maybe the self-winding watch): a handy item to be sure, but hardly the ticket to prosperity.

It should be noted at the beginning that all three are responsible for both product and process innovation. We are all familiar with the role played by the automobile industry in the emergence of the assembly line. The counterpart of the assembly line for the production of homogeneous output is continuous (as opposed to batch) processing. It is from the chemical industry that this concept spreads to food, oil, paper, and textiles in the interwar period (Mowery, 1983b: 966).[20] We have already had much occasion to see the far-reaching impact that electrical machinery had on methods of production. It would be both simplistic and wrong to argue along simple Schumpeterian lines (see below) that product improvements naturally precede and engender process innovations. This simple approach is most clearly mistaken in the case of electricity, for electrical machinery has its greatest effects from the beginning outside of the electrical products sector. All three forms of process innovation came to maturity in the interwar period.

The Second Industrial Revolution had diverse effects, many of which served to further enhance productivity. The industrial research laboratory itself first emerged in the electrical field in the United States, first rose to prominence in the world in chemicals (Mowery, 1988: 966), and through this century most industrial research has been done in chemicals, electronics, automobiles, pharmaceuticals, and oil (Reich, 1985: 256). Some of the developments in factory organization, as we have already seen, were due to the advent of electricity. Much of the study of management was devoted to the problems of our key sectors; the key breakthrough in industrial sociology in the 1920s resulted from a lengthy study undertaken at a General Electric plant.[21] The effects extended beyond manufacturing as well; tractors, chemical fertilizers, and electrification changed agriculture radically. The service sector too was transformed by electronic gadgetry. The automobile created new suburbs and gas stations, while almost destroying home delivery.

All three technological strands produced major new product lines long before 1930: automobiles, rayon, and radio for example, and long after the onset of the Depression: airplanes, pharmaceuticals, nylon, television and computers, to name a few. Given the much greater complexity of the airplane than the automobile, nylon than rayon, and television than radio, it is not at all surprising that decades would elapse between the one

The New Theory: Technological Change

and the other. The misfortune for the American economy was that gaps in product innovation occurred in all three cases. If the product innovations of the 1930s had come a little earlier, they could have absorbed much of the labor cast aside by process innovation.

Why then was the timing such as it was? Usher warned us long ago (1954) that while the order of stages of technological innovation is logical, the timing of any innovation is indeterminate. There is a substantial stochastic element to technological change. Even a heavily government-funded space program can not guarantee when a particular breakthrough will occur. And who knows when a cure for cancer will be found? The world of privately-funded commercially-oriented innovation is certainly no more predictable. If we can not predict accurately ex ante when a particular innovation will occur, we can hardly hope to completely explain ex post the course of technological change.

The situation is far from hopeless. We can isolate some of the determinants: the degree of technical difficulty, the amount of effort put forward, and so on. We can, especially, recognize that each innovation is a synthesis of pre-existing ideas. "Invention is a new combination of the prior art" (Gilfillan, 1935: 6). Technology develops through physical artifacts (Basalla, 1988: 30), and thus there is considerable logic to the order in which innovations occur. Rather than springing from the hands of some heroic inventor full grown, particular innovations generally can be traced to the efforts of many researchers over a lengthy period of time. The problem must first be recognized as worthy of solution, existing knowledge must be synthesized to create something new, and then the invention must be refined to allow commercialization (Usher, 1954). This does not happen instantaneously. Having identified three broad strands of technology, we can hope to be able to explain the relative timing of various innovations. Innovations can not emerge until the innovations they themselves are based on are brought into the world. Thus, we can hope in Part Two to provide reasonable explanations of why key innovations occurred when they did.

As well, as Chandler himself recognized, mass production itself encouraged a focus on process innovation. In the early days of assembly lines and continuous processing, the inflexibility of production itself discouraged major product innovation which would require redesigning the whole production process. At the same time, assembly lines and continuous processing could be improved almost endlessly.

Griliches (1990) has noted that while many have criticized Schmookler's hypothesis that it is demand-side forces which determine the rate of innovation, and while the evidence for it is weak, the hypothesis has not been overturned. Schmookler himself, though, recognized both that demand may have little influence on radical innovations, and that demand is at its most important in the short-run while supply-side forces are likely the key in the longer run. It is not our intention to totally denigrate the demand side. Certainly, once the downturn began, particular potential innovators may have become less keen on pursuing their dreams (or not, if Mensch is correct). Demand-side forces would also influence the direction of change. Since we have argued that the whole Depression experience can only be understood in a long-term perspective, we naturally focus on supply-side forces here. It is the technological potential gradually released as a result of path-breaking nineteenth century innovations in electricity, chemistry, and internal combustion which primarily determined the path of technological change through the interwar period.

We will see that the three strands often developed interdependently. "The three complexes were related in significant ways. The automobile (and truck) for example, depended on rapidly evolving chemistries of oil refining and rubber manufacture, as well as on the battery, spark plugs, and other light electrical gear. Electric power played a large role in the production of many of the new chemicals" (Rostow, 1978: 209). The discipline of History of Technology had its beginnings as a field in what Staudenmaier (1989) has termed internalist history, the tradition of detailed description of the evolution of techniques within specific fields. While the discipline has increasingly focussed on the social determinants of innovation, the essential interconnectedness of technological advance may not be fully appreciated.

The importance of cross fertilization can not be overemphasized. Within any narrowly defined area of research, we would expect that the rate of advance would tail off over time, as the boundary of technological potential is approached. Norton (1986), following Freeman (1982), speaks of Wolff's Law: "Every technical improvement, by lowering costs and perfecting the utilization of raw materials and of power, bars the way to further progress. There is less left to improve, and this narrowing of possibilities results in a slackening or complete cessation of technical development." This is an exaggeration; an individual research agenda may uncover unimagined problems or potentials, and thus open up new lines of inquiry. Invention is not an event but a process; its pursuit involves the creation of new ways of thinking (Aitken, 1985: 448). Since technology must advance in discrete steps, the solution to a particular problem will rarely herald the end of investigation: "A radically new technology does more than merely perform old functions better; it makes it possible to perform functions that the technology it replaces could not perform at all. It not only solves a problem; it oversolves it. It literally creates its own future." (Aitken, 11). Still, a much more prominent source of new lines of inquiry is the conjunction of two or more research agendas; radically new ideas are most likely to arise in this way (Sahal, 1985). In many cases, combinations of this type may be intentional (see Constant, 1980: 14). Research on materials may be pushed precisely because it is recognized that developments in aeronautics require lighter and stronger materials. More often, though, investigations in one field generate results which have huge but unforeseen effects elsewhere. Cross-fertilization of ideas, across and within our three technological strands, along with the injection of new scientific discoveries and interaction with longstanding technological traditions, therefore has allowed technological investigation to expand its scope for over a century.[22] For our present purposes, the interrelations between our three strands renders the unfortunate interwar timing of innovation more comprehensible.

A Further Step Back

It might be wondered if our three technological strands could be traced to some common ancestor. That question is largely beyond the scope of this work. It could well be, though, that the expansion of education, especially of a scientific/technological sort, during the nineteenth century was important for all three. As I have argued elsewhere (1991) the innovations of the English Industrial Revolution owed very little to scientific enquiry; even advances in metallurgy and chemistry were due to trial and error experimentation without knowledge of the scientific principles involved (i.e. making steel without knowing that this requires a 2% carbon content). The late nineteenth century innovations could not have progressed anywhere near as rapidly as they did without a greater respect for and application of scientific and engineering principles.[23] Certainly many authors have attempted to attribute much of Britain's relative decline late in that century to her failure to develop a technically-oriented educational system (for example, Layton, 1978). The installation of such a system in other countries may be due in part to a political desire earlier in the century to catch up to Britain.

For whatever reasons, the second half of the nineteenth century does witness a flowering of scientific knowledge. Mowery and Rosenberg (1989: 22-3) suggest that 1859-74 may be the period with the greatest concentration of scientific breakthroughs in history: Darwin, Mendel's discoveries in genetics (ignored for three decades), Pasteur, Kekule's discovery of the benzene molecule, Mendeleyev's periodic table, Maxwell's work in electricity, Gibbs' work in thermodynamics. They feel that technological development had greatly stimulated scientific inquiry in this period (1989: 35). By following this argument, one can trace the causes of the Second Industrial Revolution back to the First. While this is a demand-driven explanation of scientific advance, it could be combined with supply-side explanations of the expansion of scientific capability. Their line of argument does fit well with the idea that particular discoveries can have unforeseen effects. As well, we can note that railroads were a late development of the first Industrial Revolution, following earlier developments in steam engines, ironmaking, coal etc. They in turn greatly stimulated the development of electrification (telegraphs) and internal combustion (the example of the steam engine, plus the clear attraction of personal travel), not to mention steelmaking. Developments in textiles greatly stimulated experimentation in chemistry; research on dyestuffs provided the basis for most chemical discovery in the twentieth century.

Before moving on we should briefly reconsider the supposed "Great Depression" of the late nineteenth century. We saw in the Chapter 1 that this was nowhere near as severe as that of the 1930s. Saul (1985), indeed, concluded that the use of the term depression was inappropriate. We would not, then, wish to pursue a Menschian explanation of the rise of our new technologies as a result of hard economic times. However, one need not embrace the idea of regular long cycles in economic activity in order to look for forces behind decades-long periods of prosperity or stagnation.

We might suspect, then, that if so much of the twentieth century's technological advance is due to late nineteenth century innovations, perhaps the slowdown of the 1870s and 1880s might be due to some degree to the fact that the technological potential of the innovations of the (first) Industrial Revolution had largely been realized. Saul's (1985) review of the literature on that period provides some evidence in support of the idea, though Saul himself is sceptical.[24] Musson's research showed that the expansion of steam power and the switch from iron to steel were both occurring at their most rapid rate in the 1870s and 1880s, but this did not prevent him from favoring a Schumpeterian explanation (Saul, 44). As we would expect, it was England, the technological leader through the early nineteenth century, that suffered the most during this period, at least in terms of industrial production; latecomers like Germany and the United States fared much better as they were still catching up. It does appear that the key industries of the Industrial Revolution, textiles, iron, and coal, faced the greatest difficulty. Booms and slumps in housing activity appear to have been an important determinant of unemployment. At least in England, the decline in agricultural employment exacerbated the situation. The Depression might have been much more severe in England if not for the rising importance of the service sector. Unemployment would have likely been higher if the rate of increase in labor productivity hadn't decreased a little in the 1870s and 1880s. It is worth noting that there is a considerable debate over the degree to which English industry was hampered in this period by capital market constraints. While this evidence is only indicative, it does seem at least possible that some of this earlier and much less severe "Great Depression" might be due to the same sort of forces we argue are behind the real Great Depression of the 1930s.

Technological Determinism? The Role of Industrial Research

I have argued elsewhere (1991) against the use of technology as an exogenous explanation of historical transformation. It is too easy to attribute otherwise difficult-to-handle developments to the mystical forces of new technology. Noble has been rightly critical of those who have ascribed various socio-economic changes, from the rise of the corporation to the expansion of education, to simple technological forces (1977: xxiv). Ellul's warnings that technological evolution was determining social evolution in the modern world can also be criticized for neglecting the possibility of causation in the other direction. Technology is clearly shaped by society as well as shaping it. The pace and direction of change depend on societal forces. One should not go too far, though. "It is much less common to discuss the impact of new technology on its ambience than the other way round." (Staudenmaier, 1989: 39). This, he notes, creates a bias toward seeing innovation as the end of a process rather than the beginning. Technology is both determined and determining. Our contention here is simply that the types of technology being developed had an important effect on the timing of product and process innovation, and that this in turn had a huge impact on the functioning of the economy.

Sahal (1985) and others have characterized technological advance as a series of branches. At various points, investigators face a decision about which path of inquiry to follow. By their very nature, these decisions are surrounded by uncertainty. Some paths may lead to dead ends, others to varying degrees of success. Obviously, the choice of which path to follow is not (entirely) technically determined. Still, there is not an infinity of paths to choose from; technical considerations do constrain the decision set (Dosi, 1988: 1145). This is even more true ex post than ex ante due to the existence of dead ends. "The stock of scientific and technological knowledge is not a kind of putty out of which almost anything can be shaped. There are important supply-side constraints on what is possible" (Aitken, 1985: 16). Moreover, once a path is chosen, the direction of advance is then technically determined until the next branching point is arrived at (Sahal, 1985). Given that innovation is a long-term investment, the pace of change, if not the direction, will still be influenced by non-technical considerations. When we reach the next branching point, we generally find new possibilities have been opened up by our research which we had not foreseen. The tree analogy can not be taken too far; as we have noted earlier, important new branches often arise only as a result of some combination of others.

Choices made at one point in time tend to be self-perpetuating (unless they lead to a complete dead end). Research itself, as well as commercial use of discoveries, causes learning, and this knowledge makes it easier to continue along a given path than to explore elsewhere. Agents with specialized knowledge have an incentive to see that the existing line of inquiry is continued. There is no mechanism to ensure that the socially — or privately — optimal path is chosen. Even if a path were originally chosen due to some temporary influence, such as labor shortage, the process of learning is likely to produce an end result which is superior to the old technology under all circumstances; the possibility that a different path might have yielded an even better result may never be considered. The potential of the path not followed can never be known. The historian can not easily posit a counterfactual: what would have happened if a different choice had been made. But research is full of choices made in uncertainty, and with unimagined effects on the future pace and direction of technological change.

A line of inquiry need not continually yield positive results in order to dominate. Indeed, Maidique (1985) has argued that success too easily breeds complacency while failure is the ultimate teacher. Failure in this regard could be of a purely technical sort. As long as the technical problem can be overcome, it serves as a valuable focussing device for future research. This can be seen as technological determinism to some extent. Failure can also occur in the marketplace. Social forces would then shape the direction of research. In both cases, the important fact is that effort be put forth to solve the dilemma. If this is to occur within the same firm, a long-term view, and long-term employment ties, are important attributes. Even a purely technical problem will only be overcome in the right environment.

Recognizing the potential for environmental influence is much easier than identifying particular cases. Innovation at all times has been an activity which involves a relatively small proportion of the population. There is no need, therefore, that the motives of researchers and investors need match those of society as a whole. There is, in fact, no need that innovation be socially beneficial. One is hard-pressed to name more than a handful of twentieth century innovations which have not had a negative impact on some segment of the population. If the losers have access neither to economic nor political means to penalize the winners, changes with a net negative impact may occur. Few would go so far as to claim that the whole sweep of twentieth century technology has not been beneficial, but many would have suspicions about particular innovations. If we are to search for environmental forces which may have affected the course of innovation, then, we are well advised to focus on the environment in which researchers and investors operated.

An economist must naturally wonder about the incentives facing innovators, and look for a comprehensive theory of innovation. We can not neatly explain technological evolution in terms of the maximizing behavior of individual agents. Still, while a variety of motivations, such as fame, service to humanity, and curiosity drive individuals to perform research, the time and energy involved ensure that economic motives are generally of critical importance. As to how those financing research determine which paths of inquiry are expected to be profitable, there is still much that we do not know. However, we do not need a precise theory of innovative effort in order to investigate the social forces which impinged on such decision-making.[25]

It would be sheer folly to ignore the rise of the industrial research lab. We have already had cause to discuss this at some length. Noble does not make clear what he thinks the world would have looked like without this institution. Presumably, corporate bosses did not purposely cause the Depression. Perazick (1940: 50) argued that "the channeling of the course of technological progress [by these labs] may be an important factor in the recent failure to create new industries and thus to provide sources for new investments and new employment opportunities." Rolt recognized its two-sided character; "Given the necessary funds, it is obvious that such an organization can complete in a few years a research program which an individual would be unlikely to achieve in a long lifetime." However, "Much must inevitably be lost when inventive talent, hitherto free to flow where it will, is pooled and then canalized in this way." (1986: 206-7). We will find many innovations which seem virtually impossible without the resources of a large lab, and must at the very least have been greatly delayed otherwise. Arguably, the very complexity of modern technology makes the interaction of experts in different areas, plus access to expensive equipment, much more important than previously. There is no guarantee, though, that large corporations pursue the public interest. Some technologies are easier to control than others (Aitken, 28); this may affect in an important way the direction of industrial research efforts.

It is, of course, much easier to see what is gained than what is lost. Beyond possible biases within industrial research itself, it also attracted resources from alternative employment. Academics during the 1920s were suspicious of the new institution. They felt basic scientific discovery was being downplayed. "What must be done to create a science of pure physics in this country rather than to call telegraphs, electric lights, and such conveniences by the name science," wrote Henry Rowland. Nevertheless, university research was heavily influenced by industrial research (Perazick, 1940: 44). Universities came more and more to devote their greatest efforts to supplying graduates to industry; many worried that students were no longer taught to think for themselves (Wise, 1985: 27). While scientific research may have suffered as talented researchers were lured away by high salaries, it should also be noted that technological advance fuels scientific inquiry; it raises questions and it furnishes important experimental equipment. It thus affects the pace and direction of scientific advance.

Independent innovators may have found it advantageous to join a large organization. If they remained on their own, they might find themselves in patent battles. Large companies, on hearing of an independent patent, could claim prior development and tie the case up in court until the independent gave in (Wise, 99-100). Skillful lawyers in the employ of large companies inundated a small and poorly trained patent staff with these and other claims (Williamson, 1963: 150-1). Who can say how many may have been put off inventive effort entirely by such a strategy? Many interviewed by the Temporary National Economic Committee argued that large corporations abused their patents and TNEC's *Final Report* concluded that the patent privilege had been "shamelessly abused"; however Folk (1942), a former counsel to Bell and then with the National Association of Manufacturers, maintained that the case was not clear and recommended only minor changes to patent law.[26] Small firms may lose some of their incentive to innovate by the very existence of large powerful firms. RC Cola was the first to introduce canned soft drinks, but the market power of Coke and Pepsi ensured that these received most of the benefit. Especially for non-patentable improvements, this may be a common occurrence.

This is important for our purpose because large corporations might face different incentives concerning the relative attention to be paid to product and process innovation. Alexanderson in the mid-1920s suggested that radio technology had entered a breathing phase where the gains of the previous decades could be consolidated; such breathing spaces were necessary, he suggested, for those who had invested in early developments to recoup their expenditures (Aitken, 514). We could question the accuracy of his perception as it concerns radio. Still, it is clear that firms do have an incentive to space their innovations in such a way that they do not destroy the profitability of one discovery by introducing a yet further advance too soon. One factor which arguably slowed the development of television was that the companies concerned were involved in radio as well. Even expectations of future innovations can slow the rate of adoption of recent innovations (as anyone who has recently considered buying a personal computer or video camera can attest). As innovating firms usually earn the bulk of profits in the early years of introduction, it is important not only that further improvements be delayed but that potential adopters believe this to be the case. A corporation could only purposely delay innovating if it could be reasonably confident that competitors would not. Moreover, if we wished to use this line of argument as an explanation of the Depression, it would have to be coupled with some concept of corporate failure, for it was in nobody's interest that there be such a lengthy gap between market saturation by old products and introduction of new. Perhaps, one might point to inexperience in the new market setting created by the revolution in consumer durables production.

What was the balance between product and process innovation? We do know that R&D labs tend to emphasize process technology. Still, one of the main purposes of these enterprises was and is to prevent a company from seeing their product(s) superseded by a competitor. Reich speaks of creative destruction; once a new area of technology is opened, firms move rapidly to steadily improve their product line. Thus, these labs can be a powerful source of product innovation. Indeed, they often discovered whole new product lines, not dreamt of by managers (Reich, 5). Within research-prone industries, Mowery (1983: 964-66) has noted that R&D labs spread from larger to smaller firms. Perhaps, firms would become more interested in (riskier) product innovation as competitive research labs emerged. A few successful product innovations might have encouraged emulation. Reich speaks of R&D having to escape from its initial narrow focus and only then being able to create technological leaps into unimagined product areas (1985: 240). At the industry level, Williams (1982) suggests that, "On balance, the tendency is for new products to emerge outside the industries concerned, and for new processes to be internally generated." Thus, if corporate research absorbed independent innovators, we would see a shift toward process innovation.

Before World War One, research and development activities almost always involved simple tasks from the scientific point of view, such as assaying, testing, and quality control (Mowery, 1989: 37). A bias toward process technology was almost inevitable in such an environment. The development of better vacuum tubes and X-rays by General Electric, according to Wise, marked "a new epoch in American industrial research. Before 1913 research aimed at new products had mainly been done at independent labs on the model of Edison's Menlo Park. The few giant corporations that founded research labs did so to protect their established products: G.E.'s lightbulbs, AT&T's telephone service, Kodak's film and cameras" (1985: 177). Still, G.E.'s research director seems to have shied away from new fields after these early successes (Wise, 251), and a general survey of lab effort shows that even this famous lab devoted much of its effort to short-term projects which emanated from manufacturing divisions (Wise, 97-8). In 1926, Hoover wrote an article in *Mechanical Engineering* claiming that the rapid growth of industrial research was depleting the reserve of scientific knowledge on which innovation was based; this was used to convince Du Pont executives that they should support basic research (Hounshell, 1988: 223).

Arguably, labs were more focussed on process technology early in the century than today. The case should not be overdrawn; while modern labs devote two-thirds of their resources to product development (only one-third in Japan), the vast bulk of this is still minor product development.[27] Chandler's description of the evolution of the modern corporation implies a greater emphasis on process innovation early in this century. We will shed much light on the role of labs industry by industry in Part Two.

Echoes of Schumpeter?

To proffer an explanation of the Great Depression which hinges on technological change is to invite inevitable comparison with Schumpeter's *Business Cycles* (1939). Schumpeter posited that a wave of product innovations would be associated with the beginning of an upswing in economic activity. These product innovations would engender a series of process innovations that would allow the new products to be produced at lower cost. In the absence of new product innovations, this would cause increased unemployment. Beyond the superficial coincidence of stressing the importance of technology, however, there is relatively little in common between this work and Schumpeter's effort of a half-century ago.

Schumpeter's analysis hinged on his belief in long waves. He was convinced that there were at least three regular cycles of economic activity (and thought that there might be more), the Kondratieff, Juglar (9 years), and Kitchin (40 months). He felt that the ebb and flow of technological change was responsible for all of these (1939: 170-2). We have seen earlier that the vast majority of scholars have justifiably remained sceptical of the existence of long waves. Since these formed an integral part of Schumpeter's analysis, it is not that surprising that his emphasis on technological forces has received rather little attention since his time.

One can be sceptical of the existence of regular waves and still imagine that the rate of technological innovation has important impacts on economic activity. This is the position taken here. Nor does this mean, as we have seen, that we must relegate technological change to the role of exogenous force. The determinants of innovation may be difficult to discern, but we can be sure that there are reasons why some periods see more change than others. It is too simple to think that there is some symbiotic relationship between technological and economic activity. Nor is it necessary to believe this to recognize that the former exerts a major impact on the latter.

Schumpeter, in order to generate economically-induced waves of innovation, was forced to posit an artificial separation between invention and innovation. As I have argued elsewhere (1991), this distinction stands in the way of our comprehension of the process of innovation. Schumpeter imagined that there would be a pool of inventions available when times were right for entrepreneurs to want to innovate (Kuznets wondered why entrepreneurs only got excited every fifty years). He could thus ignore the question of what caused people to invent in the first place. Schumpeter had no theory of innovation. We maintain that most "inventors" only devote considerable time and energy to a new idea with an eye toward its adoption. Thus, invention and innovation are simply two aspects of the same activity; any temporal gap observed between the two is primarily due to the difficulty in making the new idea commercially feasible. Once this fact is recognized, Schumpeter's vision of cyclical behavior is exposed as simplistic. Numerous factors affect the rate of innovation, and it is too much to expect that any sort of regular cycle will naturally develop.

Though Schumpeter's work appears historical in orientation, his effort to paint a picture of regularity renders it ahistorical. In particular, the actual technology being developed is of almost no importance. Indeed he goes so far as to assert that it is a mistake to concentrate too much on the opportunity being exploited and not enough on the entrepreneurs exploiting it; if particular opportunities were not there entrepreneurs would likely find others (1939: 973-4). Periods of little innovation reflect a lack of entrepreneurial zeal rather than technology.

While in some of his historical survey Schumpeter comments on the different roles of product and process technology, no distinction is made in his theoretical discussion between the two (1939: 87-8 for example). However, we have suggested that this distinction is of crucial importance to understanding the Great Depression. In particular, the prominence of process innovation throughout the interwar period is the cause of much of the observed unemployment.

Schumpeter, by ignoring this, and the possibility of market saturation, is unable to directly explain the bottom half of the cycle. He can speak of prosperity and of recession, or a return to the natural equilibrium. The downswing may become self-perpetuating only if people believe this to be so; it all depends on expectations, morals, and the financial setup (1939: 145-50). "It may be safely asserted that in any unemployment percentage beyond 6 percent spirals and so on counted for much more than did the direct effect of innovation" (1939: 518). A downward spiral of this sort becomes more likely when all three cycles turn downward simultaneously. We have discussed above how unsatisfactory a reliance on expectations is. Schumpeter argues that mid-1930 was the point at which expectations turned sour and recession slid into depression (1939: 911-4). We have seen though, that expectations remained higher much longer.

There is a significant difference in predictions about the course of technological change between Schumpeter's theory and our own. Schumpeter would posit a flurry of innovative activity in the 1930s. We would expect that the incidence of discovery depends instead on the working out of ideas that have been the subject of inquiry for some time. Thus, when we focus on product innovations that come "too late" in the 1930s, we will be greatly interested in why they did not occur earlier. Was the invention lying around waiting for an entrepreneur as Schumpeter would have it, or do we witness a natural progression of innovative activity which extends back far beyond the beginning of the Depression? The answer, not surprisingly, will appear to be the latter.

Notes

1. New technology could save on capital as well as labor. While technically "labor-saving technology" refers to technology which saves more on labor than capital, we will use the term more generally to refer to process innovation which reduced the necessary labor input, even if in some cases it might have had an even greater impact on capital requirements. While the increase in labor productivity in the 1920s was larger, the increase in capital productivity was more impressive relative to other decades (Lorant 1966, based on figures from Kendrick). This is partly due to capital-saving technology. Further, Lorant notes that much of the investment in the 1920s replaced old plants with new, often larger factories; 56% of investment replaced old capital. The capital-deepening of the pre-World War One era had depressed output-capital ratios, and left much scope for these to increase in the 1920s. The availability of capital-saving process technology enabled manufacturers to achieve highly increased capital efficiency. Interestingly, Lorant notes that his results are driven by just five industries.

2. Even Lenin recognized this, defining, "Communism is soviet power plus electrification of the whole country" (in Duboff, 1979: 1). Electricity during the 1920s became an integral part of the production of many "new" goods, including cars, rubber, chemicals, and electrical products (Platt, 1988: 272).

3. In the examples cited of machines replacing other machines, the most important types were; replacement of steam-driven by electric or gas, use of new and better trucks, and the introduction of caterpillar treads. The *Monthly Labor Review* published a series of case studies in the 1920s of jobs lost to new technology (see Fano, 1991).

4. There was a spirited debate among machine tool producers about the effectiveness of new management techniques in the 1920s. Plans which worked in one place might not work elsewhere, due to differences in labor quality. Savings in workfloor labor were counteracted by increased managerial overhead (record-keeping, reporting, supervision, coordination); the latter would be more difficult to cut back in recession. Problems might be especially severe in machine tools with its short production runs (Wagoner, 1968: 168-75).

5. There is a tendency for productivity growth to be exaggerated in growing economies as capacity is more fully utilized and as markups increase. This is unlikely to have been large relative to the total productivity increase observed in the 1920s. Of more importance conceivably would be the opposite effect which would result from excess investment which created overcapacity.

6. Unfortunately, Bernstein's focus on a few sectors has been taken by Margo (1993) to imply that technological change — he does not distinguish between product and process — was modest and concentrated.

7. Fano (1987) argues that interwar technology was even more impressive in its ability to save on capital than on labor (This accords with Lorant's estimates). The new control devices and tungsten carbide both extended the life of machinery, for example. So did corrosion-resistant alloys and better paints and tires. She argues that the capital goods sector was considered risky by investors, due to the loss of foreign markets, and thus did not generate machinery or employment. Hansen (1938: 315) felt that the innovations of the 1920s had also required little capital, and that this boded ill for the future.

8. "Before 1918 full mechanization was already setting in, and it was by no means ending by 1939. Even within these years there are times of widely varying intensity. Yet with good warrant one may call the era between the wars the time of full mechanization. ... What the previous century and a half had initiated, and especially what had been germinating from the mid-nineteenth century on, suddenly ripens and meets life with its full impact" (Giedion, 1949: 41).

9. As well, process innovation tends to be underrepresented in patent data (Freeman 1982: 73). Griliches (1990), examining patent data through the century, finds that patents grew at about the same rate as GNP until the 1930s, fell dramatically then, and has grown much more slowly than GNP (or R&D) ever since. He rules out bureaucratic or legal explanations, as well as the possibility that research has increasingly been undertaken by non-patent-oriented industries. He notes, though, that due to the extreme skewness in the distribution of the importance of innovation, overall numbers do not tell us much.

10. Over some ranges, the relationship may be proportional though both Mansfield (1968) and Mowery (1983) have found that beyond some threshold size larger firms do not spend proportionally more on research or get proportionally better results. There is substantial cross-firm variation. Firms make different choices between cost cutting and differentiation. As well, historical choice of which lines of inquiry to pursue will influence the perceived profitability of present research. Beyond this, many firms show both a consistently higher propensity to innovate and higher profits over long periods of time (Dosi, 1988: 1151).

11. Maidique (1985) discusses the importance for successful innovation of interaction with customers during research. Contract research firms have generally supplied specific services, such as chemical analysis, to firms with in-house research capability (Mowery, 1983b).

12. Reich notes in particular that GE's domination of lighting and Bell's of telephones were likely due to their research and development activities (1985: 218). Mowery (1983: 970-7) notes that research expenditure is a significant factor in allowing firms to stay among the top 200 industrial firms (and thus creates much greater stability in the membership of that group after World War One) and appears to have had a significant effect on firm growth. Noble (1977) has provided a more sinister argument, that changes in patent practice (so that firms rather than individuals are protected), the development of industrial research, and the

professionalization of technical staff were all designed to allow corporate capitalism to bind technology to its service and support monopoly power.

13. Again, the answer depends on our view of the role of industrial organization in general. Gort notes that oligopolists were much more likely to diversify (because, he feels, of the difficulty of expanding in their own industry), and 40% of their new product additions placed them in industries where they became one of the eight largest (1962: 104, 136).

14. Mensch suggests that much new technology was available in 1929, but only the cauldron of Depression brought it into use. We will see in later industry studies that he has ignored the lengthy period of revision necessary before an idea becomes feasible. The new products of the 1930s were simply not ready in 1929.

15. "During the last decade the emphasis has been on technological changes of a labor-saving rather than plant-expansion character". Fano feels that while concentration was on cost-saving, real research expenditures rose enough that some product technology emerged. Cohen (1984) argues that in paper, at least, effort turned toward quality improvements and new product innovation in the late 1930s in order to expand sales. Observers at the time, including Schumpeter, noted that while there was some product innovation, it was not capable of providing enough employment to end the Depression (Fano, 1987).

16. Reich has described the various forces encouraging the rise and expansion of industrial research and development in the early twentieth century: the emergence of science-based technology, the education of large numbers of scientists and engineers, the emergence of large technology-based companies, the enforcement of the Sherman antitrust act after 1900 which cut off other avenues of competition, and the Progressive movement which expected professionals to aid society (1985: 12). Mowery and Rosenberg emphasize the growth of large integrated firms; they feel that large scale operation by its nature requires that more attention be paid to matters such as homogeneity and thus encourages scientific inquiry (1989: ch. 3, 71).

17. Buxton (1979: 16) notes that, for Britain, while old industries are consistently worse than new in terms of output and employment, the pattern of consistently inferior performances is broken when productivity is looked at. While in part due to shedding excess labor, most of the gains are due to technological and organizational changes.

18. Burstall (1963: 366-7) is one who feels that World War One greatly stimulated both private and public research efforts. Braun and MacDonald (1982) describe World War One as the chemist's war and World War Two as the physicist's war. It could be that the two wars had quite different effects on their respective postwar periods.

19. Sawers (1978: 28) downplays the connection between science and technological advance at least in the short term; "Where there is a link it is unusual for the time-interval between a scientific discovery and its application to be short, as it was with nuclear fission." Science, he concludes, is a permissive factor which rarely determines the rate of technological change. It may play a larger role, of course, if a longer-term view is taken. Scherer, for example, has suggested that

R&D occurs mostly where there is the greatest technological opportunity, and this in turn is determined by broad scientific technological development.

20. "During 1921-1946 the chemical industry functioned similarly to machine tools in the nineteenth century American economy, acting as a source of innovations and techniques that were adopted by a wide range of other industries" (Mowery, 1983: 966).

21. The argument could be turned around. Cross and Szostak (1995) note that the assembly line itself borrowed heavily from scientific management.

22. With each innovation, the potential for new combinations increases; this may be one reason why both research effort and innovation have increased over time. Dosi has suggested that differences in rates of innovation across countries may depend on the degree to which facilities for cross-fertilization of ideas across industries have developed (1988: 1147).

23. Certainly, the establishment of research laboratories required the existence of a large group of educated researchers. Layton (1978: 146) notes that American industry donated about $110 million dollars to university research between 1890 and 1901, in addition to donations of equipment.

24. In speaking of the long wave literature, Saul interestingly takes a position similar to our own; "We cannot try to answer such questions in full here, but it is obvious that we must spread our net beyond the confines of the 'Great Depression' years and look at those years in the context of the long development of the economy" (1985: 11).

25. In Szostak (1991), I argue that the development of an early modern transport system was the key which encouraged the rise in the rate of innovation which characterized the First Industrial Revolution. There is much more research to be done on the determinants of economy-wide research effort. In this work, we will take as given that considerable research effort was undertaken through the early twentieth century, and focus on the determinants of direction and timing of innovation. Aggregate research effort should largely be determined by the perceived profitability of specific research projects.

26. After World War Two, compulsory licensing laws forced large firms to share their discoveries with smaller firms (Reich, 1985: 257). Especially since one of the goals of R&D was to beat out small firms, it would be interesting to know what effect this had on the orientation of research efforts.

27. We noted in Chapter 3 that advertising expenditures soared in the 1920s. Many commentators have suggested that this encouraged meaningless brand proliferation and style changes, at the expense of new product development.

5

The Labor Market

"The final solution of unemployment is work."
Calvin Coolidge

We have had many occasions in the preceding chapters to hint at the causes of unemployment during the Depression. We have particularly noted the fact that product innovation will likely increase employment, and process innovation decrease employment, at least within the industry concerned. We begin by expanding on the notion of technological unemployment here and suggesting that technological forces can induce unemployment at the aggregate level. We note in particular the number of economists who shared this view in the 1930s. We then discuss an alternative technological argument: that innovation created a mismatch between job vacancies and the skills and location of workers, and thus much interwar unemployment was structural. We argue that there were in fact few job openings and thus structural unemployment is primarily symbolic of a much greater unemployment problem.

Our next task is to note how economic theory ignores the role of entrepreneurial decisions. In the absence of market opportunities in new product lines entrepreneurs could not be expected to quickly find profitable employment for displaced workers. Equilibrating mechanisms are extremely sluggish; it is not because of these, but rather because economies are usually characterized by greater equality between growing and declining sectors, that we do not normally see unemployment as severe and longlasting as during the Depression.

We then move on to discuss how the initial declines in unemployment would be augmented by both labor dishoarding and the multiplier mechanism. Finally, we look at product markets, and argue that while price inflexibility there can increase unemployment, it is not essential to the labor market analysis pursued here.

Technological Unemployment

Writers of the time noted the dangers of productivity increasing faster than effective demand. If consumer expenditure had kept pace with income (or workers were willing and able to take the benefits of increased productivity in terms of increased leisure time), the new process technology would have been a blessing which would have allowed per capita incomes to rise. In the world of the 1920s, though, released workers had nowhere to go, and process technology was no longer a blessing. A 1928 Brookings study concluded that workers displaced by technology had great difficulty in finding new jobs; one half of the workers studied were unemployed for over six months (Jacoby, 1985: 189). Re-employment often involved taking a lower wage (Jerome, 1934: 387).[1] The term for workers who lose their jobs in this way is technological unemployment. The idea has been around for a long time. Since the Industrial Revolution at least, worker organizations have been suspicious that labor-saving technology must reduce employment. At the industry level —depending on how the gains are divided by capital, labor, and consumer, and on the elasticity of demand for the product in question — this may be true. At the economy-wide level, and in the long run, we might expect full employment to be restored. Over the period 1899-1939, most productivity increase was absorbed by increased quality and quantity of goods, and a 20% drop in the work week; the rest was dissipated in unemployment (Fabricant, 1945b). In the short and medium run workers will at least experience some frictional unemployment, and perhaps structural unemployment, and may be in for a lengthy period of unemployment if the economy is not creating job openings as fast as layoffs.

Already in 1929, Dennison had recognized that the degree of technological unemployment could depend on the relative rates of product and process innovation. Keynes (1930) said, "we are being afflicted with a new disease of which some readers may not yet have heard the name — namely technological unemployment. This means unemployment due to our discovery of means of economizing the use of labor outstripping the pace at which we can find new uses for labor." Taylor (1933: 122) reiterated this point. "There is no warrant for complacent assumptions about self-correcting adjustment mechanisms, which would lead rapidly and automatically to 'compensating' new employment for all the jobs lost through technological change" (Freeman, 1982: x).[2] Fabricant (1945) noted that over the period 1899-1939 "mature" industries tended to see productivity growth outstrip output growth, and thus employment fell; a world dominated by mature industries would be likely to see significant unemployment.

While the idea was around for a long time, the phrase technological unemployment was coined during the 1920s (of course the word unemployment itself only emerged in the late 1880s in the United States). Ricardo and Marx had both recognized the possibility (Sardoni, 1987: 3) but had used different phraseology. No less than Jean-Baptiste Say had advocated public works to combat technological unemployment (Weintraub, 1937). The Temporary National Economic Committee attributed the great interest in the phenomenon from 1927 to the large numbers of workers let go in cigar-making,[3] mining, railroads, and textiles (1941: 38). Mitchell (1929) noted that the emergence of the new phrase was not due to the fact that the problem was new but that it involved much greater numbers than before. Jerome concurred, adding, "fortunately, there appears to be a greater public consciousness of the existence and significance of the evil" (1934: 386n). Over the course of the 1920s, millions of workers were rendered redundant, and had difficulty finding new jobs. Chronically high unemployment rates were the result.[4] Concern with technological unemployment was enhanced in the late 1930s as output recovered much more quickly than employment (Temporary, 1941: 49, David Weintraub in Gourvitch, 1940).

That so many economists spoke of technological unemployment is especially striking given the limited scope which neoclassical theory gave to the concept. When Gourvitch surveyed the literature for the WPA in 1940, he found that within the mainstream he could only find justification for a bit of frictional unemployment. Classical economists had at least recognized the possibility of a declining demand for labor in the long run. Only within the nascent business cycle literature and the "much scantier" literature on long-term fluctuations could Gourvitch find indications that labor reabsorption might occur less smoothly than equilibrium theory suggested. While the situation has improved in some ways in the last half century, the fading memory of the Depression renders it similarly difficult to argue for technological unemployment.

Still, Phelps (1993: 357-8) and others have in recent years explored theoretical links between technology and unemployment. Katsoulacas (1986, 1991) has developed a model of the relationship between technology and employment. He finds that process innovation does have a tendency to reduce employment in the short run. Even in the long run this may happen, for the income elasticity of demand will fall far below unity as people become satiated. If firms do not expect to be able to increase sales in the future (due to sticky prices or inelastic demand) the result of embodied technological change is decreased investment and increased unemployment.[5] Whether the decrease in labor demand is felt in layoffs or reduced hours will depend on institutional flexibility. Product innovations, alternatively, are found to have a positive effect. Again, stickiness in wages or prices would limit this.

Structural Unemployment

We must be careful to distinguish structural and technological unemployment; the Great Depression was much more than a brief period of structural imbalance. Structural unemployment refers to workers who find that either their skills or their geographical location are inappropriate. This could result from technological change but could also be a result of other structural changes. Technological change will only cause structural unemployment if it changes either industrial location or the skills required of the labor force.[6] Thus, some technological unemployment is structural and some structural employment is technological, but neither is a subset of the other. Structural unemployment should be a medium-term phenomenon, for (most) workers will eventually be expected to acquire the necessary skills or location.

Some authors have suggested that the interwar years were a period of unusually high structural unemployment. It was common in the 1930s to view the Depression as a structural phenomenon; there had been overinvestment in some sectors and unemployment was due to the slow transfer of labor and capital elsewhere; this extreme view became untenable as the Depression deepened. Kesselman and Savin (1978) estimated rates of 6-8% throughout the 1930s.[7] In modern parlance, the natural (frictional plus structural) rate was unusually high, and thus cyclical unemployment was nowhere near as severe as casual inspection of the figures might suggest. Alternatively, Georgescu-Roegen has argued that the constant technological change since World War Two has meant steadily increasing rates of structural unemployment (Katsoulacas, 1986: 6-10). This could make us suspicious of attributing too much of the excessive unemployment of the 1930s to this cause.

Heim (1984) has argued that structural unemployment was of great importance in interwar Britain. New sectors had a strong tendency to hire young workers and/or women. She attributes this to a preference both for low-cost labor and for labor that would not have developed bad work habits in previous employment. Alternative explanations could be put forward, such as that young workers were preferred so that training costs could be recouped over time. She also finds that the new assembly-line technology was introduced outside of the old industrial areas. Again, she suggests that the desire to tap new sources of labor was the key. It is not that surprising that new technology was embodied in new facilities, and there are other possible explanations of locational choice. While we might quibble about the forces at work, Heim does succeed in showing a significant mismatch between the skills and location of the unemployed and interwar job openings. This is powerful evidence of widespread structural unemployment.

The United States is unlikely to have been spared the structural unemployment problems faced by England, though perhaps it might have suffered to a lesser degree.[8] Weintraub (1937) felt that there were two parts to the unemployment problem, one aggregate, the other structural. Gill (1940) noted that "industrial changes which alter the nature or volume of demand for certain types of labor can create special groups of workers who are as effectively stranded in a highly diversified industrial community as those residing around the tipples of shut-down coal mines." Of the weavers and loom fixers displaced by the automatic loom, 29% were unemployed in 1936; 20% of these had been unemployed for more than five years. There was also the locational element. New Bedford lost 80% of its mill jobs in the interwar period and had 30% unemployment in 1939. In some southern Illinois counties where coal mines were closed, unemployment ranged from 40-50%. As well, "as automobiles, trucks, and improved roads made large cities more easily accessible, many a rural shopping and marketing center has lost most of its trade."

The dividing line between structural and frictional unemployment is not clearcut. Workers may delay moving or retraining if they mistakenly perceive that a dying industry will recover, especially if changing jobs will involve a loss in status. In this way, the failure of real wages to decline either relatively or absolutely in declining sectors may have sent the wrong signal to workers. Various studies of migration in the modern period have shown that the timing of migration is governed more by pull factors from growing industries than by push factors (Casson, 1983: 219). Extensive and longstanding structural unemployment is likely only to be observed, then, when there is a scarcity of job openings elsewhere. Indeed, Simon (1992) finds that structural unemployment was actually higher in industrially-diverse centres than one-industry towns, and hypothesizes that this was because firms in such centres preferred to lay off some workers, rather than cut wages and risk losing (their best) workers to others. We will not make structural unemployment a focal point of our case studies, but will accept its existence as evidence of a much larger unemployment problem.

If interwar unemployment were entirely structural, there should have been millions of job openings which nobody could fill. Our contention in this work is that there was simply a severe shortage of jobs of any type, anywhere. One of the purposes of our case studies will be to show that firms in general had no interest in hiring more workers, rather than that they simply could not find appropriate workers.

Clearing the Labor Market

It should by now be clear how the initial decrease in employment in the 1930s is to be understood. A fall in investment and consumption demand hit a number of industries at roughly the same time and caused them to lay off workers. New process technology exacerbated this effect. The sharpness of the fall was due to the unprecedented rise of new industries in the previous decade, in concert with the rapid advance in process technology. It is worth noting that layoffs were more common in the earlier twentieth century than they might be today. Large seasonal fluctuations in employment were the rule; Kuznets estimated that the labor force in automobiles was four times higher at some times of the year than others. Firms were beginning to form long-term ties with workers, but this process still had far to go. Across a number of sectors, then, large numbers of workers were released from employment, to find themselves in a world with few vacancies.

We would normally expect full employment to be restored as wage demands fell. However, at any wage above subsistence, the number of job openings was much less than the size of the unemployed labor force. This is not hard to understand. A unit of labor (a worker), unassociated with any capital, and possessing no access to wholesale raw materials, nor wide markets, is inherently unproductive. Left to their own devices, the millions of unemployed could scarcely be expected to earn their own subsistence. Begging, standing in soup lines, doing odd jobs, and accepting government relief would be the available alternatives. Given the quality of these alternatives, the fact that they did not do better is powerful evidence that they could not (and the unemployed certainly had time and motive to consider entrepreneurial solutions).

Falling wage demands were not enough to guarantee full employment. This is especially the case given that investment must by its very nature be based on long-term considerations, rather than what might be considered a temporary decline in wages. Economic theory depersonalizes the economic process, and does not describe in detail dynamic processes of adjustment. Entrepreneurs had to come forth who were willing to hire these workers. The path to equilibrium is clear. Some entrepreneurs hire cheap labor to produce goods aimed at the market provided by the two-thirds of the labor force still employed. By doing so, they widen the market and create opportunities for further increases in production. Eventually we return to full employment. The speed with which we do so will depend on the degree to which entrepreneurs see opportunities in the marketplace worthy of investment.

We have seen that aggregate expectations remained high in the 1930s. Entrepreneurs, though, must find a particular product for which they see

The Labor Market

an opportunity of making an expected profit worthy of the degree of risk they are undertaking.[9] National income accounts can blind us to the fact that individual entrepreneurs make much more narrowly defined decisions than to expand output. Feeling sure that the economy is on the verge of takeoff is nice, but unless this is coupled with perceived opportunities in particular product lines it will not yield a surge in investment. Aggregate approaches will inevitably fail to comprehend why high unemployment rates were sustained so long. In Part Two we will show that very few sectors had a motive for expansion.

Expanded production of existing products would be the simplest response. Indeed, then limited investment might be necessary, for existing plant could be double or triple-shifted. With saturated markets, output could only be increased to a limited degree without lowering prices so far that subsistence wages could no longer be paid. In such a world, new workers could only replace old workers. We cannot casually assume that there were infinite opportunities to add to aggregate consumption. Only as new consumer desires were created would employment increase.[10]

Matthews (1968) has attempted to draw a dichotomy between the pre- and post-World War One eras in Britain. Before the war, Britain was like a developing country; unemployment was due to a deficient capital stock (the existence of excess labor served to encourage the employment of casual labor which could be let go in recession). After the war, this problem was solved and unemployment could only be due to (trade-induced) deficient demand. The deficient demand problem had now been solved and low unemployment "can confidently be expected to persist" (1968: 568). He failed to recognize that due to the specific nature of capital, even advanced countries can suffer from a capital deficiency. Structural change is like a war in that it wipes out real productive capacity which can only be used to produce the goods of dying sectors, but is harder to recover from, for instead of rebuilding along the same lines the economy must develop a whole new industrial structure.

The new goods which would support postwar expansion were not available. The industries which did achieve increases in output in the 1930s simply did not employ enough workers to solve the problem. The big employers, such as textiles, lumber, primary metals, and transport equipment, engaged in very little expansion, while those industries where investment recovered quickly, such as chemicals and petroleum, employed relatively few workers (Bernstein, 1987: 32). To some extent, this was due to the fact that these industries were relatively capital-intensive. As well, they themselves experienced an especially pronounced degree of process innovation in the 1930s (Bernstein, 1987: 140). Overall, though, it was the lack of sufficient new product technology which caused sluggish employment growth. In a world of limited vacancies, the wage demands of the unemployed were severely under-represented in wage rate data, and the observation that real wages rose is no longer bizarre.

Second and Third Degree Unemployment

We have so far dealt mostly with what might be termed first-degree unemployment. We have at times alluded to what we can here term second and third degree unemployment. Each time a worker lost a job, they were forced (fairly quickly as we have seen, for life savings were small) to adjust downward their consumption stream. Thus, other workers in turn lost their jobs as a result, and further workers when these cut back on spending. The sequence will be familiar to any economics student. The exact size of the multiplier should be one divided by the marginal propensity to save; that is, the only thing that keeps it from driving employment to zero is that workers do not consume their entire income.

In practice, the size of the multiplier effect on employment in any period is impossible to ascertain. In the normal course of events upswings and downswings follow each other too closely for the multiplier mechanism to work through too many iterations. Faced with a decline in demand, employers must judge how longlasting it is likely to be. As we noted earlier, the Great Depression may have encouraged layoffs disproportionately due to its depth and length. This straightforward multiplier effect creates third degree unemployment. Normally, a dynamic economy has both industries which are expanding and others which are contracting. Forces causing a negative multiplier effect are counterbalanced to some degree. In the Depression, the multiplier could work at full force.

Recessions, if they serve any purpose, cleanse the economy of its weakest firms. Leonard (1988: 48-9) studied the experience of Wisconsin in the early 1980s, and noted that, "The employment growth rate declines not because all establishments are growing slower (they are not) but rather because shrinking establishments shrink faster and because at about the middle of the distribution establishments that were growing start to shrink. It is primarily this shift of only 5% of the establishments that lowers aggregate employment growth." This would seem to indicate that the proportion of troubled firms at the start of the recession largely determines the level of unemployment which results.

The argument can be extended to the industry level. We have dealt so far mainly with industries that could be responsible for a large autonomous fall in employment. There are other industries, suffering a slow and painful decline, that are full of weak firms, and others with bloated workforces. A severe downturn could cause these to fold and/or disgorge their excess labor.[11] (As Irving Fisher argued, once the economic "ship" tips too far, further disequilibrating forces come into play.) There are four reasons to expect a mass of layoffs from firms who stay in business. First, as layoffs are observed elsewhere, the social objections and public relations costs of firing workers are seriously diminished. Managers do not optimize on a daily basis as our textbooks imply, but tend rather to put off distasteful decisions (see Sharpe, 1994). With profits falling, and other firms firing, the decision becomes easier. Second, many firms suffer heavy losses in the early 1930s, and thus are "forced" to make tough decisions they had avoided in the 1920s. Third, the fact that others are letting their workers go may indicate that they know something about the future; even while expectations remained high firms would be wary of not following the lead set by others. Fourth, one cost of firing is that associated with hiring again if the situation changes. As unemployment rates rise, firms can feel more comfortable with the idea of shedding excess workers. Thus, to the extent we can detect industries which appear to be carrying excess labor in the late 1920s, we can expect a substantial employment decrease in the early 1930s. The workers involved will scarcely care that they are only second degree unemployed, or that they in turn will induce further third degree unemployment.

We will find great diversity across industries in terms of labor dishoarding in Part Two. Some actually hoard labor (employment falls less than output, and thus productivity falls). Bresnahan (1991) has studied the interwar auto industry and observed two conflicting forces at work on productivity, and thus hoarding/dishoarding. On one hand, the least efficient plants/firms were most likely to close. On the other, plants which remained in business usually hoarded some labor. The net industry-wide effect thus depends on the proportion of firms pushed to the point where they must close down (or at least disgorge massive numbers of workers). Industries like textiles, with widespread bankruptcies, inevitably dishoarded hundreds of thousands.

By 1933 net investment was negative and consumer durables purchases extremely low (and would have been even lower if not for the fact that some durables were new enough to still have growth potential). Non-durable consumption and government spending comprised 90% of output. Labor dishoarding and the multiplier mechanism were largely spent forces (the sectors leading the decline had laid off the vast bulk of their employees by 1931). Decline would go no further. The meager expansionary forces at work in the economy could now operate in isolation.

Price Rigidity Revisited

Economists have not yet completely integrated the results of macroeconomic and microeconomic theory. Within the structure of the latter, demand can only not equal supply, both at the level of individual goods and in aggregate, if prices are unable to adjust (fall if there is excess supply, rise if there is excess demand). It might be thought, then, that saturation in some markets can only be a problem to the aggregate economy if the demand released there can not be accommodated by price adjustments across sectors (falling in the saturated sector, generally rising elsewhere and thus encouraging increased production). In a general equilibrium framework, with freely adjusting prices and wages, the rise of new products is not necessary to restore full employment, whether job losses come from saturation or process innovation.[12]

Keynesian macroeconomic theory tells a different story. If desired saving exceeds desired investment, economic activity will decrease. No matter how smoothly prices adjusted in the late 1920s, many consumers, with inelastic demand for most goods, were going to increase their savings rate. Viable investment opportunities which could absorb these savings did not exist. Static micro theory hides from us the hardship which can be encountered on the journey from one stable equilibrium to the next. We have never claimed that full employment (perhaps only unemployment levels we have recently become complacent about) would eventually have been restored even in the absence of new products; just that this could not happen overnight. As savings exceeded desired investment, output and employment fell, and the multiplier mechanism went to work. Only the continual rise of new industries prevents us from more regularly seeing this process in such stark outline.

While price inflexibility is not necessary for widespread unemployment to occur, it can exacerbate the situation.[13] Especially in those sectors in which demand was elastic, failure of prices to adjust rapidly could add to the total of economy-wide excess demand. Thus, while price flexibility will not be a major focus of the case studies to follow, we will have cause to discuss its importance in a handful of cases.

Notes

1. A survey of long-term unemployed in the 1980s found that those who were re-employed in another sector suffered a much greater wage loss than those able to find re-employment in the same sector. Only between 20% (mining) and 55% (services) were able to find re-employment in the same sector (Podgursky 1988: 28-32).

2. "The conventional assumption that productivity-increasing investments reduce employment is certainly too simple. But it is equally insufficient to argue that growth in aggregate demand can simply compensate for the employment-reducing impacts of technological change. Rather, we must understand the economic and technical forces that shape the pace and scope of aggregate demand itself." (Hirschhorn, 1988: 377) "Mechanization has reduced employment di-

The Labor Market 139

rectly and trade indirectly. These are not the necessary effects of mechanization, but occur because production and consumption are out of balance" (Taylor, 1933: iii).

3. The cigar machine reduced labor requirements by over 50%. Jerome (1934: 368) estimated it had replaced 16,800 workers by the end of 1929.

4. This is likely an acceleration of a previous trend. The census of 1900 had argued that new machinery was causing unemployment and a reduction in skill levels. The potential long-run benefits were ignored. The N.I.C.B., fearful of government regulation, argued in 1935 that mechanization was not the problem; it was sluggish output growth. TNEC, at least, was not persuaded that mechanization was not part of the problem.

5. Neisser (1942) has suggested that technological unemployment could occur in the long run as well. He notes, following John Stuart Mill, that "Demand for Commodities is Not Demand for Labor." Thus, while there may be long-run equilibrium in the goods market, this need not lead to full employment. Neisser's basis for this result is excessive scepticism about the substitutability between labor and capital. Decreased wages may not then increase employment. He is left with a Marx-like description where two largely independent forces, capital accumulation versus labor-saving technological change, determine long-run unemployment. While arguing for the theoretical possibility of long run technological unemployment, Neisser doubts that it has been empirically significant.

6. If new technology requires unavailable skills this will cause entrepreneurs to be unwilling to invest in either the old or new technology (Katsoulacas, 1986). Jensen (1989) feels there was a technology-induced human capital shortage by the late 1930s. We should note that while some new process technology may replace skilled work with a machine, it can also be skill-enhancing, especially since machine making and repair tends to be skilled. Mechanization of handling, in particular, is likely to increase average skill requirements (Jerome, 1934: 391-403).

7. Mills, from census data, calculated that 4.9% of the workforce were forced out or withdrew from the industry in which they were working every two years between 1923 and 1929. Before World War One, the comparable figure was 2.1% every *five* years. All figures describe net movement, not gross (in Jerome, 1934: 384).

8. The United States was not only less unionized but also had fewer large moribund industries. Labor mobility was likely higher in the U.S., which had seen large interregional flows for decades. The slowdown in population growth would exacerbate any structural problem, as in a growing economy new workers will start in the growth industries.

9. Schultz (1990) has argued for the importance of entrepreneurial activity in moving to a new equilibrium in a modernizing economy. Dixit (1992) has argued that entrepreneurs will not invest the moment expected return exceeds the cost of investment, but will wait for more information. Uncertainty slows investment further.

10. This was not just a problem in 1933. Cornwall (1972: 222) attributes the end of the recovery in 1937 to the fact that consumption had not risen as fast as output, and thus investment naturally tailed off.

11. Burnside (1993) finds evidence that labor hoarding is an important characteristic of postwar business cycle behavior, though it is impossible to reliably quantify this role.

12. Sinclair (1981) develops a model and argues that if prices and wages are perfectly flexible, technological unemployment is impossible (Venables (1985) reaches a similar result). He notes that in the real world the degree of unemployment depends on the price elasticity of aggregate demand, the elasticity of substitution between capital and labor, the share of profits, and the elasticity of the work week to changes in real wages.

13. Mills (1932) noted that price declines in response to productivity advances would have little employment effect if demand was inelastic; there could thus be difficulties in equating "purchasing power created" and "productive energy released." Both price inflexibility and unequal income distribution could make matters worse. Wied-Nebbeling (1993) develops a model in which labor-saving innovation reduces employment no matter the market structure if demand is inelastic; this result may obtain with elastic demand too in the presence of market power.

PART TWO

Sectoral Analysis

6

The Procedure for Sectoral Analysis

In order to validate our approach, we must look closely at major sectors of the American economy. Existing aggregate-level analyses act as if the Depression left all sectors equally prostrate. Even if a different approach were taken to the Depression, the different effects felt by different agents would seem a natural focus for study.[1] Yet these are virtually ignored. "The differential effects of the slump on regions, on industries, and on classes of income recipients, are but a few of the missing parts in the received scenario" (Bernstein, 1987: 48). "Given the overwhelming importance of the Great Depression in the economic history of the United States, one might suspect the microeconomics of interwar unemployment to be old hat, but the opposite is true: the issues have scarcely been addressed" (Margo, 1987: 327). Some of the lack of interest in disaggregation is due to the belief that a limited number of factors should be at work in any plausible explanation of the Depression. Brunner has criticized non-monetarists for coming up with a number of little factors to explain the autonomous fall in spending rather than one big one (see Warsh, 1984). While it is extreme to argue that there should only be one cause of the Great Depression, it is nevertheless true that a large number of independent causes would be unsatisfactory. We have outlined above an approach that can potentially explain much of the technological experience of the period. We have also suggested a rationale for market saturation. We will, in what follows, attempt to show that these forces were of key importance across sectors. We will not, though, refrain from pointing out other forces at work at the industry level which served to exacerbate the Depression experience.

In this chapter, we discuss the approach which will be taken in the case studies which comprise Chapters 7 through 11. This allows us to forego lengthy introductions in each of these chapters and proceed directly to the evidence. We begin by discussing the necessity of studying a wide range of sectors. We must look not only at industries which release workers but those which were unable to absorb them. We then review the information which we hope to obtain from these sectoral studies. That is, we review the hypotheses implicit in the analysis of Part One, and discuss how these can be validated. We close the chapter with a brief survey of the focus of each of the next six chapters, and a methodological note.

Extent of Sectoral Survey

There is a limit to how far we should disaggregate. We want to know how exogenous forces affected industrial growth rates. Particular firms, due to luck or good management, handled the situation better than others. Generally, such success came at the expense of others. While it is interesting to investigate why some firms did better than others, our dominant focus must be on how the industry as a whole fared.

We must analyze a wide range of sectors. Bernstein (1987) focuses on sectors employing less than one quarter of the labor force. This will not do. In the 1930s, national income could be divided as follows: 26.5% manufacturing; 9.3% agriculture; 2.1% mining; 16.1% distribution; 13.1% financial services; 12.1% government; and 11.1% other services (Bernstein, 1987: 50). Table 6.1 shows the employment loss suffered by these sectors 1929-1933. The mining sector was very small, and could perhaps be ignored, but the changes within it, and the impact on certain localities, were so dramatic as to make it worthy of study (in Chapter 10). The other broad divisions were all too large a portion of the total economy to be safely excluded from study (though we need not examine every subsector). It is essential, after all, that we understand not only why some sectors released substantial numbers of workers from employment, but why others were unable to absorb them. Finance, government and other services did account for a greater increase in national income in the 1930s than manufacturing (Gordon, 1949).

Within manufacturing, nevertheless, we would naturally wish to devote special attention to those industries which released the largest absolute numbers of workers in the early 1930s. Table 6.2 lists manufacturing sectors in order of the decrease in employment suffered between 1929 and 1933 (Table 6.3 gives a more detailed employment breakdown). Along with the much studied automobile and iron and steel sectors, we find that lumber, textiles, stone and clay, electrical and non-electrical machinery, non-ferrous metals, printing and publishing, furniture, other transport equipment, apparel, and petroleum each released over 60,000 workers. Together, they accounted for well over two million job losses. This list includes almost all of the most labor-intensive industries, those where wages amounted to over 40% of value added.[2] It also includes those in the top ten by value added for both the pre-World War One and early postwar eras. We must add to this list for study[3] the postwar glory industries, chemi-

TABLE 6.1 Sectoral Employment Changes, 1920-1929 and 1929-1933 (1000s)

Sector	Change, 1920-1929	Change, 1929-1933	Total Employment, 1929
Agriculture	-453	-451	10,450
Mining	-151	-340	1,102
Construction	913	-830	2,319
Manufacturing	13	-3,335	10,662
Transport Public Utilities	-122	-1,242	4,071
Wholesale and Retail Trade	1,858	-1,418	8,376
Finance, Insurance, Real Estate	253	-212	1,591
Selected Services	1,113	-529	4,499
Government	463	101	3,066
TOTAL	3,987	-8,161	46,136

Note: these figures understate employment growth 1920-1929.

Source: Historical Statistics of the United States (126, 137) for sectors covered there, Henderson, 1961: 12.

TABLE 6.2 Major Employment Losses Within Manufacturing, 1929-1933 (1000s)

Lumber	412	Electrical Machinery[a]	150	Petroleum	62
Non-Electrical Machinery	250	Stone & Clay	133	Engines	47
Textiles	231	Printing & Publishing	95	Rubber	43
Iron & Steel	225	Furniture	88	Industrial Chemicals	39
Automobiles	204	Other Transport Equipment	79	Paper	37
				Leather	36
Non-Ferrous Materials	169	Apparel	64	Glass	18

[a]This figure has been adjusted to reflect the fact that radio apparatus were included through 1931 but not after.

Source: United States Bureau of the Census: *Census of Manufactures*, 1929, 1933.

TABLE 6.3 Manufacturing Employment: Selected Years

Industry	1923	1925	1927	1929	1931	1933	1935	1950	1960
Food and Kindred	1,715,298	665,279	737,931	635,359	666,237	797,442	888,298	2,396,152	2,036,810
Textiles	931,816	1,628,283	1,694,416	1,707,798	1,420,808	1,476,794	1,687,737	750,499	590,735
Lumber and prods.	219,955	921,266	862,667	866,599	509,665	454,171	579,012	477,752	579,269
Paper and Allied	307,064	222,745	225,506	233,393	194,581	196,380	235,665	762,716	909,091
Printing & Publish.	220,208	314,021	327,534	357,988	316,769	262,993	304,842	307,509	413,473
Ind. Chems & Fert.	26,319	200,325	228,227	259,159	215,176	220,507	258,922	103,762	
Pharmaceuticals	66,717	24,487	25,209	27,122	23,673	19,683	22,128	139,925	134,868
Oil Refining	71,613	65,324	71,234	80,596	68,824	69,047	77,402	68,385	
Other oil-related	137,868	71,042	70,421	66,620	51,322	15,156	18,968	238,671	244,071
Rubber	344,545	141,869	141,977	149,148	99,259	106,283	114,681	385,123	357,682
Leather & prods.	73,335	314,025	315,991	318,472	272,757	282,000	310,755	134,391	
Glass	278,357	69,371	65,825	67,527	49,917	49,797	67,138		
Stone and Clay	892,660	283,665	275,419	250,079	165,656	116,682	155,380	782,953	817,575
Iron and Steel	296,911	851,270	835,501	880,882	626,331	655,536	879,089	346,087	362,705
Non-ferrous Metals	907,707	275,292	270,545	314,741	180,477	155,969	214,986		
Machinery (Electrical Machinery)	234,892	858,843	893,059	1,101,053	693,834	470,580	687,465	765,613	1,340,272
Motor Vehicles	404,886	239,921	241,566	328,722	214,734	130,857	179,641	756,937	694,542
Aircraft	2,901	426,110	369,399	447,448	285,575	243,614	387,801	289,545	680,988
Ships and Boats	62,287	2,701	4,422	14,710	9,870	7,816	11,384	86,642	134,047
Railroad Equipment	80,590	50,224	55,014	55,089	45,262	30,885	44,830	69,592	42,247
Photographic	9,905	50,393	38,031	40,015	18,785	14,266	21,481	43,198	62,420
		9,154	12,120	12,967	10,609	8,975	12,000		

Source: United States Bureau of the Census: *Census of Manufactures*.

cals, pharmaceuticals, airplanes, television, photography and computers. Some other sectors, such as rubber, will be treated along the way, for the fortunes of the rubber industry depended in large part on the fortunes of the auto industry.

The list for study contains all of the industries in which the most famous basic innovations of the interwar period occurred. While we should not lose sight of the fact that the cumulative effect of numerous small innovations is greater than the effect of the handful of major innovations which get all of the attention (this was the conclusion reached by Enos (1962) in his pathbreaking study of the oil industry), we must of necessity focus on the most important innovations, for more is generally known about them. This is not a grave problem, for the small advances are generally derivative of the major breakthroughs.

Not all sectors are equally concerned with research. Recent statistics on the proportion of value added devoted to R&D show that the most research-oriented sectors in the U.S. are aerospace (32.6%), office machines and computers (21.7%), instruments (20.5%), electronics (12.7%), automobiles (12.6%), pharmaceuticals (12.1%), oil (6.4%), and chemicals (4.3%) (Dosi, 1988: 1124). Technical potential is limited in any line of inquiry. Thus, some sectors may find themselves facing several promising lines of exploration, while other sectors face few. There are, presumably, more possibilities worth exploring in electronics than in leather. Small firm size also limits research effort in agriculture, textiles, leather, printing and publishing, and wood; these sectors therefore borrow both product and process technology from other sectors, often from specialized machine and instrument manufacturers.

Table 6.4 provides the most recent estimates of hourly labor productivity growth across a number of manufacturing sectors. While incomplete, these tables give us a starting point from which to analyze the structural and technological forces at work on various sectors. We should note how great the cross-industry variation in productivity growth was. Relative trends in output and employment thus differed markedly across sectors. This is further evidence that a broad sectoral survey is essential.

Outline of Sectoral Studies

It is advisable to outline in some detail how the industry-level studies will proceed, so that we can avoid needless repetition later on. We must understand the intensity of the structural forces at work in a particular sector. However, structural change is generally defined as a change in the relative importance of various sectors. Thus, for every loss there is an opposing gain in some other sector. We wish, though, to understand why the number of sectors experiencing absolute decline far exceeded those

TABLE 6.4 Labor Productivity Growth in Manufacturing 1919-1929

Industry	Productivity Growth Per Year	Change in Rate Versus 1909-1919	Industry	Productivity Growth Per Year	Change in Rate Versus 1909-1919
Food	2.94	1.6	Rubber	6.69	0.5
Textiles	4.08	3.09	Leather	3.25	4.3
Apparel	4.0	1.3	Stone	3.27	4.1
Lumber	2.5	3.7	Iron and Steel	4.37	1.6
Furniture	4.2	4.7	Nonferrous Metals	2.69	0.6
Paper	5.74	4.4	Machinery	3.25	1.8
Printing	1.54	0.3	Transport Equipment	3.04	-0.6
Chemicals	4.04	3.8			
Oil	5.87	3.2	Miscellaneous	5.15	5.4

Source: Woolf (1984), except apparel, lumber, and furniture from Kendrick (1964).

experiencing growth. Some help is provided by looking at the periods before and after the two world wars. This tells us what the long run trend for that industry was. Neither Schumpeter in the 1930s nor R.A. Gordon in the 1940s could know what the long run trends would be. We must, of course, look beyond the numbers to see what is causing the trend. It is important to emphasize that postwar experience can not be used to "explain" sectoral shifts, but only to provide verification of changes which can be attributed to the forces we have spoken of.[4] In a number of cases, this involves an attempt to describe the degree of market saturation, which is defined as an (imperfectly) inelastic demand curve which may be shifting to the left.

The computation of technological unemployment is not without difficulty. Figures are often available on output per hour. If we can describe the trends in output and hours worked, we will know how many workers must be let go or hired. However, productivity itself depends on the cyclical behavior of the economy. "To appraise the causes of observed unemployment is a task rendered difficult by the fact that it is hard to distinguish between the labor-saving consequences of increased productivity and declines in employment that arise from changes in marketing conditions" (Jerome, 1934: 6). We must therefore separate the cyclical element from the technological element. Weintraub (1937: 43) noted that it was

The Procedure for Sectoral Analysis

impossible to completely isolate the technological basis of changes in productivity. We will thus endeavor in what follows to establish the technological roots of productivity advance.

We are interested in technological development in its own right in any case. In particular we wish to discern similarities across industries in technological evolution. We will spend much time tracing the course of innovation. We will highlight the long-term nature of research programs. We will explore the technical and social determinants of the timing of innovation, especially when we discuss those product innovations which came "too late."

Throughout these five chapters, a handful of characteristics of technological progress will repeatedly be stressed. Innovations take time and involve a great deal of uncertainty (thus we should be sceptical of simple short-term feedbacks from economic activity to innovation). Technological advance does not occur evenly through time, but occurs rather in discrete jumps. We will stress the degree to which innovations of all types build upon previous discoveries, and also the economy-wide impact of assembly lines, continuous processing, and electrical machinery.

The key variable in our analysis is employment, for it is the incredible unemployment experience of the 1930s which has drawn our attention. This is a natural focus, and in line with macro-level analysis. "Whatever the differences among them, the various macroeconomic theories share a common goal: the explanation of aggregate unemployment rates" (Margo, 1987). We want to be able sector by sector to determine how much of the changing levels of employment were a result of aggregate forces, and how much a cause.[5] We have argued that the problems of the 1930s can also be discerned in the 1920s. Total employment grew by 10.5% 1919-1929. The labor force grew even faster. 4.5 million jobs were created (net), but these were not enough to prevent chronic unemployment. Sectors in which employment grew less fast than average in the 1920s were already contributing to the unemployment problem. For example, the agricultural sector would have created 1,100,000 jobs if it had an employment growth rate of 10% in the 1920s, but instead 300,000 jobs were lost. We would wish to know to what extent this shortfall was due to the structural and technological forces we have outlined.[6] Then, we would extend this analysis to the 1929-1933 period, to see why another 350,000 jobs were lost in agriculture. In total, there were 13 million unemployed in 1933. We want in the end to be able to utilize the most powerful of mathematical techniques, simple addition. We will, in Chapter 12, sum across industries to see how much of the observed increase in unemployment can be attributed to structural change and technology.

We must look behind the productivity figures (which are weighted by the entire workforce, no matter how active these might be) to see how much labor hoarding there was in 1929. Only then can we discuss how many of these were later let go. After calculating first and second degree unemployment, we can discuss appropriate multiplier effects, for the initial employment changes surely cost many additional workers their jobs.

We can not focus exclusively on employment; the movements in output require explanation as well. While employment fell only a little more than output (33% versus 29%), if we add on the reduction in hours of 21% experienced by the remaining two thirds of the labor force, we see that hours worked fell 18% farther than output between 1929 and 1933. This must represent two forces at work, continued productivity increase in the early 1930s, and the removal of redundant workers that firms had kept on during the 1920s. It is worth recalling the correlation between productivity and employment found by Real Business Cycle proponents. Productivity usually falls in a recession. Recall also that Hamermesh (1989) has suggested that we will see labor-hoarding in a recession but layoffs in a depression.

This large differential between the shifts in output and hours worked has received little notice in the literature (note that output and employment also diverge markedly in the late 1920s). It is important to recognize that much of the unemployment experience must be attributed to forces other than aggregate demand, which could only affect employment levels through affecting output decisions. Moreover, once we recognize that both hours reduction, and especially layoffs, would automatically decrease consumption, by using the multiplier mechanism we can attribute much of the fall in output to these forces as well. It should be noted that over 60% of the decline in both hours and employment had happened by 1931. The same is true for the decline in (private non-farm) output. Given that unemployed workers had to quickly adjust their spending, the contemporaneous occurrence of these declines does not rule out the importance of the multiplier mechanism.

Beyond the extent to which the multiplier mechanism might explain the decline in output (or the degree to which in an era of falling prices, the decline in output may be exaggerated),[7] we have to recognize that neither technological unemployment nor simple market saturation (just an inelastic demand curve) can explain why output falls. Three possibilities remain. It is necessary not only that we have a very inelastic demand curve but that this demand curve be shifting to the left in some sectors. We have seen earlier that this was likely the case, notably for automobiles. This, coupled with a lack of expanding sectors, could be responsible for a significant decline in output. The second possibility is that declining industries which were hoarding labor in the 1920s had built up inventories of finished goods in that period (inventories did fall in the early 1930s). The third possibility involves the role of investment. A world of few rising but many declining

The Procedure for Sectoral Analysis

sectors was destined to be a world of declining investment, and therefore output. While output fell 29% to 1933, consumption expenditure fell only 18%. We have seen before, and will want to expand on in what follows, how productivity was able to improve markedly in the 1930s without requiring much investment. We will also take special note of the experience of capital goods industries.

We should emphasize that the case studies to follow are not carbon copies of each other. As different elements of our argument are naturally of varying importance, each case study emphasizes a different set of points. Some sectors are important because they declined, others because they failed to grow. Moreover, elements not mentioned above are found to be important in particular cases. In some sectors, the technological history is traced back much farther than in others. While the interrelatedness of technological evolution is emphasized throughout, we have avoided discussing the same developments in different sections. The full flavor of the connections can only be obtained by reading various case studies. The sectoral histories are purposely descriptive; we have made our position explicit above, and can let the evidence speak for itself in what follows.

One variable which will only receive occasional mention is profits. We have no general prediction about the level of profits. While market saturation should generate the falling profits which Marxian or Keynesian approaches would expect (Epstein (1934: 41, 46) does suggest that profit rates peaked in 1926, but detects a very slight decline to 1929), productivity increases should have increased profits in other sectors. Epstein found that in the 1920s one fifth of manufacturing firms saw a steady increase in profits, two fifths saw a steady decline, and two fifths witnessed either stagnation or fluctuations without trend. Moreover he suggests that the link between profits and investment was weak.[8] At times, such as when we discuss textiles, profits can be taken as an indication of industrial decline, but for most sectors they need not be a focus of attention.

The main purpose of Chapter 7 is to show that most chemical advances before 1930 were labor-saving (producing low-cost substitutes for existing products), while the key product innovations, though based on research projects begun much earlier, only hit the market well after 1930. Chapter 8 performs a similar function for electrical products, with some additions. The timing of development and the labor-saving effects of electrical machinery are discussed. Several electrical products emerged before or in the early 1920s; the reasons for the temporal gap between these and later product innovations is outlined. Market saturation in radios and appliances is discussed. In Chapter 9 the direct and indirect impacts of internal combustion are assessed. The automobile affected American society in myriad ways; it is important not only to assess its own rise and fall in the 1920s, but the various ripple effects in road construction, rubber, steel, etc. We discuss the timing of product and process innovation in these other sectors, as well as in airplane production. Chapters 7 through 9 each begin by tracing the course of evolution of our three core technologies from the Second Industrial Revolution.

In Chapter 10, we look at manufacturing sectors which appear less closely related to the three key innovations of the Second Industrial Revolution but are nevertheless affected by the same forces. We examine the secular forces which caused employment in energy production to fall through the interwar period. We discuss the product innovations which allowed food processing to fare well in the 1930s — too many of these came later in the decade (while continuous processing steadily displaced labor). In lumber, the rise of cheaper and better substitutes caused secular decline in output, which was aggravated by substantial process innovation. Textiles saw both these forces at work, with the addition of some market saturation in clothing.

Chapter 11 surveys the non-manufacturing sectors of the economy (mining is discussed in Chapter 10), to see why these, far from absorbing displaced labor, actually exacerbated the interwar unemployment problem. In construction, a secular decline was severely exacerbated by an automobile-induced boom in the mid-1920s; output and employment fell sharply from 1927. The transport sector naturally suffered as output fell elsewhere; employment fell much more rapidly than output due to process innovation. Construction activity in railroads, pipelines, and shipbuilding would have fallen in the 1930s even without a Depression. In agriculture, innovations flowing from all three elements of the Second Industrial Revolution depressed incomes and employment in a sector facing inelastic demand and many forces causing demand to decrease. Services is a diverse sector: some elements are closely associated with goods production and thus naturally suffered as it did; secular forces were not acting to increase government spending until long after the Depression began and producer services until well into the postwar period; business machines replaced clerical workers; labor-saving appliances severely limited the scope for domestic service; chain stores and supermarkets dependent on the automobile and refrigeration replaced workers in retailing; and the automobile caused housewives to replace the labor of delivery personnel. With all of these negative forces at work, services could absorb very few of the unemployed.

Chapter 12 will briefly summarize the major results from our case studies. The major task of this chapter is to add up the employment losses which can be attributed to the forces we have described. We will find that with a reasonable multiplier effect we could potentially account for all of the unemployment experience of the 1930s.

A Note on Methodology

The spirit of the Cliometric Revolution of the 1950s was that that which could be quantified should be. Much valuable research has been undertaken under the banner of that revolution in the succeeding decades. Yet even many of it leaders have warned in recent years that, like any revolution, it is subject to excess. There are limits to the degree of exactitude with which we can ever hope to treat the past. Lance Davis has justifiably denigrated the "I would rather be fancy than right" school. There are, quite simply, no statistical means by which the employment and output effects of market saturation or technological change can be measured precisely. The attempts of past writers to measure saturation in the automobile and construction sectors attest to this. This bleak fact would prevent many who style themselves cliometricians from pursuing a project such as this. However, in the social sciences knowledge is advanced not by conclusive proof but through interplay among scholars. (This is fortunate, for Eichengreen and others have recognized that time series econometrics will never tell us what caused the Depression.) Only when a particular result has been replicated, ideally by different means, should it be accepted. I will pursue in what follows the simplest of methodologies. Sector by sector, we will describe the various forces at work before, during, and after the interwar period. We will quantify these forces as much as possible. Then, using common sense we will establish ranges for the likely employment and output effects. My greatest hope is that scholars with expertise in particular sectors will refine these estimates (in what follows, we will quote at length from the authors of secondary works, to establish that writers with a different focus have generally reached conclusions in line with our arguments). Much more could be said if industry histories in the future could deal more explicitly with the Depression-related questions raised here. Only sector by sector can the quantitative importance of this line of argument be denigrated.

Notes

1. Buxton, in the introduction to a collection of essays on British industries between the wars, notes that "in an analysis of the instability of the economy after 1918 the studies presented here concentrate on a vital area. They provide the groundwork upon which general theories of the relationship between secular and cyclical growth during the interwar period must ultimately depend" (1979: 10). Along with change in technology, industrial organization, and residential construction intensity, Buxton feels that trends in per capita income and the marginal propensity to consume have important explanatory power. "Several important conclusions emerge, above all the danger of making broad generalizations" (1979: 14).

2. Bernstein (1987: 131) lists these as cotton (51.1%), non-automotive transport (50.7%), leather (50.4%), lumber (50.2%), textiles (49.9%), automobiles (48.9%), apparel (47.4%), iron and steel (44.4%), and furniture (43.8%). He notes that the relatively successful sectors such as chemicals (19%) and electrical machinery (35.6%) had a more limited role for labor.

3. Stone and clay, printing and publishing, furniture, and non-ferrous metals will only be discussed in passing in related sectors. There has been little research performed in these areas.

4. It could, after all, be claimed that the postwar boom was due to pent-up demand from the Depression and war. Employment and output might have been squeezed in some sectors by rapid growth elsewhere. Thus, we would have to be careful in treating the postwar world as the natural outcome toward which the interwar economy was moving. Still, if we find that a sector does poorly in the postwar period, we will be less willing to downplay the role of secular forces in poor interwar performance.

5. On the basis of the 1940 census returns, Margo (1987) has ascertained that unemployment was quite unevenly distributed sectorally, even at that late date. Construction workers had a 25% unemployment rate, miners 8%, and workers in manufacturing just under 6%. Duration of unemployment was also quite different for mining, construction, heavy industry and services. By occupation, rates ranged from 1.3% for managerial and 2.5% for professional to 5% for clerical, 6.4% for service, 9.1% for skilled, and 13.9% for unskilled labor.

6. David Weintraub estimated that productivity increase between 1920 and 1929 had made redundant 32 out of every 100 workers in manufacturing, but that increased output had replaced 27 of these, for a net employment loss of 418,000. Extending his analysis to railroads and mining, he arrives at a net employment loss of 1,003,000 (in Jerome, 1934: 382).

7. There is, after all, an impression to be gathered from those who were employed through the 1930s that life was good. If this is so, and if the unemployed had tended to have lower average incomes beforehand, we would not expect output to fall anywhere near as much as employment. There is, as always, some difficulty in accounting for the changing quality of goods.

8. He found, for example, that one sixth of industries witnessed investment increases greater than sales growth, despite the fact that profits were falling. Kennedy (1939) suggested that within industries profits were becoming concentrated in fewer companies. Thus, along the lines of Bresnahan (1991), the link between profits and such variables as investment and employment would depend on the rate at which inefficient firms/plants were squeezed out of the industry.

7

Chemicals

The chemical industry has existed as long as there have been dyestuffs, cleaning agents, and medicines. Improvements in bleaching and dyeing made an important contribution to the English Industrial Revolution. Yet the chemistry of that era must look primitive to modern eyes. Experimental method was employed in laboratories, but, "The old industries were arts, conducted by skill of eye and hand, acquired by long practice of the craft; for this skill there was no substitute. The new industry has steadily discarded judgement and skill in favor of analytic control, calculation of the optimum conditions, and the check of experiment" (Taylor, 1957: 12). The same chemical of different degrees of purity was often called by different names, and different chemicals known by the same name. The eighteenth century did see the scientific explanation of combustion, and elucidation of the relationship between salts, bases, and acids. The nineteenth century, though, would see much greater advance in chemical theory. In the first decade, Dalton applied atomic theory to chemistry. He assumed that all compounds combined elements in a simple numerical proportion. Compounds were thus limited in number, and of fixed composition. If there were different compounds of the same elements the elements must occur in numerical proportion: there was twice the oxygen/carbon ratio in carbonic acid as in carbonic oxide. Though there was no way of measuring these proportions at the time, chemical equations were of great use in understanding chemical reactions. By the middle of the century most of the industrial processes in use could be given an explanation in chemical terms; nonessential ingredients were identified, and the best temperature and volume for a reaction ascertained. About 1860, Canizarro in Italy extended Avogadro's work to develop a theory for the establishment of atomic weights. Formulae of all substances which could be turned to vapor, including most carbon compounds, were thus ascertained. In 1882, Raoult developed a method to establish the formulae of all solubles (only in the 1920s, using X-rays, could the molecular weights of silicates be discovered). Also about 1860, valency theory was developed; this showed that atoms of particular elements always combine with a certain number of other atoms. This greatly facilitated the establishment of formulae for compounds.

Organic chemistry (traditionally viewed as dealing with living organisms but which generally revolves around the manipulation of compounds of carbon) scarcely existed before the mid-nineteenth century. Yet the modern dyestuff, pharmaceutical, plastic, and other industries are dependent on organic chemistry. At mid-century, Berthollet established that organic compounds could be synthesized. Often, very precise conditions of temperature and pressure were required for successful reaction. The training received by German chemists in the 1840s and 1850s paved the way for future developments. "The methods of industrial organic chemistry were worked out in the establishments of the synthetic dye industry, between 1856 and 1880" (Taylor, 1957: 233). The plant and intermediate substances used for dyes could be applied to other organic chemicals. In the 1860s the main constituents of natural dyes such as madder and indigo were synthesized. By the 1870s trial and error research gave way to research planned on scientific principles. There were 15,000 known organic compounds in 1880, 150,000 in 1910, and over 300,000 in 1940 (Haber, 1971: 7).

The pace of chemical change accelerated through the late nineteenth and early twentieth centuries. Haber identifies three key differences between the two centuries. Equipment was increasingly standardized (the requirements of organic chemistry having previously encouraged the development of chemical engineering) and layout rationalized. Methods for continuous, rather than batch, processing were introduced. And, with the development of electricity production, electrolysis became an increasingly common chemical method. All of these developments owed something to science (Haber, 1971: 3). The most significant change at the turn of the century was the intermingling of the physical sciences. One result was the laws of thermodynamics, which helped establish the time and temperature conditions for chemical reactions. The role of catalysts, which aid the process of reaction without reacting themselves, was also understood only in the early twentieth century, though they had been used for years.

Between 1900 and 1930, the American chemical sector averaged annual growth of 6.3% compared to overall industrial growth of just 4.6% per year. As in Europe, the chemical sector had become one of the growth sectors of the economy. Its effects spread far. In the postwar era, it is hard to find a product which has not undergone chemical alteration at some point. In the early decades of the century, the industry was gradually extending its tentacles into a variety of new products. Most new chemical products of the early decades were substitutes for existing products. The Temporary National Economic Committee noted the labor-reducing effects of such diverse products as rayon, synthetic chemicals, Duco lacquer, and paper bags.

Within a generation of 1900, the American chemical industry had become the largest and most diverse in the world. As in Europe, the educational system had an important role to play. Free secondary education had provided an enormous pool for nineteenth century universities to draw upon. Universities were not as distant from industry as in Britain nor as academically oriented as in Germany. The research-oriented graduate education of Germany was adopted in the late nineteenth century. Training in the early twentieth century leaned heavily toward applied work and had a strong practical bias (Haber, 1971: 61-8).

The First World War exerted perhaps its greatest effect on the chemical industry. It encouraged the growth in scale and size of firms. Allied governments (as well as market shortages) encouraged developments in areas such as dyestuffs and pharmaceuticals which had previously been dominated by German industry. The immediate postwar period would see tariffs to protect these "new" industries. The allied governments also took over German patents as part of the war indemnity and made these available to domestic producers.

As the twentieth century progressed, large firms increasingly dominated the process of innovation. The chemical industry was one of the first to embrace the concept of industrial research laboratories in the United States. Du Pont opened its laboratory in 1902. It could draw on the precedent set by German dyestuff firms from the 1880s. Du Pont's policy of product diversification was achieved for the first decades by obtaining rights to new technology which was being rapidly developed in Europe, and using its labs to commercialize it. By the late 1920s, though, its central research lab had come to perform basic research which could affect a number of its manufacturing divisions. Still, the focus was on development through the 1920s, the major research effort being that which resulted in neoprene, the synthetic rubber, in 1930 (Hounshell, 1988: 150). Du Pont's success with this and with early efforts toward synthetic fibers, along with continued fears of antitrust, and perhaps also the economic situation itself, encouraged greater devotion to fundamental research in the 1930s.[1] As the Depression worsened, the new research director tightened the reins on fundamental research and emphasized applications more; the fundamental effort relied heavily on a financial reserve built up in the late 1920s (Hounshell, 1988: 239-42). Overall, though, Du Pont research spending would double in the 1930s; there was a decision at the top that research to be successful had to be continuous and that they should not give in to short-term concerns. Other chemical firms followed Du Pont's lead in research. The number of research labs rose from 550 in 1921 to 1000 in 1926. The number of professionally trained researchers more than doubled 1919-1929. Contacts between industry and university research developed slowly, and this meant there was a gap between pure and applied research. In the 1930s this gap would be closed (Haber, 1971: 352-72).

The advance of chemicals was inextricably tied to advance in other sectors. The boom in radio "miraculously transformed our plastics industry" (Haynes, 1948: v.4, 5). Ignition trouble in early cars was solved by plastic distributor caps; chemistry also produced anti-knock gas, better lubricants, antifreeze, coated fabrics, exterior lacquers, and better alloys (chemists had long been involved in steel mills). The movie industry was the third which had a large impact on chemical production in the 1920s (Haynes, 1948: v.4, 6). The textile sector also played a role, as a receptor both of dyes and synthetic fabrics. Soap, paint, glass, and tanning were other important outlets.

The 1920s would see a number of new chemical products. Viscose rayon appeared in 1920 and acetate rayon in 1928, cellophane in 1923, synthetic ammonia in 1924, improved photographic film in 1924, and tetraethyl lead for gasoline in 1924. Yet the 1920s would pale in comparison with what would be achieved in the 1930s. Haynes, indeed, in his multi-volume history of the chemical industry, would subtitle the volume on the 1930s, "the decade of new products."[2] In 1937, Du Pont could claim in its annual report that 40% of its sales came from products relatively unknown in 1929 (Fano, 1987). Scientific advances in understanding the atom, and the effects of temperature and pressure, had increased the potential for new product creation (Haynes, 1948: v.5, 37). We can list only a few of the 1930s product innovations: freeon in 1930, better paint pigments from 1931, lucite plastic from 1936, nylon from 1939, vitamins and a whole range of new pharmaceuticals.

Efforts at product creation had not totally detracted chemical firms from process technology, to say the least. Du Pont had from the outset encouraged the study of chemical engineering, recognizing the problems inherent in scaling up laboratory discoveries into commercial propositions. The 1930s would witness the introduction of a variety of new apparatus to achieve automatic control of temperature, pressure, weight and volume of flow, specific gravity, and humidity (Haynes, 1948: v.5, 37).

The Second World War had a mixed impact on the chemical industry. Much wartime research had little civilian spillover. Du Pont concentrated on the Manhattan Project, but decided to leave the nuclear field after the war. As in other industries, fundamental research not linked to the war effort ground to a halt. One glimmer of light was the fact that the wartime excess profits tax meant that many firms only paid 38¢ for each dollar of research undertaken. Moreover, "World War Two revolutionized the chemical industry by greatly accelerating the rate and extent of the displacement of natural substances by synthetic polymers" (Hounshell, 1988: 474).

Synthetic fibers, rubber, and plastics would be major areas of postwar growth.

Du Pont executives in 1945 predicted a postwar boom in research both in industry and in universities. The next five years saw Du Pont alone spend $50 million on new research facilities, and the five years following even more. The company saw its future success depending on long term research projects. Their success with nylon at the end of the 1930s provided the prototype which they and other chemical firms were to follow for the next decade. The result, not surprisingly, was that a range of new products emerged.

It would be impossible to trace the development of all the chemicals which emerged during the period of our interest. Even those who have written books on the industry have been forced to be selective in their coverage. We have tried to sketch the big picture above, and will highlight what seem to have been the developments with the greatest economic significance in what follows. Some are final products whose role is readily apparent; others are inputs into a wide variety of other industrial processes. Hopefully, a reasonable picture of the forces at work will emerge. Unfortunately, statistics are often unavailable on the output of individual product lines (Haber, 1971: 4). Ballpark estimates, especially in areas of rapid growth, are usually available.

Electrochemicals

Valency theory had given chemists great facility in handling organic chemicals, because of the behavior of carbon, hydrogen, and oxygen. It had been difficult to apply the theory to inorganic chemicals, however. Chemical reactions there could only be understood with the recognition that molecules were electrically charged. Ionic theory emerged in the 1880s, and became widely accepted in the 1890s, though it did not become common in textbooks until the 1920s. It resulted from collaboration between physicists and chemists. Finally, a unified theory could explain the behavior of both organic and inorganic chemicals.

Electricity had been used for electroplating from the 1840s, copper refining from the 1860s (electric wiring providing the greatest market for copper), and aluminium smelting from the 1880s. The scientific understanding of electrolytic disassociation from the 1880s was to open up a number of new possibilities. Electrolysis had previously been a tricky art; ionic theory explained why it worked and forecast the best conditions. Inexpensive electricity also supported growth. "Electrochemicals made no headway until the dynamo had been perfected and the rotary converter introduced so that direct current for the cell systems became available" (Haber, 1971: 31). By 1900, along with electrothermal methods of producing phosphorus, carbide, and graphite, there were electrolytic methods for manufacturing aluminium, caustic soda, chlorine, caustic potash, chlorates, hydrogen, and hydrogen peroxide. Developments were heavily dependent upon advances in chemical engineering, for continuous processing was the key to commercial production.

Much of this must be considered process innovation — lower cost production of existing chemicals. However, these advances also facilitated the mass production of automobiles and machine tools. Aluminium was the most far-reaching success story. Electrolytic methods were devised in 1886 which would form the basis of aluminium production for decades. They resulted in a tremendous drop in price and increase in purity, which allowed aluminium to enter a steadily increasing range of uses. Inexpensive electric power after Niagara in 1895 provided a massive boost to aluminium production. Aluminium, under the right conditions, reacts violently with oxides of other metals. It thus became the key to the isolation of manganese, tungsten, chromium, molybdenum, and combinations thereof. These were used to manufacture steel alloys of special toughness, heat resistance, and hardness. Among the other uses which would open up would be tungsten cutting tools in the 1910s and tungsten carbide cutting tools in the 1930s, which would greatly increase labor productivity across industry. Manganese and aluminium itself were combined around 1910 to produce the very light but strong duralum.

Electrolytic production of alkali also emerged in the late nineteenth century. The Dow chemical company got its start in this line of production. While overshadowed by advances in organic chemicals, these inorganic chemicals had important roles to play as well. Their main uses at the time were for fertilizers and explosives. At the end of World War One, indeed, producers feared they would suffer from overcapacity. In the next decade, though, soda ash would become an important input for a range of chemicals, glass, soap, and paper, while caustic soda would be used widely in the oil industry and rayon production. The alkali revolution was one of the most profound changes in the chemical industry in the 1920s (Haynes, 1948: v.4, 112). Demand for soda purity from rayon producers induced a number of technological improvements.

There was competition between different electrolytic processes using different inputs. This resulted in a number of other chemicals emerging as byproducts. Chlorine, while not an alkali, was commonly produced at the same time. The lower prices which resulted caused chlorine to replace bleaching powders in textiles, paper, and other industries, and aided its introduction into water treatment in the mid-1920s. Researchers soon sought out still other uses for chlorine. Sulphuric acid was another byproduct; lower price and higher purity suited it to the manufacture of dyes and drugs, as well as petroleum refining.

Among other results of the combination of electricity and chemistry was acetylene (from 1892), which had uses in lighting, welding, and fertilizers, and would become an important input into the manufacture of vinyl, synthetic rubber, and a variety of plastics. Carborundum, or silicon carbide (from 1893) was an important abrasive (for sanding equipment etc) and furnace lining. Graphite (from 1893) was used for electrodes which greatly enhanced the quality of electronic equipment.

Petrochemicals

The manufacture of organic chemicals, with a few exceptions, continued to rely on coal tar derivatives through the early decades of the twentieth century. Only as petroleum production expanded so that petroleum derivatives were readily available did this change, and even then the familiarity with coal tar had to be overcome. Some products were produced from petroleum during the war in both Europe and America. The United States was to take the lead after that. Three different companies established large works to produce a range of chemicals — isoproponal, ethanol, acetic acid, hydrogen — from oil and natural gas. "Petroleum chemicals, which now account for the bulk of chemical output, date from the 1920s" (Haber, 1971: 373). They were not new products though; all had been previously prepared from coal tar, fermentation, or wood distillation. The only advantage was cost. As oil production expanded in the United States, byproducts could be obtained at virtually no cost (some had been burnt as waste). In volume terms, the requirements of the chemical industry were minuscule compared to oil output. The use of oil involved much higher temperatures and pressures than had production based on coal tar; thus intricate chemical engineering had to be devised (Haynes, 1948: v.5, 208). Average price per ton fell as a result from 43¢ in 1921 to 13¢ in 1939.

While the advent of petrochemicals in its early stages must be viewed as process technology, the dramatic fall in cost through the 1930s did induce the creation of hundreds of new chemicals (see Haynes, 1948: v.5, 208). Moreover, the carbon in oil has different properties from that in coal. It turned out to be superior for the manufacture of antifreeze and a variety of solvents. Most importantly, polymers, the building blocks for plastics and synthetic fibers, require oil-based carbon, for there the atoms are organized in short chains rather than rings.

Plastics and Synthetic Fibers

Plastics[3] and synthetic fibers would be two of the most important growth sectors in the postwar period, but had shown little of this potential in the interwar period. They are both based on polymers: large molecules made up of thousands of smaller ones joined in a chain. They both, therefore, share a common ancestry. As a byproduct of the search for a filament for electric lamps, Swan in England and Chardonnet in France both experimented with cellulose in the 1880s. At least three different fibers were created within the next decade, though these were all surpassed by metallic filaments. None of the cellulose-based products can be viewed as truly synthetic, as they all began with polymer-based plant matter. Nevertheless, the methods employed in their production over the next decades would prove to be of great value when true synthetics appeared.

A handful of firms in the United States and elsewhere spun cellulose fibers for such products as ribbons and tassels in the early twentieth century, but their output was minuscule. As well, there was small scale plastic production. Celluloid had first been produced in 1869, and had come to be used in such products as combs, cigar cases, and billiard balls. Another important product was film for movies; celluloid would be the mainstay through the 1920s, after which its flammability would lead to replacement. All told, only some tens of thousands of tons of plastic were made on the eve of World War One.

Another plastic, bakelite, was patented in 1907. Its inventor, Baekeland, had been searching for a better varnish at the time of discovery. Simultaneous research effort in three countries resulted in much patent litigation. Production by a handful of firms of this phenol-based substance remained small before the war though it did find uses in car parts (especially brake linings), wiring, and as insulation. In the 1920s, expanded use in automobiles plus its appearance in radio components caused production to rise from 2600 tons in 1922 to 14,700 tons in 1929. It was providing by that time keen competition to celluloid. Bakelite established the possibility of making strong, hard articles of complex irregular shapes, and thus spurred the search for other plastics (Taylor, 1957: 258). Through the 1920s, though, plastic manufacture remained in the hands of small firms. The large chemical companies which would come to dominate the field were only beginning to show interest.

With hindsight, we can see signs of promise in the 1920s. "Diversification of plastics materials, the salient development of the next decade, started during the 1920s" (Haynes, 1948: v.4, 354).[4] Casein plastics were used to make buttons. Synthetic lacquers and resins were developed and improved. The development of the radio and the end of Baekeland's patent provided important spurs to research effort.

Much of the research effort in plastics in the interwar period took place in Germany. This was in large part due to German government interest in synthetic materials. German industry also was very supportive of scientific inquiry in the area, and due to the complexity of plastics, scientific and technological advance were closely tied. Other countries often did not adapt German innovations of the 1920s and 1930s until the postwar period (Freeman, 1982: 88-95).

As the name cellophane suggests, this transparent film was first developed in France. Du Pont obtained the American rights from the French firm in 1924. They financed one researcher at a cost of $5-10,000, who succeeded in developing a moisture-proofing for cellophane from 1927. This, not surprisingly, had a huge impact on sales. Sales tripled 1928-1930, and increased another 60% in 1931. Cellophane, then, is a small example of a radically new product whose output grew despite the Depression. The expansion was aided by changes in methods of distribution toward self-service and unit sales (customers wanted to see the meat they were buying, for example, and cellophane allowed cuts to be individually wrapped). As with many new products, Du Pont horribly underestimated the prospective market; its 1928 forecast was out by 200% by 1931 despite the Depression. Though the 1930s, and especially 1940s, Du Pont devoted millions to researching methods of reducing production costs of cellophane (Mueller, 1962: 328-9).

The most impressive development of the 1920s, though, was in the area of synthetic fibers. Their production had been given a big boost by World War One. Still, it was developments right after the war that were the most astonishing. World production of synthetic fibers was 11,000 tons in 1919, 61,00 tons in 1924, and 197,000 tons in 1929. Rayon (a cellulose-based fiber) suddenly appeared as a potential rival to cotton, though total production remained small in comparison. Rayon also expanded at the expense of lisle and silk (rayon became the middle class silk) and a mass market in hosiery emerged (hemlines were on the rise in the 1920s). Rayon was the fastest growing chemical sector through the interwar period, averaging about 19% per year. Rayon producers earned high profits through the 1920s. The early product tended to be weak, brittle, and irregular, and disintegrated when wet. Acetate rayon, from 1929, proved greatly superior to the previous viscose rayon. A series of small product innovations were essential to the expansion in sales, allowing rayon to invade more and more textile markets (see Hounshell, 1988: 167). From 1934, for example, rayon was used in tire cord, though its advantages were only fully recognized with synthetic rubber tires in the 1940s.

Process innovation also played a key role in rayon expansion. Not only did it replace more expensive materials, but its cost of production steadily fell. Hollander (1965: 192-9) has examined in detail the course of productivity improvement in rayon. For the 1930s and after, he finds productivity growth rates of over 4% per year on average for his selected plants. These increases came not from a handful of breakthroughs but from a series of minor technological improvements (increases in scale, and improvements in labor quality and organization, make only minor contributions). These appear to have developed on the shop floor rather than in the laboratory. Much of this new technology did not require investment (or could be easily facilitated with replacement investment); thus productivity rose 22.2% in the 1930s, about as fast as the 26.6% recorded in the 1920s, despite net investment being eight times higher in the earlier period. It appears that both product and process technology spread rapidly across firms.

The 1930s would see many important developments in both plastics and synthetic fibers. The largest markets for plastics continued to be cars and radios, though novelty items in five and ten cent stores quickly appeared (Haynes, 1948: v.5, 329). The first new plastic of the 1930s was urea-formaldehyde. This had a variety of uses, from light fixtures to plywood to an antiwrinkle agent in fabrics. In 1934, new resins allowed plywood to be set in minutes rather than days. Acrylic (from petroleum) was commercially produced from 1931. By 1935, it had been used to create Plexiglass. There were numerous uses for this transparent moulded substance. It was quickly adopted for lenses, reflectors, and watch crystals, and was used extensively in airplane windows during the war. The patent for lucite, another acrylic derivative, was obtained from the English producers by Du Pont in 1936. The third new plastic was vinyl. This allowed more grooves on records, and was used in toothbrushes and dentures. As new potential markets were found for the new plastics (the process being aided by the importation of the first injection molding machine from Germany in 1935; this greatly expanded the color range available, and made transparent and translucent moulding possible), a plastics boom emerged in the late 1930s. Costume jewellery, buttons, knobs, gadgets, and novelties were among the results.

Costs of plastics production fell steadily. The price of acetate, used in steering wheels and control knobs, for example, fell from $1.25 in 1930 to 35¢ in 1939 (Haynes, 1948: v.5, 331). Plastic still remained more expensive than wood or metal. It thus had to obtain its market by either being of better quality for the purpose, or serving a purpose previously impracticable with traditional materials.

The molecular chains in rubber are kinked, and this allows rubber to expand and contract. The key to producing synthetic rubber was to imitate these kinked chains. Du Pont had been developing chemicals in the laboratory for the rubber industry since the early 1920s. Toward the end of that decade the firm launched a basic research effort to find a synthetic rubber; they succeeded, partly by accident, within a couple of years. The discovery of neoprene (the name came later) occurred in 1930, was announced in 1931, and commercial production began in 1932. There were problems in controlling the chemical reaction; explosions caused three deaths. The company had to devise new methods of polarization. The problem was solved by the end of the decade, in part through licensing German technology. The sales campaign focussed at first on tasks which natural rubber performed poorly. Neoprene was more resistant to weather, oil, and gas. The development of odorless neoprene in 1937 opened markets in gloves, shoes, heels, etc. Still Du Pont lost money through the 1930s on neoprene; it was only with the onset of war and thereafter (rapid growth occurred through the 1940s and 1950s) that it recouped its heavy research expenditure.

The most famous of Du Pont's, and the industry's discoveries during the 1930s was in the synthetic fiber area. About the same time it began working on neoprene, the firm started basic research on synthetic fibers. In just over a quarter of a century it would engineer a revolution in fibers (Hounshell, 1988: 384). The fiber revolution, which was to have wide-ranging effects on both input and output industries, required the creation of a whole new body of knowledge (Hounshell, 1988: 384). The efforts of the late 1920s were only one step in the process; the first truly synthetic fiber in 1930 was unable to withstand either washing or drycleaning. Though executives continued to believe in fiber research, the research lab devoted its efforts elsewhere for three years. The 1930 discovery might well have been lost but elements in the company, notably the rayon department, pushed for the research to continue. The firm hired Wallace Carothers to engage in basic polymer research (he was only interested in such), but then induced him to extend his research toward application. The first nylon fiber was produced in 1934. Nylon, like neoprene, required the development of sophisticated production technology, as the reaction had to be strictly controlled. Du Pont launched a crash five-year program to bring nylon to the market, spurred on by the impregnable patent position they had achieved. While they were able to borrow expertise from various company divisions, numerous entirely new developments in processing and testing were required. Nylon manufacture required a number of stages which had to be designed from scratch. Du Pont also had to be able to supply advice to those firms which would produce finished goods with nylon. Du Pont aimed its research at the stocking market. During the war, though, nylon output would be diverted to parachutes, tire cords, and glider tow ropes. A pent-up demand was therefore ready at war's end for not only lingerie but tire cord and carpets as well. Nylon would become in the postwar period the greatest moneymaker in Du Pont history (Hounshell, 1988: 237-72).

The success of nylon fired considerable research effort in the postwar period. Major developments in the 1950s included Orlon acrylic (discovered 1948) and Dacron polyester (1949). The development costs for these were even greater than for nylon, in part as resilience and water-resistance had to be established for these to gain their market niches (and resilience had to be studied in its own right before it could be achieved). The achievement of water resistance in turn created problems with dyeing. In the end, though, polyester was destined to become the most important synthetic in the U.S. market (Hounshell, 1988: 386-408).

Plastic production would expand 600% in the decade after 1939, largely due to wartime necessity (the government had proclaimed in 1941 that plastic should be used wherever possible in place of materials such as aluminium which were in short supply). The war did not so much see the introduction of new products as the large scale production of products introduced in the 1930s. In the postwar period, producers could tailor plastics to the needs of buyers, with minor adjustments in production processes. Intense competition would lead to the synthesis of literally thousands of new polymers.

The most important new plastic was polystyrene, destined to become the first billion pound per year plastic in the 1950s. Discovered in England in 1933, the lack of a viable production process delayed its commercialization till 1940. After military application during the war, such as in radar, it emerged in the postwar era as a major input in film, coated paper, moulded and extruded articles, wire and cables, bottles, and pipes. Output would expand 50% per year 1948-1953, and by 40% per year through the rest of the 1950s.

Teflon was another important postwar development. Discovered in 1938, its development was aided by the Manhattan project. Still, it was only in the mid-1950s that all of the production problems were worked out. The very properties that made Teflon so useful, such as resistance to heat, rendered it difficult to manufacture. Sales expanded steadily through the 1950s and 1960s. Only in 1961, after feedback from salespeople, and with the threat of French exports, did Du Pont overcome its fear of lawsuits and apply Teflon to coated cookware.

The polymer revolution was not without limitation. Du Pont, which spent $150 million on research in the 1950s, increasingly focussed on the production of special purpose chemicals for yet unsatisfied needs. The new products of the 1950s and 1960s had saturated most markets (Freeman, 1982: 96-7). "The polymer bonanza was over" (Hounshell, 1988: 501). While it lasted, though, it had provided not just chemical companies but the economy as a whole with a major source of growth.

Photography and Motion Pictures

We can conveniently date the era of modern photography from the late nineteenth century when celluloid film replaced glass plates. The mass market for photographic materials and apparatus was opened after 1889 with the advent of roll film (Jenkins, 1975: 4). Lenses free of aberration were developed at the same time (hand cameras had become available in 1880). Output expanded by 11% per year for three decades. By 1900 one in ten in both England and the United States possessed cameras. Chemical research improved film sensitivity and decreased the necessary exposure time. In 1906 a panchromatic emulsion sensitive to all colors of the spectrum was developed; this was only possible due to recent developments in dyestuffs (Gernsheim, 1978: 1282). Eastman Kodak, which dominated the American industry, increasingly moved toward continuous processing in film production. Years of German competition (as well as the prestige associated with science) encouraged the firm to devote greater efforts to product innovation (Jenkins, 1975: 181-5).

The first decades of this century produced few real breakthroughs. "Even if the inventions that transformed the technological evolution of photography in the first five decades of the twentieth century seem less startling than those of the previous six, they nevertheless provided the final and logical solution to a number of important problems that had engaged scientific minds for a long time" (Gernsheim, 1978: 1282). Better films and lenses allowed continued reduction in exposures. However, solutions which could significantly expand the market for cameras generally came after 1930. Even allowing for a propensity to exaggerate, the opinion of Sheldon and Arens (1932: 67) is revealing: "As offered to the public today, cameras of all kinds are too complicated for the general run of consumers. As a result of unsatisfactory prints many people who could easily afford expensive cameras prefer the cheap box affair with its fixed focus. Others have become disgusted with poor results in the finished prints and have given up picture taking entirely." The development of 35mm film for movies stimulated the development of "miniature" cameras; the first successful one was the German Leica in 1925. American models, as well as new and improved enlargers, emerged through the 1930s. Pictures were very grainy until fine grain film was developed after 1933. After various methods had been experimented with for decades, the first successful color films — Kodachrome and Agfacolor — emerged in the mid-1930s for slides; extension to prints awaited the 1940s (widespread amateur use also awaited the postwar period). Film of all types became less expensive with synthetic production of acetic acid in the 1930s. The first light meters and important shutter improvements also emerged in the 1930s (Gernsheim, 1978: 1284-90). Nye ends his study in 1930 because "New reproductive technologies emerged in the 1930s, changing the photographer's work and the range of possible images" (1985: xi). He speaks of smaller cameras, better color film and better flash bulbs. While the postwar expansion in output and employment in photographic equipment and supplies may reflect various social developments (including rising incomes), it must in large part be seen as the result of technology that had not been there a couple of decades before.

The social repercussions of the onset of motion pictures likely far exceeded the strictly economic impact; however the employment fortunes of tens of thousands both inside and outside the industry would be affected. The first experiments with movies had occurred at the very end of the nineteenth century. It had long been recognized that the eye could be fooled into perceiving motion when presented with a rapid sequence of still pictures; flash cards which achieved this illusion had been a popular novelty for some time. The technical difficulties with taking and showing a rapid sequence of photos were gradually solved. The first decade of the twentieth century saw the emergence of thousands of storefront nickelodeons which would create shows out of several one-reel (several minute) films. By 1909, production of movie film had become of greater importance than film for still photography (Jenkins, 1975: 4). The audience for these films was largely drawn from the working class (especially immigrants) and the films were often produced by working class members to highlight everyday problems of the wage-earning urban masses. Before World War One, multi-reel feature films (the word feature, and other terminology, being borrowed from vaudeville) were introduced. The previous opposition to the existing social order disappeared. By the end of the decade movies had won a large audience among the middle class, where they could be viewed as a substitute for playgoing. The patent monopoly which had previously dominated film making was broken up in 1915. Major Hollywood studios were established, and tens of thousands of movie palaces were constructed. In the 1920s the industry continued to grow, as movies became a national obsession. By the end of the decade, virtually every town in the industrialized countries had a theatre (Thomas, 1978: 1302).

The idea of sound had been there from the beginning; Edison had experimented with movie-making with an eye toward expanding the phonograph market. Studios showed relative disinterest through the early 1920s, perhaps conscious of the cost involved in switching to sound. The problem was synchronization; a method had to be developed of imprinting the sound track on the film itself (and reading same) to ensure that sound and picture moved together. This was done in the early part of the decade by borrowing from radio, phonograph, vacuum tube, and photoelectric cell technology. The Jazz Singer in 1927 launched the new era of sound. Studios borrowed $350 million to finance the conversion to sound. While there were only about 100 theatres wired for sound when the Jazz Singer premiered, there were 8700 by the start of 1930 and 13,500 by the end (Williams, 1982: 1313). It took three to five years for all of the technical difficulties of sound to be worked out, but attendance and studio profits jumped from the start.

The movie industry, for largely technological reasons, boomed in the 1920s. Sound movies were considerably more expensive to make than silent films, so that there was a positive employment effect at the production end. In theatres, "the rapid spread of the `talkies' has resulted in a sharp decline in demand for theatre musicians" (Jerome, 1934: 32). Ten thousand, or about half the total, had been dismissed by mid-1931 as a result (Jerome, 170-1). This number was balanced by an increased demand for projectionists as attendance rose in the late 1920s. In terms of structural unemployment, of course, the tradeoff may be quite negative. Kraft (1994) discusses how displaced musicians flooded the Los Angeles market for studio musicians in the 1930s; excess supply prevented the musicians' union from wielding much bargaining power. The biggest employment effect, though, was on vaudeville. Most of the 1000 vaudeville houses were turned into full or part time movie houses. Centralizing the production of entertainment in Los Angeles proved to require less labor than the previously decentralized approach.

Attendance did fall off in the 1930s (some of vaudeville's decline might also be partly attributed to the Depression). Many contemporaries felt that radio was an important challenge to movie attendance. The average American still saw one movie per week. The studio response to the Depression was to introduce the double feature with "B" movies – low budget films which were leased to theatres at a flat rate rather than on a per-patron basis. Serials, which would induce patrons to return again the following week, were also a big part of Depression fare. It was after 1933 that the real decline in attendance set in, indicating that people had come to view movies as almost a necessity by 1929, and that perhaps radio was an important cause of decline.

Attendance might have held up somewhat better if the next technological advance had occurred earlier. Methods of creating color film by tinting the film had been developed near the beginning of film making. Among various problems with this technique, including poor color quality, was the fact that it could not be applied to sound film. Filming twice through different filters was tried for short films in the 1920s. The modern color movie process was developed by Technicolor Corporation in 1935. Their exercise of patent rights limited the spread of the technology. Even in the late 1940s less than 12% of Hollywood features were in color. Due to the end of the Technicolor monopoly, and the rise of competition from television, this figure rose to 50% in 1954 and 94% in 1970.

Television would provide damaging competition for movies in the postwar era. Employment in movies would fall from 250,000 in 1947 to 187,000 in 1959 (almost the same level as 1939). The movie industry thus had room to grow in the interwar period that it would not have possessed if television had been developed earlier.

Pharmaceuticals

Some man-made substitutes for natural drugs, such as ether, had been used since ancient times, and others, such as chloroform, were developed in the early nineteenth century; still, one can date the beginning of a pharmaceutical industry to the late nineteenth century (Taylor, 1957: 243). Drug production in the first half of that century was still largely undertaken by pharmacists themselves. Advances in chemical understanding changed this. Chemical firms previously focussed on dyes began working on drugs. The acceptance of bacterial theory about 1880, along with the success of quinine against malaria and mercury against syphilis, encouraged the belief that drugs could be found to combat other diseases. Antitoxins, especially for diphtheria, created a large scope of operation for drug companies, and this was aided by the demand for vaccines from emerging public health departments (Liebenau, 1987: 3). Army demand during World War One spurred growth worldwide. The war also caused American companies to produce drugs previously supplied by Germany.

While output was still small by modern standards, the pharmaceutical industry took its modern form 1890-1930 (Liebenau, 1987: 1).[5] Expansion in size of drug companies can be traced to mechanization (tablet-making machines from 1890s), demand for standardized products from health agencies, possibilities of mass advertizing and distribution, and the increasingly supportive attitude of the medical community. Research effort was both a cause and result of growing firm size. The first American research lab was that of Parke-Davis in 1902. Others soon followed. Work over the next decades, though, was concentrated in testing, measuring, standardizing, and general quality control (plus providing advertizing copy); only a few companies later moved beyond this to devote some portion of their effort to original product research (Liebenau, 1987: 6; Swann, 1988: 1). The interwar period would witness both an impressive increase in scale of research, and a reorientation toward basic product-oriented investigation. Merck expanded its employment of scientists from a handful in 1920 to over 70 in 1930, and multiplied its budget ten times. Squibb filed one patent in 1920, 21 in 1930, and 164 in 1940. Lilly, Merck, Wyeth, Abbott, Ciba, Squibb, and a number of others opened fine new well-equipped labs in the mid-to-late-1930s (Haynes, 1948: v.4, 246). Moreover, much stronger ties were established with university academics; "Cooperative biomedical research did not emerge on a systemic basis in America until the 1920s and 1930s" (Swann, 1988: 24). Many university fellowships were established by drug companies in the 1920s.

The 1920s emerge as a period of consolidation. Research was undertaken on a variety of fronts, as we shall see, but there were no major breakthroughs. The number of products expanded from 58 to 82, 1923-1929, the major new product being mercurochrome the antiseptic (Haynes, 1948: v.3, 245). Domestic manufacturers continued to slowly win loyalty away from German brand names, after taking their plants and patents after the war. "Processes were overhauled and prices brought down from wartime levels" (Haynes, 1948: v.3, 245). The number of firms fell from 36 to 30. Beyond labor-saving within the industry, the emergence of standard brand name products packaged for individual sale saved much labor among pharmacists and doctors; the switch from apothecary to drug store begun at the turn of the century was largely completed in the 1920s, and these were being superseded by chains (Haynes, 1948: v.3, 274). Chains in turn encouraged concentration and specialization in pharmaceuticals (Liebenau, 1987: 133).

The 1930s would stand out in any chronicle of pharmaceuticals for the development of new products; many emerged in that decade, while important steps were taken toward others which came onto the market in the 1940s (Haynes, 1948: v.4, 245). Process improvement continued as well – the price of aspirin for example fell 50% 1935-1938[6] (see Haynes, 1948: v.4, 251-7) – but this was outshone by product improvement along four different lines.

The nineteenth century promise of anti-bacterial agents was only to be realized in the 1930s. As in most other areas of drug research, inquiry was informed by experience gained in the dyemaking industry. Dyes had the valuable quality of adhering to some substances but not others. Drugs which could fight disease while not harming the host must share this property. We have still an imperfect understanding of chemical reactions within the body. There has always been a strong empirical bias to drug research, therefore. Once it is established that certain (often natural) compounds have a specific physiological effect (often through chance discovery at first – e.g. the sleep-inducing property of sulphonal), similar compounds are synthesized and their effects tested (molecular theory is obviously important here). In the first decade of this century, Ehrlich in Germany, following the observation that certain dyes killed bacteria, isolated arsphenamise (Salvarsan), which had some success against syphilis. "Despite this promising start, a quarter of a century had to pass before the next major pharmaceutical discovery" (Reekie, 1975: 4). There was considerable research effort in the 1920s to follow up on Ehrlich's discovery, but little success (Haynes, 1948: v.3, 248-9).[7] I.G. Farben tested every dye in their possession for therapeutic properties and only in 1932 discovered one red dye which had some success against streptococci. It was later found that the dye worked better in mice than in test tubes. At mid-decade, a French firm discovered that the dye was metabolized into sulphanomide, and that this latter was the key anti-bacterial agent. This discovery also overcame the Farben patent, and the sulphanomides revolution had begun. Numerous sulpha compounds would be produced commercially by the end of the decade.[8] Reekie graphs pharmaceutical discovery, and notes a turning point in 1935 which he credits to the sulpha breakthrough (1975: 4).

Fleming, building on work which had evolved gradually since the time of Pasteur, had isolated penicillin in 1922 (in England). This "did not lead immediately to the development of a useful drug" (Schwartzmann, 1976: 41). Decades of research at Oxford and elsewhere were necessary to establish its therapeutic value. Moreover, great difficulties had to be overcome in at first increasing the rate of natural production in mold, and after 1944 in synthetic production. Penicillin only became practically available in 1940. World War Two provided a big boost to both production and research.

Histamines, which cause inflammation or allergic reactions, were isolated only in the 1910s. Decades of research followed before successful antihistamines were discovered in the late 1930s. Schwartzmann, (1976: 43) asserts that the high rates of innovation observed in the 1940s and 1950s (including the development of antibiotics) were a direct result of the previous developments in sulphanomides, penicillin, and antihistamines. Success in these lines provided both the experimental basis and the incentive to look for other therapeutic agents.

The 1930s also saw the emergence of vitamins. Only in 1911 had biologists arrived at the notion that consumption of some substances in doses too small to have any significant impact on energy could still be essential for various body functions. Casual empiricism, such as the suspicion of a role for some fruits in combating scurvy, could only provide the starting point for research. A ton of rice bran contains only five grams of thiamine. 30,000 eggs yield one tenth of a gram of riboflavin (Haynes, 1948: v.4, 266). Trial and error animal tests under strict controls were necessary to determine the role of trace amounts of various chemical compounds. The 1920s and early 1930s would see the identification of the major vitamins: thiamine in 1926, C in 1928, A in 1930. From isolation to synthesis generally took several years: thiamine in 1936, C in 1933, A in 1937. Vitamins have complex chemical formulas; these were difficult first to establish and then reproduce. Given this the pace of the process could be viewed as impressive (Haynes, 1948: v.4, 267). Commercial production followed shortly after, and was associated with steady improvements in production technology. The price per ounce of C fell from $213 in 1934 to $34 in 1936 to $2.75 in 1939. By the end of the Second World War, it was common to add vitamins to food products. This use, plus production of vitamin pills, would expand steadily thereafter.

Concluding Remarks

The early products of chemical innovation almost all replaced natural products. Competing on the basis of price, their net employment effect must have been negative (depending, of course, on the elasticity of demand). Of the really new products available in 1929, cellophane emerges as the most important. (Motion pictures with sound proved to be one of those product innovations which actually had a negative employment impact.) After that point, while replacing natural substances remained a key focus of chemical research, truly new products become more common. Plastics fulfilled roles which natural materials could perform poorly at best. So, later, did synthetic fibers. The new pharmaceuticals also were truly new products. It was these sectors that were to dominate the postwar chemical industry.

Notes

1. Du Pont hoped that pure research would gain them prestige, aid their recruitment efforts, allow them to trade information with others, and might have practical application (Hounshell, 1988: 221-3). "In general, however, the commercial developments in the period [1900-1930] did not stem from fundamental work, but from the need for improving the quality of the existing range of products". Whereas in 1930 the chemical industry was producing much the same product mix as in 1900, it would come over the next three decades to produce a much wider range of products (Haber, 1971: 372).

2. He argues that before World War One, the tendency was to concentrate on process technology. In part due to overproduction, the emphasis shifted to product innovation after. He notes that chemical prices fell through the 1920s (1948: v.4, xvii-xviii). Haber (1971) feels that the slump, by exposing the danger of relying on only a few markets, greatly encouraged the search for new products.

3. Plastic simply means that the substance can be formed when heated, and will retain its shape when cooled. There are some natural plastics, such as rubber.

4. Four pages earlier, he notes, "The rush of new types of plastics did not come until the 1930s: nevertheless the period was signalized by significant innovation in the old phenolic and cellulose groups."

5. Despite its obvious importance to human wellbeing, and the fact that it has tended for a century to be the most profitable industry in the country, there has been virtually no historical research on pharmaceutical companies (Liebenau, 1987: 1). Liebenau is justifiably critical of accounts which credit the post-World War Two development of antibiotics and psychoactive drugs with 'making' the industry. While these promoted remarkable growth, they were based on a structure long in place.

6. Liebenau feels that non-price competition was becoming more important through the interwar period as the industry developed (1987: 126).

7. Between 1910 and 1935, various tropical diseases were treated with drugs. These were, however, protozoal diseases quite different from the bacterial diseases of temperate climates.

8. Schwartzmann (1976: 38) speaks of these as a family of hypotensive, diuretic, anticonvulsant, uricosuric, and antidiabetic drugs.

8

Electrical Products

Scientific experiment with electricity had been occurring for centuries, but we can conveniently date the modern era as beginning with Faraday in 1821. He discovered electromagnetic induction, and thus demonstrated how an electric current could be mechanically produced. It would be almost half a century before the mechanical problems in electricity generation could be worked out. Some motors (for transforming electrical to mechanical energy) were constructed in the 1830s, but they were so inefficient that electricity was almost solely used for communication for the next decade (there was also some electroplating of gold and silver jewellery and utensils). Maxwell in 1856 provided a mathematical interpretation of Faraday's theory and this spurred further inquiry. Still, the development of dynamos (for translating mechanical to electrical energy) was hindered for decades by a lack of scientific understanding (Sharlin, 1963: 140-1). Armature design was improved, and electromagnets, which Faraday had used but not recognized the importance of, were applied.

The 1870s saw two important innovations. The development of an improved dynamo stimulated the invention of the electric lightbulb, which provided the first large non-communications market for electricity. For decades, lighting would be the wedge with which electricity service would enter new markets (Nye, 1985: 20). A gradual process of generator improvement (in turn stimulated by the development of electric lighting and advances in the science of electromagnetism) and development of new markets proceeded over the next decades. Over 180 cities introduced electric streetcars within two years in the late 1880s.

Some of the potential of electric lighting had been demonstrated from the 1850s with the carbon arc lamp in lighthouses. Many had been working on the electric lightbulb in the 1860s and 1870s. A key facilitating development was a better vacuum pump. Edison, impressed with advances in generators, proceeded to synthesize existing ideas to create the lightbulb. Before he was commercially successful he needed a better vacuum, better filament type (a result of chemical research) and shape, and a whole system for large scale electricity generation and measurement. (This proved to be especially important as development of improved gas mantles 1885-1893 rendered gas temporarily cheaper than light in some localities). He succeeded due to financial backing from J.P. Morgan and other noted financiers (Nye, 1985: 11).

Edison developed his system based on direct current. The ensuing decades would see a battle with advocates of alternating current; the latter would facilitate long distance transmission and would allow different customers to receive different doses. Transformers (Faraday had discovered the principle; it was developed as long distance transmission loomed as a possibility) and alternators were developed in the late 1880s. In 1888, Tesla developed the first electric motor using AC; this promised to solve the peak load problem raised by evening use of lighting. The victory of AC followed in the 1890s; the first big AC generation scheme was Niagara in 1895.

"After a gradual start (1889-1905) electric power capacity grew so fast that it effectively replaced other power sources within two generations" (Du Boff, 1979: 57). Light and communications are a very small part of present day electricity consumption. The growth in electricity generation depended on its application to mechanical power. "It seems hardly too strong to state that the primary electric motor has been the chief innovation in manufacturing technology in this century" (Du Boff, 57). Electricity as a proportion of primary power in industry rose from 4.4% in 1899 to 22.6% in 1909, 51.6% in 1919, 77.4% in 1929 and 85.5% in 1939 (Du Boff, 64).[1] Hughes (1983) has provided us with a detailed description of the development of the electricity generating sector to 1929. The small generators and limited distribution systems of the late nineteenth century had begun before 1900 to give way to large steam or water driven turbines and high-voltage long-distance transmission. Larger turbines decreased production costs significantly, in part as they could take advantage of cheap energy sources, and also as the heat and magnetic field losses associated with generation depend on the surface area of the turbine rather than the volume. Equally important, large regional systems could serve users with differing temporal energy requirements and thus achieve greater capacity utilization. Linking together different generators led to savings in that the amount of reserve capacity needed to deal with emergencies naturally declined relative to the size of the system. Such regional power systems came to dominate in the 1920s; by 1930 they had matured and later changes were more regular and predictable. The decade thus witnessed both a dramatic extension in service and decrease in costs.[2] In the late nineteenth century electric motors were used to drive belts to power many machines, much like steam engines; this gave way through the early twentieth century to individually powered machines. This improved material flow (power stoppage was localized) and gave workers control of their own machines. The associated reorganization of factory layout and design was completed by about 1930 (Du Boff, 1979: 146-7).[3]

A steady stream of new machines naturally emerged for home, office, and industry as the potential of the new power source was recognized. "Once the small motor reached a point of practical perfection in the late 1890s, the application of electricity to mechanical devices was limited only by the imagination" (Platt, 1988: 262). We will see the ramifications of these developments in many areas, notably appliances, service employment (Chapter 11), and machinery.

Edison set up what was arguably the first research laboratory in the United States, and the development of electronics was to be closely tied to the efforts of such laboratories. These labs tended to be narrowly oriented at first. Bell Telephone's research, for example, focussed on a series of pressing problems with telephone technology until the research branch was given more autonomy in 1925. Though the company recognized the importance of obtaining patents, and felt that researching new products, even if unprofitable, could aid them in their goal of obtaining a national network, the goal-oriented research required in this dynamic era in telephone technology meant that Bell did not create a range of new products. The biggest innovation was likely the repeater which made transcontinental transmissions possible and was developed 1912-1925 (Reich, 1985). New management, the threat and opportunity provided by radio, fear of antitrust, recognition of the public relations benefit, and the example of General Electric, all played a role in the establishment of Bell Labs in 1925.

General Electric's early research also tended to be goal oriented. By 1900, the firm had captured 90% of the lightbulb market; the possibility that Europeans or Westinghouse might challenge their position with improved technology motivated the firm to establish its laboratory (Wise, 1985: 74). Outside experts were often hired on a short term basis to pursue particular projects. A patent agreement with Westinghouse removed much of GE's incentive to innovate.[4] Filaments were a continuing concern, especially in the face of European developments. An improved carbon filament was developed in 1905. In 1912 success was achieved with tungsten (an element which chemists had only recently isolated), which proved to be the rugged and highly efficient filament for which scientists and engineers had been searching for decades.[5] As the twentieth century began, the scope of GE's research expanded. It achieved important results in power transmission and radio, and played a key role in the development of X-rays. Radio and X-rays both took GE research increasingly into new product lines. In the 1920s, among the new products which they would introduce was the tungsten-carbide cutting tool, which would have far-reaching effects in industry.

The history of electronics can be defined as the development and manufacture of components; it could be said that the 1920s saw the birth of the component industry (Dummer, 1978: 3). The simple constituent parts of basic electrical devices, such as switches, fuses, wiring, and lamp sockets, had been developed in the 1880s (Bakelite, the early plastic, was widely used in these). More sophisticated electronics would require devices for processing electrical signals. The diode of 1904 (an adaptation of lightbulb technology) was soon superseded by the triode of 1906, which after improvements in 1912 proved to be able to amplify signals. These two tubes were both developed for use in radio, and improvements to the triode through the next decades were driven by the requirements of radio (Tillman, 1978: 1093-5). While military application in World War Two (notably radar) would encourage standardization, reliability, capacity to withstand extreme temperature, and miniaturization, the success of components manufacturers in the early postwar period was still dependent on radio and television (Dumner, 1978: 7).

Interwar Employment Trends

Electric light and power production continued to increase throughout the interwar period. Beyond the effects this had on employment in electricity-using sectors, it also tended to raise employment within electric utilities themselves (both electric and gas output were, though, growing at a decreasing rate). Against this trend, however, were substantial productivity increases as early difficulties with electricity generators were alleviated. Since electricity competed with gas in many important uses, it is best to consider electric and gas utilities jointly. Gould (1946) estimated that output per worker expanded by 18% 1919-1929, and a further 47% 1929-1939. Utility output actually rose 2.9% in 1930 and was still 1% higher than the 1929 level in 1931. Due to 2.5.% and 5% labor productivity advances in those two years, employment barely expanded in 1930 and fell by over 27,000 in 1931. Output then fell for a couple of years while productivity continued to expand; the result was that 82,000 fewer were employed in 1933 than 1929. Since output had fallen 8.5% while productivity had expanded 13.3%, the great part of this employment loss must be attributed to the latter. Even though output was 37% greater in 1939 than a decade before, employment never reached its 1930 peak again in the 1930s.

Telephone

The telephone was largely the creature of tinkerers rather than scientists. Alexander Graham Bell, indeed, had been interested in the study of voice rather than electronics. The potential of an innovation which seems so obvious today was scarcely recognized by the inventor. Bell's successors viewed the telephone as a mere substitute for the telegraph; socializing by phone was actually discouraged for decades (Fischer, 1988). Costs of telephone service remained high until Bell patents expired in the 1890s. Telephone rental fell from 10% of average income to 4% by the early 1900s. After that, rates changed little till after World War Two. Per capita telephone use expanded by nine times 1893-1902, and by a less startling 4% per year till 1929. By that date 42% of households possessed one, though this fell to 31% in 1933 and 37% in 1940 (Fischer, 1988).

The first manual switchboard was established in New Haven in 1878. It was the automatic exchange, though, which made possible the universal use of the telephone. This was first patented in 1889 (by an undertaker with little training in electronics), and many improvements were made to it before 1900, but only by the onset of the 1920s did engineers and the phone company feel confident that it was superior to manual switchboards (Sharlin, 1963: 61-3). By 1929, over 4 million stations were supplied with automatic switchboards, though toll and long distance calls still required an operator. Dial phones had increased from 1.7% of the total in 1919 to 31.9% in 1930.

The number of operators did increase by 25,000 1921-1930, but "it is estimated that 69,421 additional operators would have been required to handle the 1930 volume of business at the 1921 output per operator rate" (Jerome, 1934: 380).[6] While presumably the rate of increase would have been slower, the net employment impact of technological change in the 1920s was likely negative. A further big expansion of telephone services, especially inexpensive long-distance, would come in the late 1930s. The adoption of frequency modulation from radio would allow one cable to carry many conversations simultaneously (Williams, 1982: 306-7). Once again, this would come a little late.

Radio

Wolman (1929) speaks of radio as "Probably the outstanding development in consumption in these past years." The press used words like "sudden", "rapid", "fever", and "outstanding" to describe its rise in the 1920s. (Douglas, 1987: xv). Copeland concurs; "In the enumeration of changes in demand which occurred during the period covered by the survey, radio sets furnish the outstanding example of a new type of merchandise placed on the market with a short phenomenally rapid increase in demand" (1929: 322). There were only 60,000 radios in use as late as 1922, but 7.5 million in 1928 and 10 million in 1929. Demand had quickly spread from rich through middle class to poor (75% of sales were by instalment). The growth was even more phenomenal than that within automobiles.

The beginnings of radio can clearly be traced to scientific inquiry in the nineteenth century. While the telephone can be viewed as a result of trial and error tinkering (dependent, of course, on the previous discovery of electricity), the radio was not. As Nelson (1959) has noted, it is doubtful that anyone looking for radio would have discovered it. Neither Hertz nor Maxwell were; they studied the properties of electro-magnetic waves in order to understand nature, and particularly electricity. Since at least Faraday's dynamo the relationship between magnetic forces and electricity had been a subject of inquiry by physicists. The existence of waves was posited and then shown by experimentation. During the second half of the nineteenth century, many physicists experimented with the transmission and reception of electro-magnetic waves. In this way greatly increased knowledge of the properties of such waves was gained.

Hertz himself thought that wireless transmission of sound would be impossible. Oliver Lodge demonstrated wireless as a scientific curiosity in 1894. Many inventors, notably Marconi (also Teale and Fessenden), began searching for commercial applications of this scientific knowledge. Marconi saw the greatest potential in ship-to-shore communication: wires were naturally impossible, great gains in safety could be achieved, and interference would be minimized. Decades of effort to expand the range of wireless transmission would be financed by revenue from ship-to-shore communications.

Modern radio requires the instantaneous translation of sound to waves and vice versa. In the beginning, only intermittent transmission such as Morse Code was possible. Marconi generated waves by creating electric sparks. While numerous valuable improvements were made during the era of spark transmitters — the best wave lengths were investigated, and antennas were developed — transmission could at best be intermittent. As well, transmission could not be limited to one wavelength; such transmitters are banned today for they would cause widespread interference.

Technological inquiry in the early days was naturally dominated by attempts to improve sparks. Only in the closing years of the nineteenth century did a handful of people start to recognize that a technology beyond sparks was desirable. Like those who would later start work on jet engines while piston engines were still being improved, these people saw some potential for voice transmission which sparks could never satisfy (Aitken, 1985: 6-10). If wireless had only been viewed as a substitute for the telegraph, such a need might never have been felt.

While the limitations of the spark were obvious to some, the solution to the problem of continuous transmission was not. Three competing technologies emerged in the early twentieth century. The first voice broadcast in 1906 used the alternator. Neither this nor the arc transmitter had a lasting role in radio, but both were important for early experimentation with voice transmission, and served to establish its viability. Both suffered from excessive power usage and frequent breakdown (Sturmey, 1958: 22). De Forest had worked briefly on improving the arc while pursuing the triode.

Aitken (1985) has, laudably, examined the development of these "failures" along with that of the finally successful device. He notes that both resulted from intensive research efforts focussed on improved transmission, few surprises were encountered in either effort, and the end results fulfilled the basic expectations of those who had undertaken the research efforts. The expertise of GE in power generation ensured that the full potential of the alternator was realized before it was superseded. The vacuum tube, on the other hand, emerged from an attempt to improve radio receivers, and turned out to have capabilities its inventor had not imagined (Aitken, 1989: 549-552).

The vacuum tube was a result of scientific research performed over a period of decades with no eye toward practical application (MacLaurin, 1949: 241-2). Fleming (in Britain), the inventor of the first vacuum tube, had begun with curiosity about the "Edison effect"; the pattern of discoloration of lightbulbs which had intrigued scientists for decades (it provided evidence that charged particles moved away from the filament). Edison himself had experimented with the insertion of a second electrode in the lightbulb. Fleming was the first to notice that particles only moved from the hot to cold electrodes, and recognized the potential for an electronic valve in this result. His combined knowledge of electron theory (a recent development in physics), lightbulbs, and the needs of the wireless industry (he had consulted for both industries) allowed him to develop the diode (Aitken, 1989: 210).

The diode had received little interest. The Marconi company used it sparingly, and had not licensed others to use it. De Forest, in 1906, discovered that adding a third electrode and thus creating the triode greatly increased sensitivity. The triode quickly ousted competition after World War One. De Forest maintained that his invention owed little to that of Fleming. Aitken has recently analyzed this question in detail (1985: 205-15). He concludes that it is unlikely that De Forest did not learn much from the diode; while he had been working along similar lines before, his first vacuum tube looks identical to that of Fleming. On the other hand, his conception of the tube was entirely different. Fleming's view of how the device worked precluded the addition of a third element.

De Forest had long been interested in radio, but had only begun work on vacuum tubes after over a decade of working with other devices (and some shady financial dealings). His main motivation was to develop a receiver which would not be dependent on the previous patents of Fessenden. A mistaken conception that ionized gases, not electrons, moved between electrodes did not hamper his investigation. The science of electrons was so poorly understood at the time that advances naturally occurred due to trial and error experimentation. The idea of inserting a control electrode was not entirely novel; De Forest was familiar with their use in cathode ray tubes. Still, inserting the control grid appears to have been just another thing to try (Aitken, 1989: 550).

The triode was little used for five years. It was unreliable, and could not yet outperform competing devices. De Forest eventually abandoned attempts at self-finance and associated himself with Bell Labs. General Electric launched a large research effort in 1913; recognizing the role of electrons (versus gases), they achieved a much better vacuum and thus a much better tube. Only from 1912 was it recognized, by those whose previous telephone experience made them familiar with questions of amplification, that the vacuum tube could be used for transmission. While not originally designed for transmission its usefulness in that regard had inevitably to be discovered (Aitken, 1989: 546-7).

Until the postwar development of the transistor, the triode was to be a basic building block of not just radio but all electronic equipment; very few developments in electronics over the next decades would have been possible without it. (MacLaurin, 1949: 244). "The origin of the electronics industry is associated with the development of the radio, and until World War Two radios and related equipment accounted for nearly all of the industry's output" (Tilton, 1971: 7). Bell Labs from 1913 devoted considerable research effort to all aspects of radio, mistakenly viewing it as a complement to telephone transmission. A major development by Bell Labs in 1920 was the copper glass tube which allowed greater power output without the danger of the glass cracking; this was especially important for radio transmitters.

World War One saw greatly increased (and improved) production of transmitters and receivers, and the training of a mass of radio operators. The Navy, due to its dislike of having had to rely on British Marconi during the war, encouraged American companies thereafter. GE established the Radio Corporation of America in 1920, and this subsidiary quickly reached cross-patent agreements with the other electrical giant Westinghouse and with Bell. The terms of agreement indicate that none of the principals, even at that late date, recognized the potential of the new technology. Early radio sales depended on amateur transmission.[7] In 1920, Westinghouse started the first commercial station in Philadelphia in order to expand its sales of receivers. Nobody had foreseen the success this station achieved. A host of new producers of receivers entered the market; a new agreement in 1924 allowed the licensing of RCA patents to a limited group of fairly large firms.

By 1922, there were already over 600 radio stations. Bell continued to assert its monopoly rights (under the 1920 agreement) over transmission and advertizing revenue, and refused for a while to allow the creation of networks through it phone lines. Threatened by court order Bell agreed to leave broadcasting and concentrate on ship-to-shore communication in 1926. One result of arbitration of this issue was the creation by GE/RCA/ Westinghouse of NBC in 1926; competitors launched CBS the following year.[8]

The phenomenal growth of receiver sales in the 1920s did not exhaust the market. There were 10 million sets in existence in 1929 (one third of households), 19 million in 1933 (two thirds of households) and 28.5 million in 1940 (80%). While the sales peak of 1929 was not to be matched until 1935 in terms of number of sets, sales through the early 1930s remained at or above the levels of 1928, and well above those of any previous year. In terms of value, though, the 1929 figure was never approached through the interwar period. Unit prices fell much faster than the price level due to increased productivity. Total sales were $506,000 in 1926, $690,000 in 1928, $842,000 in 1929, $496,000 in 1930, $200,000 in 1932, $300,000 in 1933, and peaked during the 1930s at $537,000 in 1937 (MacLaurin, 1949: 139).

MacLaurin proffers a Schumpeterian explanation of the course of technological change in radio in the 1920s; a period of product innovation gave way to a period of process innovation (1949: 132). It is true that a number of new firms (with considerable turnover; 250 firms died in each of 1925 and 1926) followed the early innovators into radio production in the 1920s. It seems, though, that the extensive cross-licensing agreements had a major negative influence on research by removing the incentive to "get around" existing patents. The royalty fees which the newcomers had to pay (7.5%) may have strapped their overall ability to finance research. (The large firms certainly continued; Bell spent over $21 million on radio research 1925-1940, developing improved tubes and antennas as a result.) More centrally, the arrangement limited the value of product innovation (for the agreements covered new as well as existing patents) while the competitive environment encouraged process innovation. Competitors would not be able to look at a radio set and see how it was made. Product innovation had been necessary in the early 1920s in order to replace home-made with store-bought receivers; by 1925 process innovation was clearly ascendant.

One major innovation was the development of the thoriated tungsten tube by Longmuir at GE; the filament was less subject to oxidization and thus lasted longer. Even in the postwar period, replacement tubes were 40% of tube sales (MacLaurin, 1949: 139). Hull (1923-1932) developed a new tube which provided less feedback, provided more amplification, and was smaller. This paved the way for smaller radios. Emerson introduced the low-priced table radio in 1932; they had captured 80% of the market by 1941. Even product innovation could lower costs. Sales departments in the 1930s pushed for only minor product innovations (MacLaurin, 1949: 144). Still, Philco did open new markets in the 1930s with battery-operated units, portable radios, and car radios.

Sturmey notes that consumers tended to keep radio receivers for many years. In the absence of radical change in receiver quality this is not surprising; incremental product change was not enough to galvanize the market.[9] FM would have provided a boast to receiver sales if it had been developed earlier, as it removed the static and noise which bedeviled AM broadcasting. The large producers, notably RCA, showed little interest in the new technology. This was to some extent because FM had to compete with another nascent technology, television, for frequency space and research funds, but was mostly due to the fact that RCA had already heavily invested in AM (Udelson, 1982: 92). FM would be developed in the mid-1930s by Armstrong, a university professor who financed his research with funds from previous radio patents. He would claim that RCA stood in the way of FM (Sturmey, 1958: 183-90).

While radio does not provide a classic example of market saturation, we have seen that the particular timing of product and process innovation led to a decline in the value of industry output in the early 1930s. The picture in terms of investment is even more clear.[10] By 1929, the major investment in the radio industry had taken place. The first period of broadcasting-station construction was largely over, and the set-manufacturing industry had reached a capacity which was capable not only of filling replacement needs but of taking care of considerable expansion in demand. New investment at this stage was likely to come only from changing over to new types of transmission stations and from the reconstruction and relocation of manufacturing plants. Such a wave of industrial investment did not come until the development of FM and television in the late 1930s.

The history of radio highlights the role of surprises in the evolution of technology. It is not just that radio itself would not have been discovered by people looking for it. The vacuum tube opened up entire new possibilities that its inventors had never imagined. "The moral is that sometimes a technology seems to take charge of events and exercise what is almost a legislative power of its own" (Aitken, 1985: 16). While social forces had a role to play in the development of the vacuum tube itself, once the triode appeared it determined to a large extent what came next.

The picture became even more complicated in the 1920s. The vacuum tube made radio technology accessible to an ever-widening group of amateurs. Major manufacturers, who had thought their control of patents would guarantee dominance of the new technology, found the situation beyond their control. Amateurs proceeded to experiment with small-scale broadcasting. Only at this point did the market potential for radio broadcasting as we know it become apparent. The industrial research labs had produced the technology, but the corporation could not control its use (Douglas, 1988: 89). On a broader view, the emergence of modern radio was conditioned by both the potential of the triode and the cultural values and systems of regulation which encouraged the expansion of amateur radio.

Television

While many view television as having sprung to life fully developed in the postwar period, and others see it as developing naturally from the 1920s, there were a number of fake starts and dead ends along the way. A number of different elements, engineering, marketing, and programming, had to be brought together. The road to television can be said to begin with the scientific discovery in 1883 that the resistance of selenium varied with light. This, along with the introduction of the telephone, encouraged a surge of research in the 1880s (Burns, 1991: 183). In the Nipkow disk of 1884, light hit selenium through a rotating spiral of holes. Thus, a moving picture could potentially be transmitted as electronic signals. The selenium reacted much too slowly, however. After problems of amplification and the slow reaction time of the selenium cell were worked out, the cathode ray tube was developed in 1891. These tubes had a variety of uses (the development of electronics itself required a device capable of displaying the waveform of electronic signals) and spawned another epoch of technical enquiry. In 1905 the photo-electrical cell was devised, based on the work of Hertz; it provided much quicker reaction than selenium. In the same year, Einstein used quantum theory to explain the photoelectric effect.

The development of the radiotelegraph from the 1890s created the possibility of television broadcasting. In the pre-World War One era, the entertainment potential of TV was not recognized (as with radio); scientific, military, and industrial motives dominated. A facsimile machine was available shortly after World War One; moving pictures were much more difficult. The technical difficulties to be overcome were much greater than those required for radio. Cameras had to be developed, the cathode ray tube had to be improved, the signal had to be amplified, transmission and reception needed to be synchronized, and methods of broadcasting developed. The development of radio itself in the 1920s solved some of television's problems. Still, the further development costs involved were so large that radio companies were very wary of financing research in the 1920s, especially as radio profits were so high (MacLaurin, 1949: 196).[11]

Still, the first efforts at commercial television emerged in 1925 in both the United States and Europe. The fact that inventors in England and the United States unveiled different techniques within months is evidence of the considerable incremental research occurring at the time (Udelson, 1982: 27-8). There was a miniature boom in television after that. In 1928, there were about 15 television stations in the United States, most operated by manufacturers. They broadcast experimental-type pictures. Most receivers could only handle video; it was hoped that television would stimulate radio sales. Christmas shoppers were assured that year that television would not render radio obsolete. Programming included singing, vaudeville, sports, news, and movies. By 1933 this boomlet had collapsed, but not, it should be noted, due to any financial difficulties. The reason was the poor picture quality compared to that of motion pictures. The main drawback was mechanical scanning. The essence of television is that the picture be scanned rapidly (usually in lines, from left to right). The modern television camera operates by electronic scanning. In the 1920s the scanner would mechanically cross the picture. As a result, the picture had to be scanned in large segments. The result was an extremely fuzzy picture (see Udelson, 1982: 61). The technology of transmission also had to be improved. In 1931, it was estimated that 7 million pictures elements per second had to be transmitted. At the time, they were experiencing difficulty transmitting 4000 (Burns, 1991: 197).[12] Still, the practicality of television had to some degree been demonstrated.

Government could be accused of contributing to the decline of mechanical television by refusing to allow advertizing. However, it recognized that television could not be introduced like radio and gradually improved over time. Picture quality had to become acceptable, and standards developed concerning the length and number of lines to be scanned and the number of frames per second, so that transmitters and especially receivers could be standardized. As well, television used up very large band widths and thus space had to be found in increasingly crowded airwaves for the new technology. The use of Very High Frequencies (VHF) had the unfortunate side effect of limiting transmission distance (and thus encouraging network formation), as VHF waves do not bounce off the ionosphere but travel along the ground.

From 1908, it had been recognized that if the cathode ray tube was employed in the transmitter as well as the receiver, the system could be rendered fully electronic. Only in the 1930s were CRTs improved to the extent that electronic scanning became feasible. Zworykin had filed a patent for electronic TV as early as 1923, but the pictures he had obtained by that time were very poor. He was employed at Westinghouse at the time, and they were relatively disinterested in the project. He had greater success after becoming head of RCA's Electronic Research Lab from 1930 (by

1932 he had 60 people working for him, though this was cut to 10 when it was recognized commercialization was years away). The image was improved and the degree of flicker reduced. He would not be awarded a patent until the late 1930s due largely to the conflicting claim of Philo Farnsworth. The latter had developed an electronic television in the 1920s which could only transmit very bright scenes, but embodied a superior scanning system. The patent battles between these two and others in the 1930s, while they may have slowed the development of television, indicate how active research was in that period (Udelson, 1982: 84). RCA and Farnsworth finally agreed to cross-licensing in 1939, for each had discovered important complementary elements of the modern TV.

When RCA announced a million-dollar plan to develop commercial television in 1935, the New York Times bemoaned the five year delay. Industry spokesmen responded by pointing to the expense involved. A survey in 1939 found that the five leading firms had spent at least $13 million without any financial return. The size of investment required for commercialization: transmitters, receivers, and programming, was simply immense.

The late 1930s saw not just technical experiment but programming experience as well, as RCA ran a station in New York 1936-1938. By 1938, a handful of companies indicated a willingness to produce receivers (though the technology was still far from perfect). They could not agree on the standards to be utilized. As well, FM proponents were competing for the same wave lengths. Only in 1941 was the Federal Communications Commission able to license commercial television, settling on a 525 line screen, 30 frames per second, and a six megahertz band width. Thirty-two stations were licensed by year end.

Not surprisingly, "The outbreak of the war delayed development of the service on both sides of the Atlantic" (Williams, 1982: 318). Progress was rapid afterward. Synthesis of four different picture tubes resulted in the "perfect" picture tube in 1946 (Udelson, 1982: 116-7). Receiver sales had been sluggish in 1941 due to high prices (from $150 to $650 and up), poor programming and lack of standardization, and a fear of obsolescence. Programming changes, especially in sports and special events, played an important role in enhancing sales thereafter. Television sales took off from 1947. Half of American households owned a television by the mid-1950s.

Arguably, television could have been developed earlier. "I believe that, if we had decided in 1928 to press forward in television research and development with the same all-out energy that we put into radar, it would have been technically possible to advance as far in two or three years as we have gone in twenty years" (MacLaurin, 1949: 265). If we accept that government subsidy was not feasible politically at the time, we could only blame private enterprise for what the New York Times viewed as delay.

The research costs involved were such that they could really only be shouldered by corporate research labs (MacLaurin, 1949: 191). Many of the companies involved were also in radio, and many may have been less than enthusiastic about competing with themselves, especially since much research was going on in radio itself through the interwar period. There was some feeling in 1938 that manufacturers were deliberately delaying to guarantee high profits on introduction (Udelson, 1982: 96).

However, the extreme expense involved in establishing reliable commercial television indicates that "delay" is hardly the appropriate description. To a large extent, television built upon radio technology; there was thus a logical temporal sequence. Heavy expenditure on electronic scanning was naturally unlikely until mechanical scanning had proven its inability to provide entertainment. Mechanical scanning also provided publicity for television which encouraged research over the next decade. The experience with mechanical scanning also indicates that television should not be viewed as a Menschian reaction to the Depression. Rather, it appears as a difficult outgrowth of a number of strands of long-standing scientific and technical inquiry.

Appliances

For Giedion (1941), the interwar period was the point at which mechanization hit the domestic sphere. Most appliances in use in the first half of the twentieth century were improvements on nineteenth century devices, the big change being the application of electric motors (Williams, 1982: 387). The electric iron emerged in the 1890s. So did the electric kettle, though kettles with internal elements were only introduced in the early 1920s. The vacuum cleaner was invented in 1901, but the earliest ones were extremely bulky. The Chicago fair of 1893 had sparked great interest in the United States in the potential future of electrical appliances. The manpower shortage of World War One had given a further boost to efforts to develop appliances.

While advances in the harnessing of electricity must be viewed as the key proximate cause of the rise of home appliances, we should not lose sight of the broader social context. Cowan (1983) has provided an overview of the impact of technological change on housework since the mid-nineteenth century. The most interesting result is that time spent by women on housework has changed little, though the composition of that time has been greatly transformed. Major nineteenth century improvements such as running water and the replacement of wood with coal stoves tended to economize on the labor of men and children (and was one factor encouraging men to work outside the home). In the interwar period, it would be the labor of domestic servants which would be replaced (itself an important source of unemployment). The housewife not only absorbed duties from displaced men and servants but took on new functions. The availability of hot running water encouraged new standards of cleanliness. The automobile brought a number of new transport functions (while throwing thousands of delivery people out of work). Women were encouraged to devote more time to cooking and childrearing. Technological potential interacted in a complex fashion with social forces to introduce new appliances to the home.[13] It was not General Electric but a number of small competitors which introduced a range of appliances after World War One; GE reacted by buying these companies (including Hotpoint), and improving their products through research (Nye, 1985: 24). Most appliances spread rapidly, once their utility was clear. "Some terribly burdensome chores were either eased or totally eliminated when the new kitchen came into being, which is no doubt why most people seem to have acquired conveniences and appliances as soon as they were able to pay for them and, given the prevalence of instalment credit plans after 1918, sometimes even sooner than that" (Cowan, 1983: 97).

We have discussed previously the rapid extension of electricity into urban homes in the 1920s.[14] By the end of that decade 68% of all households had electric power (versus 33% in 1920). This guaranteed that there was a growing market through the 1920s for the new appliances. As late as 1923, a survey found that 80% of household electricity was used for lighting and 15% for ironing (Platt, 1988: 268). While the advent of electric lighting had encouraged the development of a number of appliances in the 1890s, the dominance of lighting meant that these had generally to be plugged into light sockets (or wired directly). The cost and inconvenience of this was a major deterrent to appliance utilization. The modern two-prong plug and wall receptacle were creations of the early twentieth century; manufacturers only agreed to standardize these in 1917; the process of standardization was completed by the 1930s (Schroeder, 1986). Complete market saturation was not possible in the 1920s in any case; rural electrification especially would bring new households into the market (though this would come late in the 1930s). Appliances fare better than most industries in the 1930s. Sales per household expanded 3.2% per year, compared with 4.5% per year in the 1920s (Olney, 1989b). Falling electricity costs, improved mechanical reliability, minor design refinements, and, of course, style changes aided market growth (Hogan, 1971: 1403). As well, the government encouraged public utilities to lengthen the terms of instalment contacts (the average term rose from 12 months in 1929 to 22 months in 1938) and even provided loans itself for appliance purchase (Holthausen, 1940: 14).

We should not paint too bright a picture. The experience of particular consumer durables varied considerably through the interwar period. Many branches experienced the massive sales in the 1920s and decline in the 1930s of the auto industry. "It is clear that the rise in output of durable goods in 1928-1929 was too rapid to be long maintained. Excess capacity was developing in a number of lines, and this meant a decline in demand for further capital goods. As a matter of fact, new orders for some types of durable goods declined fairly early in 1929" (Gordon, 1961). Based on prewar data, one would have significantly underpredicted sales in the 1920s and overpredicted sales in the 1930s (Olney, 1990).

The smaller appliances, irons, toasters, percolators, hotplates, and waffle irons, had saturated their markets in the 1920s, and thus output declined sharply in the 1930s (Hogan, 1079: 1412). Over three million irons were sold per year in the late 1920s. By the end of the decade 94% of wired homes possessed one and could scarcely be expected to buy another. Vacuum cleaner sales also expanded rapidly after Hoover introduced the one-person vacuum suited for private use in 1908. By 1929, 43% of wired homes possessed one. This may seem a far cry from market saturation, but already in 1926, "80% of all the affluent households that were studied by market researchers in 36 American cities had vacuum cleaners and washing machines" (Cowan, 1983: 173). Further expansion in vacuum sales could only come after the working class was convinced of its utility. "With the Depression of the 1930s [vacuum] sales dropped even more than might have otherwise been the case as its market was much more thoroughly saturated than was the market for most other appliances" (Hogan, 1971: 1078). The washing machine (33% of wired homes possessed one in 1929) would be gradually introduced into working class homes in the 1930s through the market for used machines. While some improvements were made in the machine in the 1930s (notably the automatic in 1935) it had achieved substantially its modern form (water-heating capability came only in the 1950s) by the end of the 1920s, following a series of minor improvements. Sales rose through the 1920s, fell in the early 1930s, and rose to new heights after 1934.

The emergence of the electric stove/oven would be delayed by the rivalry of gas. As gas companies lost their lighting market to electricity, they pushed strenuously for the adoption of gas ovens and furnaces. The 1920s, while seeing a hint of the future of electric cooking, would be dominated by the replacement of coal, oil, and wood by gas. This transformation provided a much greater potential saving in home labor than would the later rise of electricity. The technology behind gas ovens had reached an advanced state in the late nineteenth century, but only in the 1920s did it become the dominant technology (Hogan, 1971: 1070-1). Sales first exceeded those of coal and wood stoves in 1923. By 1930 there were 14 million gas ovens, 7.7 million coal or wood, 6.4 million oil, and less than one million electric. While gas would remain dominant through the 1930s, it could not be expected to continue its previous rate of expansion.[15]

The electric stove made great strides in the 1930s. Thermostats to control oven temperature were developed. The first chrome alloy range element had been developed in 1906, nickel-chromium alloys which would not oxidize had followed in 1908; these were safer, quicker, and more durable, and were in common use by the 1920s (Wise, 1985: 171-2). In the 1930s, the one-piece all-steel body was introduced, and various problems of stress and enamelling were solved. Electric oven sales would not surpass 200,000 until 1935, and 700,000 only in 1941. From the point of view of the Great Depression, it could be argued that the gas oven came too soon and the electric too late.

The electric refrigerator was the appliance which fared the best in the 1930s. While refrigeration had been applied on a large scale on ships and in meat stores since the 1850s, the problems of small scale refrigeration were only solved in the interwar period.[16] One part of the solution came from chemical inquiry; synthetic refrigerants replaced the toxic ammonia in the late 1920s. One hundred had died in a Cleveland hospital in 1929 because of exposure to toxic refrigerants. Hounshell and Smith feel that toxicity was the major reason why only 15% of wired homes had refrigerators in 1929 (1988: 155). The theory of atomic structure pointed to fluorine as having minimal toxicity and flammability; experiments with the physical properties of its most promising compound led to the first "Freon" in 1930. In 1927, GE developed the hermetically sealed motor and box. Wood, which rotted, was replaced by steel. Better thermostats were developed. The 1930s would witness numerous small improvements[17] including a series of better freezer units. The enamelling problem was solved in 1933 with "Dulux" a relative of "Duco" which was used on automobiles. (Du Pont created "Dulux" in 1928, but its application was delayed by a patent battle with General Electric.) The evaporator was separated from the food so that the latter would not dry out. By the middle of the Depression, most of the fundamental innovations, except the much later automatic defrost, had been made (Cowan, 1983: 133).

General Electric released the Monitor Top in 1927 and hoped to sell 10,000. It sold 50,000. The millionth Monitor Top was sold in 1931. Having not had time to saturate the market in the 1920s, and with technology continuing to improve, manufacturers could expand sales in the 1930s; 800,000 were sold in 1930, 3.85 million in 1941. Overall, refrigerator sales expanded steadily to 1937 as consumer health concerns were overcome and as a network of sales and service facilities were established. A that point, half of wired homes possessed one. The experience of the refrigerator clearly indicates that new products could fare well during the Depression.

Bowden (1994) has studied the diffusion of household appliances over the last century. Appliances which save on housework time are shown to diffuse more slowly than those which increase the quality of leisure time (such as radios or television) in all time periods. Interwar appliances such as the washing machine, vacuum cleaner, and refrigerator are not unusual; indeed they diffuse much more rapidly than earlier or later products such as the telephone or clothes dryer. This is further confirmation that sluggish appliance sales in the 1930s were not due in large part to the Depression.

Widening our focus beyond merely appliances, we can note that consumer durables absorbed 11% of household income in the 1920s, versus 9% in both the 1910s and 1930s (14% in 1948-1982). There had been a substantial debate in the 1960s as to whether the so-called "consumer durables revolution" of the 1920s had been a result of changing relative prices or reflected a change in tastes. The relative price of durables seems in fact to have risen in the 1920s (Olney, 1990); such a measure, of course, does not account for changes in quality. Olney (1989b, 1990) has estimated demand functions and found that there is a structural break about World War One. That is, the increased consumption of consumer durables in the 1920s can not be attributed to changes in prices or incomes. Olney suspects the increased ease of instalment credit purchase may provide the explanation, but recognizes that such factors as suburbanization and electrification also played a role. We could suspect that the creation of a new generation of electrical appliances had much to do with this transformation, especially as Olney (1990) finds that it is major durables expenditure which rises while minor durables spending falls.

Machinery

From 1830 onwards, engineers in various countries had been working on the problem of shaping finished steel. "No matter how accurately such parts were machined in their soft state, some distortion inevitably occurred during the subsequent hardening process" (Rolt, 1986: 188). The key was the grinding machine. Natural materials such as emery, clay, and feldspar were the major grinding materials. Whereas Watt had marvelled at precision of within the width of a sixpence, by the 1880s machinists talked in terms of a thousandth of an inch, and a generation later one tenth of that. While there was a prejudice against the use of the grinding machine in general production through the nineteenth century, it was invaluable in making economical a variety of other precision machinery. Metal cutting was greatly improved, and a superior gear-cutting machine emerged in 1897 (Rolt, 1986: ch.9).

Electrical Products

In 1895, commercial production of silicon carbide (carborundun) was begun. While this material had been produced as a laboratory curiosity earlier in the century (and was rediscovered by accident in 1891 while looking for synthetic diamonds), its production required an immense expense of power in an electric arc furnace. With the opening of the Niagara power station, this material, whose hardness was surpassed only by diamonds, quickly replaced previous natural grinding materials (Rolt, 1986: 219). Grinding machines were improved to achieve the potential supplied by this new material in the first decades of the twentieth century. The planetary spindle cylinder grinder revolutionized the production of internal-combustion engine cylinders in 1905; automatic, high-speed grinding had been achieved by 1930 (Woodbury, 1978: 1044). One worker could grind 90 tappets per hour before 1920, 150 after 1923, and (by handling three machines) 1350 by 1929 (Burstall, 1963: 383-4). By the 1920s, the grinding machine was widely used in precision engineering, could produce crankshafts in a fraction of the previous time, and was valuable for making gears for bicycles, cars and airplanes (Williams, 1982: 182).

While the steam engine had been a major driving force behind the improvement of machine tools in the late eighteenth and early nineteenth centuries, it was the internal-combustion engine automobile that would provide the dominant incentive from 1895 (Rolt, 1986, Woodbury, 1978: 1035). Improved tools allowed automakers to produce better gears and bearings which were themselves incorporated in better tools. At the same time, the adaptation of electric power to machinery was of crucial significance. Even more fundamental than these two influences, though, was the improvement in cutting materials themselves, for a machine can only be as good as the cutting tool employed (Rolt, 1986: 202-3).

Alloy steels had been experimented with for decades, but as late as 1868 their properties were little understood, and their potential for improving cutting tool performance unrecognized. In that year, Mushet in England manufactured, after years of experiment, the world's first improved tool steel from manganese-rich ore. It was used for machining manganese steel, but did not find its way into ordinary machine tools for some time. In the last decades of the century, Frederick Taylor (before he became famous for scientific management) performed over 50,000 experiments to show that rounded cutting tools were superior to the previous pointed tools, that directing a constant stream of water (saturated with carbonate of soda to prevent rust) at the point of chip removal increased output 30-40%, and that the new tools worked even better on soft metals (90% improvement) than on the hard metal (a mere 45%) to which engineers had thought them most suited. Finally, he experimented with various alloys, showing first that it was the manganese that gave Mushet steel its character, finding next that chromium was superior, and then making the astonishing discovery that superheated tungsten allowed cutting speeds of four to five times the previous level. This discovery swept the machine tool world; cutting machines had to be totally redesigned to handle the new cutting tool. In 1917, a cobalt-chromium-tungsten alloy was developed which further doubled machine speeds to 150 feet per minute.

The new cutting, grinding, and other machines[18] spread slowly in the early twentieth century. Machine shop operators were loathe to abandon their entire capital stock; replacing one machine without replacing all was likely to create production bottlenecks. The rising automobile industry, however, had no sunk costs to worry about. Proper grinding of pistons and cylinders, a much more refined task than large steam engine cylinders had ever required, was essential to automobile performance. Numerous specialized tools were developed for particular car parts. In all, the modern machine tools "were as essential to the success of the modern high-speed internal combustion engine as was Wilkinson's bracing bar to that of the steam engine" (Rolt, 1986: 233). From the 1930s, the aircraft industry would rise to a position second only to automobiles in driving machine tool improvements.

"The new materials for cutting tools not only produced important changes in the size, rigidity, and power required to utilize them for high production rates, but they also gave rise to scientific study of the actual cutting process with the application of the theory of solid state physics" (Woodbury, 1978: 1037). We have already seen the extensive experimentation undertaken by Taylor. As higher temperatures and pressures were encountered in machine use, knowledge of mathematics and science became increasingly important; increased understanding of metal resistance to corrosion and stress, along with homogeneity and precision in manufacture, allowed engineers to reduce the margin of error in machine design (Burstall, 1963: 367-76). Koenigsberger feels that the outstanding feature of the period 1890-1960 was the trend toward the scientific or at least systematic approach to the solution of problems; standards for the evaluation of machine tools were developed in the late 1920s and saw widespread use in the 1930s (1978: 1046-52). In such an environment, a series of small improvements was inevitable.[19]

The percentage of powered machinery which was electrified grew from 4% in 1899 to 30% in 1914 to 75% in 1929. Kilowatt-hours per worker rose from 1.22 in 1920 to 3.2 in 1950.[20] After the revolution in cutting materials, electrification must be seen as the most important transformation in the early twentieth century. Though Taylor's experiments began before the advent of the electric motor, they could never have been fully exploited otherwise (Rolt, 1986: ch.10). Woodbury, discussing improvements to the milling machine and lathe, concludes that, "Above all these was the introduction of local electric drive" (1978: 1038). Commonwealth Edison had opened an electricity shop in Chicago in 1909; on one floor it displayed household appliances, and on another electrical machinery. As it was for households, the 1920s would be the era of electrification for industry. Through tailoring the power source to the job at hand, immense gains in productivity were possible. One example must suffice. Electric battery trucks for handling materials replaced about three workers on average; the 12,000 in existence in 1928 thus replaced 36,000 jobs (Jerome, 1934: 368).

The machinery sector has always tended to be highly volatile in output. Mills estimated an index of instability in growth, and found machinery to be slightly more unstable than consumer durables, and almost five times as unstable as non-durables (1934: 270-8). Industrialists can be expected to postpone costly capital expenditures when they are facing hard times themselves. As with consumer durables, though, it is likely that the decline in machinery output in the early 1930s was much more than a cyclical phenomenon. (Machine output rose from 28,000 units in 1927 to 36,000 in 1928, and 50,000 in 1929, before falling to 23,500 in 1930, 12,000 in 1931, and 5,500 in 1932). We have seen how traditional mechanical engineering, chemical experiment, the rise of the automobile, and electrification, had combined to create a new generation of machine tools in the early twentieth century. Industrialists could not delay the scrapping of obsolete equipment forever. While the general manager of the National Machine Tool Business Association bemoaned the instability in 1919, he also spoke of the inelastic demand facing the industry and the danger of overcapacity (Wagoner, 1968: 121). Aldcroft has suggested that the leading role played by machinery producers in the 1920s might have limited the amount of replacement needed in the 1930s. Along with the growth of machine-intensive industry, it would appear that the evolution of machine tool technology itself must have contributed to the investment boom.[21]

The most revolutionary changes predate 1920. "After 1920 most developments in machine tools tended to improve equipment already in operation" (Hogan, 1971: 1032). With hindsight at least, the early 1920s must appear as a good time technologically to buy new machinery. Machinery was much superior to that available mere decades before, but was not to be made obsolete in the near future. Improved lubricants and fittings could be easily absorbed. A survey in 1935 found that two thirds of installed machine tools were over ten years old (Woodbury, 1978: 141). Yet Gill found that machine productivity steadily increased through the 1930s. Throughout his study he notes the limited investment required to adopt new technology. He claims that the sluggishness of recovery was in large part due to the poor performance of capital goods, and that the sluggish demand for these were due to the course of technological change (1940: 19). Productivity-enhancing innovation which did not require investing in a new machine could only ensure that machine tool output remained low. Within the machine tool sector itself, improvements in such areas as acetylene torches and methods of steel castings ensured that labor productivity rose steadily.

Tungsten carbide machine tools were a major concern of Gill's. This material was first developed by Krupps in Germany in 1926, as an offshoot of the production of tungsten filament for lightbulbs. Within a couple of years it was marketed in the United States. By 1934, the problem of cutting the material itself had been solved with the use of diamonds. Machine tool speeds now reached 400 feet per minute. If the cutter were run at high speed, but the material to be cut fed slowly, a cut so fine that further grinding was unnecessary was achieved. The new machine was admirably suited to the new generation of tough alloys of steel and aluminium, which were of key concern to the airplane industry, and were previously unmachineable. In turn, the high speeds of the new machine were only obtained by borrowing gearing and lubricants developed for automobiles. Still, the investment required to utilize the new material was minimal in relation to the labor-saving that accompanied it.

The 1920s, not surprisingly, had seen much development of conveyor belt technology. Hydraulics would also see much improvement in the 1920s and 1930s (Hogan, 1971: 1033-4). Woodbury credits the desire for higher production rates within industry with stimulating the development of electronic and hydraulic controls and drives; one byproduct was the replacement of skilled labor (1978: 1935). These interwar innovations paved the way for postwar adaptation of computer technology and steady progress in automation. While this would act to further replace labor in industry, it would at least stimulate postwar demand for machine tools.

Computers

Babbage's design for a calculating machine in the early nineteenth century possessed the major elements of the modern computer in mechanical form, including the use of mechanical counters for memory. Thousands of mechanical calculating machines had been manufactured in the 1920s. While some of these machines were electronically driven, the calculations were all performed mechanically. Devices for reading punched cards (electrical contact occurring when the pin met a hole) had been used in the 1890 census. An increasing range of adding machines, computing devices, and desktop calculators appeared after 1900. These had important impacts on clerical employment. Numeric integration of the differential equations of planetary motion was accomplished in 1933.

While mechanical devices could achieve much, the difficulties of construction overwhelmed any attempt to create machines capable of readily handling complex calculations. Mechanical devices were also inherently slow. Mauchly, co-developer of the first modern computer, had recognized in the early 1930s the possibility that electronic devices could work thousands of times faster and handle problems of much greater complexity. His attempts to attain funding for research in that direction achieved no success until the Army became interested in 1942 (Luckoff, 1979: 29-30). Developments in other fields, such as the electronic selection devices created for radio and telephone, paved the way for computers. Still, until government became interested at the end of the 1930s, no fundamental advance had occurred in commercial calculation devices since the 1900s. (Katz, 1982: 422).

"The general view prior to 1950 was that there was no commercial demand for computers" (Katz, 1982: 420). It was felt that only a handful of government departments and even fewer private firms would have use for one. Even after the first computers appeared, they were so cumbersome and expensive that only a handful found ready use. They were often produced, in fact, on a one-of-a-kind basis for particular end-users. In the early 1950s a variety of firms had the capability of producing computers but all still believed the potential sales small.

Given the poor market predictions (aided by a failure to foresee how inexpensive computer technology could become) it is not surprising that war-induced military spending played the key role in the inception of the modern computer. With the technical antecedents long in place, the expense of the ENIAC project came not so much from difficulties in conception and design, but rather in construction and getting innumerable components, including 18,000 vacuum tubes, to work simultaneously. The final computer took up 3000 cubic feet of space. Critics of the project, indeed, had harped not on the feasibility of the concept, for the technology of electronic switching was in a sense straightforward, but doubted that everything would work long enough to get meaningful results (see Luckoff, 1979: 32-33). The critics were proven wrong. ENIAC was unveiled in 1945. The first fully electronic computer, despite its dependence on vacuum tubes, proved to be 100-500 times faster than that which had gone before.

ENIAC was still a far cry from the computers of our day. To program it, thousands of switches had to be set manually. It pointed the way, however. "A wartime project which might have ended with the war's end was instead the beginning of a technological revolution" (Mauchly in Luckoff, 1979). The next step was to develop methods of storing programs, and this meant devising an electronic memory. This was in many ways a trickier problem than the original design of the computer. Many researchers went down the wrong path, experimenting with such devices as electrostatic storage tubes. Again, government financing provided the key to the solution of the memory problem (Katz, 1982: 422-3). The first stored memory machine appeared in 1949. The early stored programs contained error-checking routines to counter the effects of vacuum tube failure.

A host of improvements were made to the computer through the 1940s and 1950s. Magnetic tape readers were designed for rapid data input and output. The transistor was introduced to replace the cumbersome, hot, and unreliable vacuum tubes from 1954 (continued reliance on unreliable tubes must always have severely limited the scope of computer technology and speed of operation). The census bureau left the first UNIVAC with the Eckert-Mauchly company for two years so that the bugs could be worked out; otherwise the development of computers might have been set back for years (Luckoff, 1979: 111). Fisher spoke of "continued, rapid, often unanticipated technical change" (1983: xi).

Eckert-Mauchly (which was bought by Rand) expected to sell fewer than ten UNIVACs but eventually sold 40. In 1953, IBM forecast few sales for its "650" model; this became the "Model T" of computers and they sold 1800. IBM had begun assembly line production of computers in 1953. The sales of the "650" established that there was a huge market for business applications, whereas it was previously thought that scientific calculations would be the main outlet (Fisher, 1983: 17). Encouraged by the size of the market, further developments were made in the late 1950s, notably magnetic core memory, disk drives, and the first computer languages. Higher speeds expanded the market further to areas such as airline reservations. The viability of the computer industry was at last assured.

As with the jet engine, the long lag between the first computer and the achievement of mass production would seem to preclude the possibility that minor changes in government policy might have engendered its widespread use before the 1930s (especially given the radical improvements possible only with the transistor). Still, the necessary technological precursors appear to have been in place for some time before funding became available to finance the first stage of innovation.

Transistors

Those who remember vacuum tubes will recall their numerous drawbacks. They were fragile, and frequently had to be replaced. They were large. They had to be heated and energized before use. Other elements had to be protected from their heat. They did fulfil the fairly low standards acceptable for radio (Dummer, 1978: 5). There were many, though, who were conscious of their disadvantages through the interwar period.

The modern transistor was a result of a research program begun at Bell Labs in 1946. There were precursors; Braun and McDonald speak of a continuous chain of innovation stretching from the crystals used in early radio receivers (1982: 1).[22] But the solid state transistor was a distant dream until Bell decided to devote considerable resources to its pursuit in the postwar period. "Three things delayed the innovation of a solid state amplifier: the success of valve amplifiers [vacuum tubes], lack of knowledge of solid state physics, and the lack of a product champion" (Braun, 1982: 4). On the other hand, the clear market for transistors which the vacuum tube had established was an important spur to innovation.

Bell Labs did not start from scratch in 1946. Enough had been learned during the 1930s and the war to be able to produce germanium and to understand its electrical properties and how to alter them. Physical experiment made it clear that extreme purity would be required; the chemistry of the 1930s was unable to supply this (Braun, 1982: 21). "Wartime concern with purifying and understanding semiconductor materials was crucial to early postwar success in obtaining the transistor" (Braun, 29; this was the opinion of the inventors of the transistor). It was the wartime research that established germanium, followed by silicon, as the prime contender.

"It was not until the 1940s that solid state physics acquired any systematic, structured, and theoretically supportive framework of knowledge" (Braun, 1982: 23). This advance occurred despite the fact that most physicists were drawn into nuclear research in the interwar period; many concepts were in fact borrowed from nuclear physics. It followed upon a series of scientific discoveries dating from the nineteenth century.

From this base, much work still had to be done. The complexity and small size of transistors, plus the need for purity, meant that idle tinkering would not succeed; development in a research laboratory was required (Braun, 1982: 6). Bell Labs was in the 1940s the largest industrial research organization in the world, and its reputation allowed it to attract the cream of the scientific and engineering community. (The previous technical advances, combined with the market for an improvement on the triode appear to have induced a number of similar efforts in the United States and Europe in the early postwar era. Its time had come.)[23] Bell encouraged the necessary cooperation among physicists, chemists, metallurgists, and engineers. It is estimated that from $140,000 to well over a million dollars was spent by Bell to develop the transistor.

The transistor required much further improvement after its initial development. It took well over a decade before it could outcompete the vacuum tube in both price and performance. It had not been realized at the time of its development that the transistor would do more than be a smaller vacuum tube (Braun, 1982: 46-7). Its most significant impact, though, would prove to be in products which were impossible or uneconomic with vacuum tubes (Nelson, 1962: 553). Computers, miniature radios, hearing aids (one of the first large markets for transistors), calculators, robots, satellites, space launches, digital watches, and a whole host of other devices for consumers, industry, and the military fall into this category. After the discovery of the transistor, R&D efforts were undertaken throughout the world which produced "a multitude of new improved devices" (Tilton, 1971: 11). The development of integrated circuits (two or more semiconductors on one silicon crystal) in 1958 opened up a host of new possibilities. These various new electronic products have played a major role in providing output and employment growth in the postwar era.[24]

Notes

1. Growth was naturally uneven. Mining, pumping, hoisting, and welding were electrified greatly in the 1890s. Industries which required much heat energy, such as paper, chemicals, and petroleum tended to lag in their adoption of electricity. Industries near water, with small plants, or which were growing fast, proved the most susceptible (Du Boff, 1979: 134-43).

2. Increased use of electricity by industry helped to balance the demands from consumers and railways. There was a built-in accelerator mechanism in the entire process; increased use lowered average cost which encouraged further use (Platt, 1988: 247, 270). The now-familiar S-shaped diffusion path of new technology was first identified by Jerome in 1934 with regard to electricity (Woolf, 1984: 177). There was much backsliding after the war as the vested interests associated with small-scale production fought these new ideas, using whatever legal and political tools they could commandeer. In the United States, the most extreme example was the Muscle Shoals hydroelectric plant (attached to an electricity-intensive nitrogen fixation plant to serve wartime armaments and fertilizer needs), the largest generating facility in the United States at war end, which became a political football through the 1920s, but the centerpiece of the Tennessee Valley Authority in the 1930s. A similar saga is observed in Germany (Hughes, 1983: ch.11).

3. Manufacturers with a heavy investment in steam engines, gears, and pulleys, were naturally resistant. "Factory owners were slow to recognize the numerous, albeit indirect, benefits of reorganizing work flows according to principles of scientific management by installing a separate motor for each machine" (Platt, 1988: 263).

4. Patent agreements can have varying effects on innovative efforts. With hindsight, much of the innovative effort of the 1920s seems designed to have something to trade with other companies who might produce something in one's own line (Reich, 1985: 237). Sharing existing, but not future, patents may be optimal from society's point of view.

5. The tungsten filament doubled energy efficiency. The combination of winding the filament into tight coils plus the use of argon and nitrogen in the bulbs (1913) increased energy efficiency a further 50%. Due in part to the increased efficiency of electricity production, the cost of lighting was one quarter of the 1910 level in 1930. Lightbulbs per capita increased by 16 times. Coiling the coils

Electrical Products   201

would achieve another significant efficiency gain in the 1930s. GE also developed automated production methods which doubled labor productivity. Jerome estimated that 22,931 fewer workers were required in 1929 to produce the 1920 output (1934: 381).

6. Half the telegraph operators (8460 of 16,926) had been replaced by the printer telegraph. 3500 jobs had been lost in wire services to machines (Jerome, 1934: 380).

7. David Sarnoff had suggested to Marconi that they set up commercial radio in 1916 but had been ignored (Udelson, 1982: 5). Sarnoff would later become the driving force behind RCA. As early as 1909, some stations were broadcasting regularly scheduled news and music

8. Radio had only gained about 1% of the national advertising market at the time, but this was enough to support these networks. Still, Copeland speaks of the role of radio in making the market for many goods homogeneous (1929: 402).

9. Style did have some role to play. Philco, a battery producer which felt it had been betrayed by RCA, entered the radio market in 1928 with the idea of a radio that would look as good as furniture. It outsold RCA by 1933 (Udelson, 1982: 120).

10. MacLaurin is sensitive to the idea that innovation might cause cyclical fluctuations: "one could argue that, had we not gone on such a temperamental building spree in the 1920's in the expansion of public-utility enterprises, and the high pressure development of such fields as radio, the 'new investment' phase would have lasted until FM and television were ready."

11. American Telephone and Telegraph devoted a mere $300,000 to television research 1925-1930 and another $500,000 1931-1934. It was eclipsed by RCA in the latter period due to patent swaps among the large electronics firms (Burns, 1991: 181).

12. Burns (1991) discusses in detail various minor improvements to various elements of television technology in the late 1920s and early 1930s.

13. Some attempts to replace household labor failed. A handful of companies set up cooked food delivery operations in the late nineteenth and early twentieth centuries. They served at most 100 people, and none survived the 1930s. Given the failure of overall household work time to fall, we might wonder if all appliances (which themselves must be bought, repaired, and cleaned) have served a labor-saving role.

14. As late as 1910, most urban residents knew of electric (or gas) service only because of its rapid introduction for central street lighting. Electric companies used the schools to advertise the cleanliness and other advantages of electricity in the early twentieth century. Many poorer people gained electricity by moving into serviced neighborhoods as the rich fled to the suburbs (Rose, 1988: 230-6).

15. By 1930, gas space and water heating, while not as prevalent as gas cooking, was also becoming commonplace (Cowan, 1983: 91). Bowden (1988) studying English adoption of electric ovens in the 1930s found that the relative price of electricity and gas was the most important factor. Instalment buying, income levels, and house construction were also significant.

16. Gas was a potential competitor here too, but the technology was never adequately worked out. A gas absorption refrigerator had been developed in Sweden in 1926 and firms produced gas-powered models into the 1950s. Cowan feels that if these producers had been given the support electrical manufacturers were given by utilities the quieter and less complicated gas machine might have competed effectively (1983: 133-42).

17. See Hogan (1971: 1405). The Westinghouse refrigerator had 12,459 moving parts in 1931 but only 714 in 1941. The later 1930s saw developments such as sliding shelves and the automatic interior light. The price of refrigerators fell steadily.

18. The lathe would see steady advances in speed, power, rigidity, and precision. By 1920, electric drive and rolling contact bearings were standard. "However, by far the most important advances in all lathes, in fact in all machine tools, was in the cutting tool itself, and in the reflection of its new possibilities in the overall design of the lathe" (Woodbury, 1978: 1036-7).

19. Much greater precision in physical measurement was made possible by Johanson's measuring blocks, developed in Sweden in 1907. From 1918, the US Bureau of Standards produced these mechanically (Woodbury, 1978: 1041). Rolt bemoans the lack of either monetary reward or fame that was accorded to machine tool innovators. It is an industry which, though small, has a far-reaching impact. He suggests that innovators have been motivated by the challenge involved, and the view of mechanical engineering as a vocation.

20. Some of the reasons for the growing popularity of electric power in factories are suggested by the following quotation from the *1910 Census of Manufactures* (VIII, 331). "Electric power is largely applied by means of relatively small motors distributed throughout the manufacturing establishment, some of which are in general use while others are required only at infrequent intervals. As the electric power can be used or cut off at will, it proves both convenient and economical, especially for the operation of machinery which is in use only a part of the time; and the cleanliness and quietness of the electric motor as compared with other sources of power also give it manifest advantages in certain industries, such as the clothing industries. ... The electric motor run by purchased current furnishes power for manufacturing with a minimum of trouble or attention on the part of the operator. ..." (in Jerome, 1934: 253).

21. As Gourvish has noted for Britain, the diversity and lack of data for machine production make it difficult to discuss with precision (in Buxton and Aldcroft, 1979: 129).

22. The principles underlying the success of crystals was not understood. However, when the triode faced increasing difficulty in satisfying later applications, attention naturally turned back to solid state (Tillman, 1978: 1116). Great numbers of crystals were used in radar during World War Two; this stimulated research (Tilton, 1971: 10).

23. "It is unrealistic to see the transistor as the product of three men, or of one lab, or of physics, or even of the forties. Rather its invention required the contributions of hundreds of scientists, working in many different places, in many dif-

Electrical Products

ferent fields, over many years" (Braun, 1982: 43). Bell began its project in 1946, after showing interest through the 1930s, with the feeling that a breakthrough was at hand (Nelson, 1962: 558).

24. "Even in the depths of recession the [semiconductors] industry is hungry for capital to expand" (Braun, 1982: 85). The full effects are often hidden; Freeman (1982: 102-3) notes that electronics had increasingly been used in a wide range of products, such as machine tools, aircraft, and toys, which are not included under electronics in national statistics.

9

Internal Combustion

Automobiles

"Every description of the state of the economy [in the 1930s] has of necessity included the motor vehicle industry, which was by the time in effect the main determinant of the nation's economic condition" (Rae, 1984: 194).

As we would expect, a variety of forces combined to create the internal combustion engine.[1] The concept dates from Huygens in the seventeenth century; it thus predates the conception of the steam engine. Huygens thought of using gunpowder. Only in 1859 did Lenoir succeed in creating a workable internal-combustion engine based on a mixture of coal gas and air. His engine was inefficient (there was no compression before ignition) and thus expensive, but its limited success spurred other innovators. Among improvements in the next few years was the development of the four-stroke cycle by Rochas. It is possible that Otto was unfamiliar with Rochas' work when he developed his engine in 1876. Through this early period, internal combustion engines borrowed much from steam engine technology. Many engines were adapted for use with petroleum based (kerosene at first) fuels. All early engines, including the diesel, were designed for factory use, were large, and worked slowly. In 1885, Daimler introduced the first high-speed engine. The rapid vaporization required was achieved through the use of gasoline and the introduction of the carburetor (a device which would see much improvement by many hands in succeeding years). Daimler's engine was also much lighter; he was looking toward markets in ships, airships, trains, and manufacturing. Electric ignition, often used in previous engines, was a key element of Daimler's design. In the same year, Benz developed the first internal combustion motor car (Field, 1958).

There are three types of internal combustion engine. The first was the gasoline engine, which required a spark to achieve intermittent combustion. The second was the diesel, which was much heavier but could achieve intermittent combustion through compression alone. It would thus see its first successes in ships and railroads, and later trucks, before its power could be harnessed to the diminutive automobile. The third was the gas turbine which did away with the need for intermittent combustion, but operates best in large units; it is used for large stationary power production and for jet engines in aircraft. While each of the three presented its own technical difficulties, they also share a common technical heritage. All can be traced to the Otto engine of 1876. Improvements in gasoline engines over the years helped pave the way for commercialization of the other two types. In this way (and also as early airplanes used gasoline engines) the development of the automobile had important derivative effects on the other major forms of transport as well.

Internal combustion is inherently more efficient than the external combustion of the steam engine which preceded it. This fact was of great importance for locomotion, where the ratio of power generation to mass was a crucial variable. Some steam-powered vehicles, notably the Stanley Steamer, were commercially produced around the turn of the century. They performed about as well as the gas-powered cars of that era. Since steam locomotion in railroads would be superseded by a form of internal combustion in the 1930s, despite a head start of over a century, we can be confident that internal combustion would have established superiority even if steam-powered autos had been developed earlier. It is worth noting, though, that experimentation with the idea of the motor car was not dependent on the development of the internal combustion engine. There were other forces at work encouraging the emergence of the horseless carriage.

A key antecedent of the automobile was the bicycle. Cardwell (1972: 199), calls it a necessary precursor. Many of the earliest automobile producers had begun as bicycle manufacturers. Wherever there was a concentration of either metal works or bicycle manufacture, at least one firm would move into car manufacture (Goodman, 1988: 147). The bicycle had been one of the first new consumer goods to emerge in the nineteenth century. As prices had dropped, the bicycle had spread through society from the 1890s and 1900s (Barker, 1982: 152). By the First World War, there were six million bicycles in the United States (four million in France, five million in Britain). This was a clear indication of the market available for a new mode of transport which could provide freedom from public transport and animal traction.[2] Bicycles (and sewing machines, another of the new products of the nineteenth century) also played a significant role in fostering the development of the technology of interchangeable parts, which was to be of great importance not just to automobiles but to a variety of twentieth century industries (Rosenberg, 1963). Light tubular frames, ball bearings, chain drive, pneumatic tires, gears, wire wheels, and brake cables were among the particular technical elements borrowed by the automobile (Ville, 1990: 174).

Only with petroleum drilling and refining in the second half of the nineteenth century had a suitable fuel for internal combustion become widely available. The development of the Spindletop field in 1901 provided the American auto industry with inexpensive fuel (Crabb, 1969: 69). Innovations in tires, gearing, and metalwork were also necessary before the earliest cars could appear in the 1880s (Goodman, 1988: 145-7). As well, means had to be devised to mix the air and fuel to handle different speeds and loads, methods of cooling the engine had to be developed, a transmission had to be designed so that the engine would not stall when the vehicle was travelling slowly, and a starting mechanism was required; these earliest difficulties were only fully overcome in the first decade of the twentieth century (Bryant, 1978: 998-9). Later, as alloys of steel and aluminum made it possible to produce lighter engines with more, smaller, cylinders, problems of balancing and vibration emerged (Burstall, 1963: 408).

The last decades of the nineteenth century and first of the twentieth thus saw innumerable attempts to improve upon the automobile and the methods of its production. Often, these innovators were unaware of the actions of others. Over fifty new car companies emerged in the United States in 1902, of which only a handful would survive a decade (Crabb, 1969: 124)[3]. Countless minor innovations ensued: the steering wheel and storage battery in 1895, precision engine boring by Cadillac. Perhaps the greatest reflection of this is the fact that in 1915 automobile producers agreed to pool all but the most major patents (Seltzer, 1928: 44).

Even the most important innovation in the industry of the time, the Ford assembly line, was in large part a result of previous improvements. Hounshell notes that it is important both to recognize previous moves toward interchangeable parts, disassembly lines (meatpacking), efficiency, and continuous flow, and to recognize Ford's revolutionary role. Conveyor belts had been used in brewing, flourmilling, meatpacking, and canning. Time and motion studies had been performed in the early days of auto manufacture. While it is not clear how aware of Taylorism Ford was, these studies were used in setting up the assembly line (Hounshell, 1988: 249). Interchangeable parts had been a goal of American industry through most of the nineteenth century; it was achieved to the degree that cars could be taken apart, and the parts mixed and reassembled, by Cadillac as early as 1908 (Crabb, 1969: 114).

Both within and without the auto industry, then, there were many precursors to the Ford assembly line. Still, putting together the pieces was a bold step forward which was to have repercussions throughout American industry. Ford was confident that he would be able to significantly lower the cost of production by simply producing more of the same car (he may have simply been thinking along lines of learning by doing). He was aided in this goal by the development of vanadium steel by the chemical and steel industries; it was the key to the production of the light and inexpensive Model T (Abernathy, 1979: 31-2). His persistence would lead in 1913 to the assembly line.

The achievement of the assembly line was contemporaneous with the emergence of the American car industry to world dominance. In 1906, the United States produced more cars than France for the first time; in 1910 it produced more than the rest of the world. Some attributed the American advance to looser government regulation. American consuls in Europe, at least, attributed it largely to differences in average incomes; there just were not as many people in Europe who could afford a car (Barker, 1982: 157-8).[4] Certainly, Canada and Australia, the only other nations with a similar car/person ratio, also possessed high average incomes. We saw in Chapter 3 that economic growth in the long run must be seen as an interplay between rising productivity (and thus incomes) and the generation of new consumer goods. We would expect an expensive new good like the automobile to fare best in a society where sizeable numbers had incomes far beyond the cost of traditional necessities.

The automobile was an innovation that would have immense social repercussions.[5] Its development therefore required a host of complementary transformations. The most central was the development of a modern road system. Following on the demands of early bicyclists, early drivers pressed for roadways more suited to the new method of locomotion. The automobile itself demanded improvements. The rate of fatalities per car was twelve times higher in 1910 than in 1940. Late nineteenth century parkways had already shown the way with grade separation, limited access, and exit ramps. As the closed car steadily replaced the open air touring car (see below), commuting became more common; the first major traffic jams occurred in many American cities in 1914. By that time, virtually every state was charging a vehicle registration fee, and had recognized gasoline taxes as a painless way to finance road construction. While political considerations often meant that rural areas received disproportionate attention, a network of roads would be achieved within a decade or two.[6] The Federal Highway Act of 1921 established that the federal government would pay for half the cost of U.S. routes. This system was largely complete by 1935, though by that time it was realized that larger capacity divided highways were needed in many areas. The 1920s saw considerable bridge-building, in part due to the introduction of the suspension bridge (Hogan, 1971: 130-1). Such bridges as the Golden Gate, San Francisco-Oakland Bay, and Triborough, would be built in the 1930s. The American road network would double in the 1920s and again in the 1930s.[7] The success of limited access highways in New York in the early 1920s encouraged replication in other cities through the 1920s and early 1930s. Traffic lights and banked curves became common in the period 1918-1930 (Allen, 1952: 118). So also did tourist cabins and roadside diners. Gas stations more than kept pace throughout with the expansion of cars and roads.

The advance of technology, and often the internal combustion engine itself, greatly aided the process of road construction. Horses had begun to be replaced by the power shovel for grading from 1910; a further great improvement occurred with the introduction of the caterpillar tread from 1920. An inspection of 1925 found that two thirds of grading jobs were mechanized. Horses were also replaced for the delivery and removal of materials. The subgrader for preparing the base for paving was rapidly adopted from 1922. The concrete mixer was applied to roadwork from 1910, and caterpillar treads added from 1918. The size of mixers steadily increased. The finishing machine for shaping concrete surfaces was introduced in 1917 and was almost universal by 1925. The centralization of concrete mixing saved 25-50% of the labor involved, while the finishing machine saved 40-60% of the labor involved in that operation (Jerome, 1934: 140-5, 367-8). In 1919 work crews averaged 4.5 feet per day; in 1929 the figure was 18. Jerome lists a variety of other innovations, and notes that state governments credited mechanized haulage as one of the major sources of productivity increase.

The introduction of the assembly line from 1913, and the collateral developments in roads and other infrastructure, paved the way for the phenomenal development of the American automobile industry in the twenties. Car production had only exceeded that of wagons and carriages in 1914. In 1919, there were 6,771,074 cars registered in the United States. By 1928, that figure had more than tripled to 21,630,000. There was an even more rapid increase in car production. The first year in which over one million cars were produced was 1916, and the two million mark was only surpassed in 1922. Annual output ranged from 2.94 million to 3.78 million 1923-1928. The level of almost 4.5 million cars produced in 1929 would not be achieved again for two decades. Output slumped to 1.1 million in 1932, before rising back to 3.93 million in 1937. Wolman, in 1929, averred that, "It would be difficult to find anywhere in economic history so swift and pervasive a revolution. In 1910 there was one automobile for every 265 persons in the country; in 1917 one to every 22; in 1919 one to every 16; and on July 1, 1928 one to every 6" (Wolman, 1929: 59). He noted that not only the aggregate numbers, but also production figures for low, medium, and high priced automobiles attest to the fact that all classes of the population participated in car ownership. In not much over a decade, this revolutionary new product went from being a rarity to being in the possession of the majority of households.

The increase in production was matched by an equally impressive increase in productivity. Some estimates show the assembly line cutting costs by a ratio of eight to one, though these neglect contemporaneous improvements in other areas (Abernathy, 1979: 24). Labor productivity expanded by 7.4% per year 1919-1929.[8] Much of this was due to learning by doing in the new assembly line mode of production. Among major improvements was the introduction of Duco lacquer (developed by Du Pont) in 1924 which reduced painting time from between ten and fifty days to two, and saved at least 15% of the labor involved in car painting.

As we might expect, there were also numerous improvements in what was still a fairly new product. As late as 1920, 85% of car bodies were made of wood, and this limited the spread of the closed body because wooden bodies could not handle rough roads; by 1926 steel was the norm for most of the body and closed cars accounted for over 70% of the market. Improved fuels and lubricants, the introduction of shock absorbers, along with better cars, tires, and roads, lowered the running costs and increased the durability of cars. This naturally made cars seem an even more desirable investment. However, the increase in durability served to make the car market that much more susceptible to violent swings in sales.

Installment buying is often credited with having a significant impact on car sales. By 1923, 75-80% of cars were being purchased in that way. "Without the forced saving imposed by an installment contract, thousands upon thousands of people would never have been able to accumulate the purchase price; their income would have been spent for more immediately gratifiable wants" (Smith, 1970: 95). Since installment contracts were rarely more than a year in length, we might wonder if the effect was that dramatic. Still, as installment buying involved an effective interest payment of about 30% (Olney, 1989), it appears many did have difficulty saving on their own. Those who may have thought that installment buying was the key to permanent prosperity miscalculated. It could only at best bring about a one-time boost in sales (Smith, 1970: 96). Some of the burst in sales which occurred in the late 1920s might have been delayed to the 1930s if installment buying had not become easier in the 1920s. Ford, indeed, blamed the downturn of the 1930s on installment buying and neglect of farmers. A conference was called at the White House in 1938 to see if installment buying had led to overselling and thus contributed to the boom and bust cycle so harmful to the economy (Smith, 1970: 133).

The most important result for our purposes was that the market for automobiles became heavily saturated by 1929.[9] With the advantage of hindsight, we can recognize the inherent danger in expanding output as rapidly as was done. The virgin market was quickly inundated, and car producers became increasingly reliant on the replacement market; replacement sales were one quarter of the total in 1913, two thirds in 1925, and three quarters in 1927. In the absence of substantial product innovation, and there was little from the mid-1920s[10], consumers were likely to hold on to their cars a long time. By the end of the 1920s, the average consumer did not replace their automobile until a little more than seven years had passed. In 1929, over 90% of the cars on the road were less than seven years old, and over 75% were less than five years old. The high levels of sales in the previous three or four years meant that most car-owners had a fairly new car, and were unlikely to need a replacement for a couple of years. That the existing cars could last many years is indicated by the fact that, though car sales plummeted in the early 1930s, registration was only 10% lower in 1933 and 2% lower in 1935 than in 1929.

There was little scope for expansion of the first-time buyers market in 1930. There were 26 million cars and 30 million households in 1929. On this basis, various writers have suggested that almost everyone had a car. Stricker (1983: 30), however, has noted that once adjustments are made for business cars, two car families, and singles not counted as households, only 46% of households had cars, and among the working class the figure may have been as low as 30%. To be sure, the ratio of people to cars would fall much further in the postwar period. However, we can not casually compare eras of substantially different average incomes.

Car prices were only about one third of the cost of operating a car (and manufacturing costs comprised only half of the retail price; Rae, 1984: 162). Thus, cutting car prices might do little to increase sales. While Ford had had great success cutting prices earlier, Maxwell found that when it cut its price by $100 in 1923 its sales did not rise at all. Even in the Depression, only a few manufacturers cut their prices. Plymouth slashed prices in 1930, but still could not stem the decline in sales. The demand curve facing the industry should be even less elastic than that facing individual producers. It would seem, then, that variations in price could lure few first-time buyers to the market.

This was especially the case given the existence of a widespread used car market. New car producers had found used cars a problem from the early 1920s. The auto industry was much less vertically integrated at the time (and vehicles simpler), and thus parts suppliers readily seized the opportunity to serve the used car market. Chevrolet bought and junked 650,000 used cars between 1927 and 1930 (Hogan, 1971: 1018). Used cars also ate substantially into the Model T market in the mid-1920s (Hogan, 1971: 1023). It was estimated that there were between one half and one million used cars for sale in 1929.

Given that market saturation occurred later in other countries than in the United States (indeed this may be one reason for Britain's relatively good performance in the 1930s), it might be thought that exports, which were 11% of output in 1929, were a potential salvation. The United States had, remarkably, produced 85% of the automobiles in the world during the 1920s. Exports, though, fell from 1927, largely due to tariffs. With one minor break, exports to Britain faced a 33% tariff through the interwar period. France and Germany likewise introduced high tariffs. This encouraged American firms to establish branch plants abroad. There was a strong fear of dumping in Britain in the 1930s, but this never materialized. Exchange rate adjustments and a drop in the horsepower tax (which had hurt the larger American cars) did lead to a brief surge in American exports there in the mid-1930s. Overall, though, American exports to the rest of the world fell off, in part as demand elsewhere shifted toward smaller cars (Miller, 1979).

It might be thought irrational for car companies to have pushed production to a point where their market would disintegrate. It was, of course, difficult in the heady days of the 1920s to accurately predict the future (though already in 1926 an article in *Motor* magazine had noted that most potential first-time buyers had bought cars, and less than two million more would do so by 1930; Hounshell, 1988: 276). Moreover, policies that may be optimal in the aggregate need not be optimal for individual firms. The late 1920s was a race for market share; concerns about the extent of the market were ignored (Smith, 1970: 100). Chevrolet had been steadily increasing its share through the decade.[11] When Ford shut down its plants to replace the Model T with the Model A, Chevrolet expanded further to take up the slack. Then, Ford brought out the A in 1927 to an amazing response. Sales of both Chevrolet and Ford remained high through 1929. Ford alone would account for the increased output in 1929. Chrysler entered the market in 1928 as well with the first Plymouth, though sales of these were moderate in comparison through 1929. Numerous other companies (independents still sold as many cars as GM in 1929) attempted to retain a niche in the market in the late 1920s.[12]

The most obvious symptom of emerging market saturation in the 1920s was the institution of yearly style changes. The automobile had reached a stage in the 1920s where the major early problems with performance had been solved through improvements to brakes, tires, valves, air cleaners, and oil purifiers. Ford remained confident throughout most of the 1920s that producing basically the same car year after year was the key to success. The success of Chevrolet over the 1920s signalled that style and appearance were of increasing importance to consumers. Ford found that sales of its Model T fell from 1923; it sold 2 million that year but only 1.4 million in 1926. General Motors had instituted yearly model changes in 1923. In Hounshell's words, the Model T was the only truly revolutionary car of the twentieth century; it made everybody want one; all subsequent revolutions were in marketing (1988: 278). General Motors realized that in order to establish a reliable replacement market it was necessary that cars at least appear to change from year to year. Such a policy would have been unnecessary in the heady early days of tapping a new market, but became increasingly important as the 1920s progressed. Even Ford, after selling 15 million Model Ts, would have to bite the bullet and retool to produce the Model A.[13]

The entire auto industry had seen inventory levels rise rapidly in 1928 (Hounshell, 1988: 295). While the overall sales figures for 1929 look good, sales had tapered from the first quarter. By May, dealers were clearly facing real difficulty, and most automakers cut output by June, though Ford waited until October. If the output level of the first five months of 1929 had been maintained, 7 million cars would have been produced that year. The auto industry went into the most precipitous decline of any industry. Output briefly stabilized in early 1930, which induced thoughts of economic recovery, but then output declined steadily to 1933. Even before sales fell off, the industry was suffering from overcapacity. "The automobile market was clearly oversold; in addition, the industry's capacity exceeded even the peak production of 1929" (Gordon, 1961: 426). Certainly there was little scope for investment in 1929. Even working just an eight-hour shift, auto makers could have turned out over 8 million cars in 1929, about double the actual output. Neither product nor process innovation pushed automakers to change their equipment. Nevertheless, Ford continued to expand its plant 1929-1930. Naturally, not all of the decline in auto sales to be observed in the next years can be attributed to market saturation. Not only would the newly unemployed have been removed from the market but those who still held jobs might be expected to feel some uncertainty about the future; they "could very easily continue to run their old cars until the skies brightened" (Smith, 1970: 114). Given that the downturn in car sales began from early 1929, and that industry pundits had been speaking of saturation for years, we can be just as confident that the incredible drop in car sales was not simply a result of the Depression.

Mercer and Morgan (1972) have attempted to quantify the degree of market saturation. This can only be done by estimating the demand function for automobiles. Such an exercise, especially when focussed on a new good, must be fraught with difficulty, and thus their results must be viewed with caution. The authors start from the premise that durable goods are subject to a natural path of growth and decline. They find that in a dynamic sense saturation had set in from 1924. That is, the actual stock of automobiles grew faster from that point than the desired stock. However, in the static sense of demand actually falling, they suggest that saturation should not have set in until 1930. Of course, if they have at all overestimated the demand function, they will predict saturation setting in later than it did. We can place more credence in the clear finding that the industry was tending toward saturation through the late 1920s. This, coupled with the fact that the downturn in car sales precedes the general economic decline by some months, would indicate that saturation had occurred in 1929.

Bresnahan (1991) has analyzed how the auto industry contracted after 1929. At the onset of Depression, the industry was comprised of a few low cost mass-production firms and a number of small firms using older technologies and experiencing higher costs. Many of these latter firms failed.[14] The low cost firms hoarded some labor, and thus saw productivity fall. At the industry level, though, productivity fell much less because of the disappearance of high cost operations. When expansion in auto production was possible later in the 1930s, little labor had to be hired because of the hoarding during the downturn.

Process technology continued to improve in the 1930s. Ford, especially, introduced much new machinery during that decade, including the transfer machine. Cams and crankshafts were cast, rather than forged, with much attendant labor-saving. Thus, even as demand faltered, progress was made in further reducing the labor required to assemble an automobile.

The 1930s do see some product innovation. The all-new Plymouth, with its widely-acclaimed smooth-running engine, was unveiled in 1931. The all-steel body (previously there had been a fabric insert in the roof) emerged in the mid-1930s, after the iron industry developed wide sheets,[15] and stamping machines were designed which were suited to these. With improvements in gearing etc., it became possible to streamline the automobile by lowering its center of gravity. Such improvements, while significant, pale beside those of the preceding decades. Moreover, they clearly grew out of previous attempts to improve the automobile. Nor would the early postwar period be highly innovative. This was widely recognized; a Ford executive in 1964 maintained that there had been no major innovation since the automatic transmission (developed in the 1930s, diffused postwar).[16]

Needless to say, car owners became even more hesitant to replace their vehicles once the Great Depression set in. Recovery of the industry was destined to be slow in such an environment; this was the penalty carmakers would have to pay for the durability achieved in the 1920s. Continued model changes did have some effect in encouraging sales. Replacement sales naturally picked up (from 1932) as the cars of the 1920s wore out. Still, recovery based on replacement and conspicuous consumption inevitably crumbled (Ginzberg, 1939: 205). The military requirements of World War Two pushed car production to zero for four years. This, combined with high wartime savings, guaranteed that there would be an active market once the automakers were able to retool for the civilian market. As the postwar boom continued, and as incomes rose steadily, car production surged to new heights in the 1950s (slowed by strikes and material shortages at war end, and also by the Korean War). It could only do so, of course, in concert with the growth of numerous other sectors of the economy. The conspicuous consumption motive in car purchase had existed since at least the introduction of annual model changes in the 1920s, but in the 1950s it became the dominant consideration in design and marketing. Automatic transmission, power steering and brakes, and air conditioning were not really new technology; they just had not been worth introducing before. By appealing to this motive, both turnover and complexity were enhanced. Car sales expanded rapidly. Saturation was still possible, but for shorter periods. Record sales in 1955 were followed by a decline in sales which contributed to the ensuing recession (Rae, 1984: 113).

If we are to credit the automobile with a significant role in fomenting the Depression, we must ask what the interwar period would have looked like without it. Interestingly, the auto industry was accused around 1910 of diverting the productive resources of the country, both labor and capital, into unproductive activity (Seltzer, 1928: 162). However, when the WPA examined the question in the late 1930s, they determined that the automobile was a technological innovation which had aided employment overall (Gill, 1940). This is what we would expect. While the automobile naturally had put some carriagemakers and buggy whip producers out of business, it was basically an entirely new product which opened up new vistas in personal transport. Even in the 1930s it employed more workers than the carriage industry ever had. The experience of Britain in the 1920s provides some guide as to what the decade might have looked like in an auto-less United States. As early as 1929, some British politicians recognized the role the auto industry could play in creating jobs (Miller, 1979). It is not the car itself which caused trouble, but its spectacular rise and fall. In a world without numerous other radically new products, entrepreneurs could not cope with the onslaught of unemployed.

The decline in auto production naturally affected a variety of other industries. It is important to recognize that auto use did not decline anywhere near as much as auto production. We have seen that car registration declined only slightly. Gasoline consumption per automobile actually rose, though much less than the heady increases of previous years. Those in oil refining, gas stations, road construction, and other sectors closely associated with car use were relatively untouched by the downturn in production. It was the industries which provided inputs into car production which were rocked severely from 1929. About one in eight workers in manufacturing were involved in auto production either directly or indirectly. In 1929 auto production absorbed 52% of the iron and 18% of the steel, 84% of the rubber, 73% of the plate glass, 58% of the upholstery leather, significant fractions of various non-ferrous metals, and even 17.7% of the hardwood lumber produced in the United States.

The Rubber Industry

Starting with bicycles, tires had come to replace footwear as the major outlet for rubber in the early years of the twentieth century. By the 1920s, with tires and tubes comprising 85% of sales, the fortunes of the rubber industry were largely tied to those of the auto industry. Of course, old cars as well as new may require new tires. As with cars, though, improvements in durability (along with better cars and roads) greatly extended the average life of tires. In 1915 the average lifespan was .77 years, in 1930 2.47. "Ironically, the 1930s penalized the industry for the excellent job it had done in developing its product in the 1920s" (Bernstein, 1987: 87). By 1932, tire producers were operating at only 49% of capacity. In part, this was because, "The tire industry had been overbuilt, and tire production had fallen sharply in the latter part of 1928" (Gordon, 1961: 426). Tire industry executives apparently expected the auto industry to continue to expand into the 1930s (Bernstein, 1987: 81).

The industry saw substantial process innovation. Labor productivity virtually tripled between 1914 and 1927. After rapid design innovation to 1920, the tire industry concentrated on using mass production to produce cheaper and more durable tires (French, 1991). The central tendency was to replace batch processing with continuous processing, and to introduce conveyor belts and other forms of mechanization. The Banbury mixer replaced half the labor in the mixing and compounding divisions. In tube manufacture, the watchcase method had four times the output per personhour of previous pot heaters. Transition to the new methods did not occur overnight due to the cost of scrapping old equipment. Still, Jerome estimated that to produce the 1929 output using 1922 methods would have required 35,888 more workers, and to produce the 1922 output with 1929 methods 13,215 fewer (1934: 369-70, 381-2).

The other "traditional" outlet for rubber, footwear, also suffered severely in the early years of the Depression. The industry, led by Firestone, did begin to diversify by 1931. Conveyors, communications and electrical equipment, sporting goods, tubing, hoses, gaskets, and engine mounts all provided alternative outlets, but all grew slowly.[17] Perhaps, if the auto industry had not blossomed so spectacularly, the rubber industry would have earlier turned its attention to other products, and not suffered as severely in the 1930s. Of course, by replacing other materials it may simply have shifted some of the suffering elsewhere. The rubber industry did significantly expand its advertizing budget in the 1930s.

Another problem faced by the industry through the interwar period was that international collusion was forcing up the price of raw rubber. While Americans had laughed at German efforts to produce synthetic rubber during World War One, the 1930s witness the introduction of synthetics by both Du Pont and Goodyear. World War Two would both open up a variety of new uses for rubber and see heightened activity in the synthetics field on both sides. Synthetics would be found to be more resistant to wear, oil, and sunlight and would thus replace natural rubber for many uses. The postwar rubber industry would thus look quite different from its interwar self.

The Glass Industry

Plate glass sales naturally tumbled with auto production, and window glass with construction activity. Efforts of the WPA to stimulate construction provided some relief in the later 1930s. Those plants focussed on electrical products such as lightbulbs or vacuum tubes fared well. So also did glass bottle producers, especially as home canning increased in popularity.

Technological change characterized all parts of the glass industry. The nature of window glass production had meant that little except the mechanization of ancillary functions had occurred before 1900. In 1903 mechanization of some parts of the glassmaking process itself had occurred with cylinder technology. Only in 1917, following decades of effort in general, and over a decade of development by two separate companies in particular, had sheet drawing (rolling) emerged; this was practically automatic, quicker, economized greatly on labor, and produced sheets of better quality than the previous method of casting on a flat iron table, grinding, and polishing. Hand production fell from 100% in 1903 to 50% in 1912, 24% in 1923, and virtually zero in 1927. Even cylinder production fell to only 20% by 1929 (Davis, 1970: 181-91). The Fourcault sheet machine, a later import from Belgium, reduced the labor cost by 62% for single strength and 56% for double strength glass (Jerome, 1934: 368-9).

Rae credits the rise of the automobile with stimulating the development of continuous processing for plate glass (1984: 67). Ford itself pioneered the mechanized continuous production of plate glass after World War One. By 1929 this method provided one half of capacity (Davis 1970: 255-9). It would only be superseded by the float glass method in 1952. Output of plate glass nearly tripled in the 1920s.

The 1920s also saw the elimination of hand blowing of glass containers. "By 1920, after thirty years of almost continual mechanical revolution, almost any bottle, whether wide mouth or narrow, could be made by automatic machine better and vastly cheaper than in 1890" (Davis, 1970: 211).

Jerome estimates that the automatic glass bottle machine resulted in a labor saving of 86-97% (1934: 368-9).[18] The next decade saw a steady process of minor improvement in these machines. Half of all bottles were produced by automatic machinery in 1917, 80% in 1927, and 90% in 1924. For lightbulbs as well, 95% of production was automated by 1927. Productivity was also enhanced by continuous processing of the molten glass; some have claimed that automatic machinery was impossible without this (Davis, 1970: 211). There were also a number of improvements in factory organization. New methods of milk and beer distribution were stimulated by these advances.

The post-World War Two period would see a burgeoning of new uses for glass. Half of Corning Glass sales in 1950 were products which did not exist in 1940 (Allen, 1952: 198). Even in what seems a "traditional" sector, then, there was potential for creating new markets. This potential, unfortunately, was not to be realized in the 1930s.

Iron and Steel

The iron industry had been transformed by the first Industrial Revolution, and was transformed again in the mid-nineteenth century by the Bessemer and Gilchrist-Thomas steelmaking processes. These methods were developed in the old trial and error fashion, but the fact that the Bessemer worked on some (non-phosphoric) ores but not on others stimulated much scientific research into the chemical processes of metallurgy. Carnegie hired a chemist, and felt that this gave him a huge advantage over his competitors (Mowery, 1989: 29-30). The new production techniques, and the new industries the iron industry came to serve, required that the industry become increasingly dependent on scientific research (Mowery, 1989: 40). While not a creature of the Second Industrial Revolution, then, the iron industry both shaped and was shaped by the new attitude toward scientific inquiry of that period.

As a producer goods industry, the interwar output experience of the iron industry was largely determined elsewhere (there is reverse causation too: improved steels increased the durability of products and thus lowered demand for steel). Blast furnace output expanded from 36.9 million tons in 1920 to 42.6 million in 1929. This reflects to a large extent increased sales to the auto industry (not only was output rising there, but steel was replacing wood). Appliances, construction, machinery, containers, and pipelines were also significant sources of increased sales. The overall decline in iron output in the 1930s masked wide differentials in firm experience based on the industries which they served.[19]

Those affected least in the 1930s were those producing for the appliance or tinplate industries. Appliances had only been recognized as an important steel market in the mid-1920s. While some consumer durables had saturated their markets in the 1920s, others, such as refrigerators, expanded through the 1930s. Sales of steel for appliances declined by one third 1930-1932 and only returned to 1929 levels after 1934.

The canning of food made tremendous strides through the interwar period. Tinplate firms had respectable investment levels and higher profits through the 1930s. Demand for canned food only dropped 13% and picked up after 1933. In part, this was due to the discovery of means by which new products such as beer, oil, and fruit cocktail could be canned.

Those producing for railroads suffered throughout the interwar period. Construction demand collapsed from 1926. Public works projects in the later 1930s would alleviate an otherwise bleak situation. Oil pipeline construction had surged in the 1920s but would contract in the 1930s by 60%. The demand for steel for use in farm equipment and oil drilling also fell in the 1930s.

The dominant firm in the iron industry through the interwar period was US Steel, a giant which was worth 6.8% of American GNP when it was formed in 1901. Unlike most firms created by merger, it did not undertake much rationalization of its operations. Instead, management appears to have devoted much of its effort to (successfully) maintaining price stability in the industry. The courts, faced with this strategy, decided that they preferred stability to the previous cutthroat competition of Carnegie (McCraw, 1989: 594). Since steel was a small part of the final price of most goods, and thus faced an inelastic demand for much of its output, this concentration on price stability likely did not contribute much to the great fluctuations in output experienced by the industry in the period.

The 1920s saw no major single technological breakthrough in the blast furnace. There was, however, a steady accretion of scientific knowledge, an increase in scale (with associated technology), increased use of exhaust gases, and mechanization of casting and charging operations. Better furnace linings of vacuum-pressed brick were also introduced. At one Pittsburgh-area furnace personhours per ton fell from 2.4 in 1920 to 1.4 in 1923 to .83 in 1929. Productivity gains such as these allowed steelworks to move from the 12-hour day to the 8-hour day in 1923-4 (Hogan, 1971: 831-4). While only thirty new furnaces were built during the decade, these were usually parts of large steelmaking complexes, and thus engendered large savings in material transport costs. Only three new furnaces were built in the 1930s, while over 70 were scrapped. Due to a slight increase in scale during relining and rebuilding operations, total capacity only declined slightly. Still, the decline in numbers can be viewed as a reaction to overcapacity, for there were a number of idle furnaces in 1929 (Hogan, 1971: 1133-4). Interestingly, personhours per ton in Hogan's representative plant rose to 1.1 in 1931 before falling to .6 in 1939. While there was continued productivity improvement through the 1930s, there seems to have been some labor hoarding in the early Depression.

The open hearth furnace had replaced the Bessemer furnace as the principal method of producing steel in 1908. The process of replacement continued through the interwar period. There was also considerable improvement in the open hearth itself. Improved linings, inclined walls, and dolomite spreaders became common in the 1920s. Scale of operation increased without any matching increase in labor. Personhours per ton fell from 3.1 in 1920 to 1.9 in 1929. As a couple of plants were under construction in 1929, capacity actually rose 1930-1931. Open hearths operated at about 20% of capacity in 1932. A number of plants were abandoned in the 1930s. Still, further process technology was introduced. Instrumentation to regulate temperature and pressure had been used in a few plants in the 1920s; it spread widely in the 1930s. An improved carbometer was developed. A continued process of gradual change, of replacing worn out with new and improved equipment, resulted in a 54% decrease in personhours per ton in the blast furnace 1923-1946 and a 62% decrease in the open hearth (Hogan, 1950: 130-7).

For producing alloy steel and ingots, the electric furnace had emerged before the war (the first electric furnace dates from 1878 in Germany). It replaced the crucible process which could only produce small batches, had high labor costs, and relied on expensive raw materials. The new alloy steels that chemistry was making possible could thus be produced with much less labor than would have been possible without the advent of electricity and its application to heating. Over 80,000 tons were still produced by the crucible method in 1920, but this figure fell to 7,400 in 1929 and 2,500 in 1930. Electric furnace production expanded from 566,000 tons in 1920 to 1,065,000 in 1929, before collapsing to 686,000 in 1930. The automobile industry was the dominant source of demand for the new alloy steels during this period. There were a number of developments in electrical equipment for the furnace, such as better transformers, electrodes, and regulators, charging was mechanized, and furnace size increased. The switch to top-loading in the mid-1930s effectively increased capacity 11% and increased labor productivity considerably (Hogan, 1971: 836-9, 1146). Further improvements in the postwar era would result in the electric furnace being used for mass production of steel.

Rolling mills provide us with an example of a major breakthrough. While there had been relatively little change in the hand processes for making sheet metal between the eighteenth century and 1924 (steel being passed back and forth through the rollers), continuous hot strip rolling was introduced at that point. This depended on previous advances in furnaces, roller bearings, and variable-speed electric motors (Hogan, 1971: 1006-9).[20] Rae has argued that pressure from the auto industry for more and better sheets induced this innovation, though canning also played a role (Williams, 1982: 178). In any case, it replaced what had previously been a highly skilled occupation (rollers were adjusted by hand to achieve the desired thickness). The victory of the new method was not complete by the end of the decade. This area would thus see construction of 28 new mills in the 1930s. Many hand mills would abandon their attempts to compete with the new technology in 1930 and 1931 (Hogan, 1971: 839-41, 1010, 1148).

Airplanes

"Aeronautics ... for practical purposes is entirely a product of the twentieth century" (Williams, 1982: 258). Clayley has estimated the power to weight ratio necessary for flight; in the nineteenth century this was far beyond the capacity of existing engines. The development of the internal-combustion engine for cars caused the kilogram per horsepower ratio to drop from 200 in 1880 to 4 in 1900. The airplane became feasible (the Wright homemade engine had a ratio of 6 kg/hp.).

Flight presented different problems, and engine design soon began to deviate from that for automobiles. By 1918 airplane engines were ten times as heavy and 100 times as costly as their automotive cousins. The problem of knock, or premature combustion, had to be overcome; an intensive research effort involving both chemistry and engine design provided the answer (Bryant, 1978: 1012-4). The supercharger was developed to compress air at altitude. Progress in both engine and aircraft design was rapid before 1914; simple improvements in structure and aerodynamics could be made cheaply. The cost of launching a new plane in this period could be counted in the hundreds of thousands of dollars; by the 1930s this would be in the millions, and by the 1950s in the hundreds of millions (Sawers, 1978: 39).

Still, by 1914 the airplane had not advanced beyond the stage of an expensive hobby. Moreover, most advances after the Wrights had occurred in Europe. "Until the first war in 1914 flying was not taken seriously except by a few enthusiasts, but after 1918 it was generally appreciated that the construction of aircraft for both military and civilian uses was likely to provide manufacturing industry with an outlet of considerable size and economic value" (Burstall, 1963: 395). The effect that war had in advancing airplane development is incalculable. In 1914 there were worldwide some 5000 airplanes (only 49 were produced in the United States that year); another 200,000 were constructed by 1918. Major improvements occurred in terms of both reliability and safety. A patent pooling arrangement similar to that in the auto industry was arrived at in the U.S. in 1917 (it would be extended in the late 1920s). Without the impetus of war, costly legal battles might have greatly delayed aircraft development (Mingos, 1968:

29). The war and its aftermath also induced European experts such as Fokker and Sikorsky to migrate to the United States.

The immediate post-World War One period naturally saw a contraction of the industry as wartime demand fell off. Auto plants and others went back to their prewar uses. As with shipbuilding, war surplus sales limited the market for new planes. Only 263 planes were produced in 1922. Still, many firms survived, and some new firms, including Douglas, entered the market. The fact that aircraft technology was steadily advancing acted to minimize the depressing effect of the war-related glut. More centrally, the military, and especially the Navy, continued to place orders for new planes.[21] The early growth of Douglas, Boeing and of most other companies of the time, was largely dependent on Navy orders (Rae, 1968: 10, 16). Rae feels that no private company could have borne the development costs of new aircraft without government contracts. We must recognize, though, that civilian aircraft required different characteristics; demand from airlines for lower maintenance thus resulted in the frequency of engine overhauls falling from once every 50 hours during World War One to every 300 in 1929 and every 500 in 1936 (Bilstein, 1984: 88).

The 1920s were a decade of steady technical progress, the acceptance of flying by the public, and the beginning of air flight as public transport (Gibbs-Smith, 1970: 180). The public had seen the military value of the plane; only with continued improvement could it be perceived as more than an agent of death. Government provided not only the major source of research funding but the development of airports, charted airways, beacons for night flying, safety laws, and a weather service. In the United States, the introduction of air mail service in 1918 and the contracting out of mail routes from 1925 provided a major impetus to the construction of large planes suited to passenger transport (Europe sees passenger routes from 1919, the United States only from 1927). Through the 1920s, passenger service only proved profitable in conjunction with lucrative mail contracts (Bilstein, 1984: 56). Though many airlines, such as American, Delta, United, and Northwest emerged during the late 1920s, passengers were considered a nuisance until into the 1930s (Rae, 1968: 53). As the decade wore on, planes were found useful for private and corporate flying, cropdusting, seeding, and photography.

The most important technological advance was the introduction (first in Europe) shortly after the war of stressed-skin construction. This can be compared to combining the chassis and body of the automobile into one unit. The skin of the plane itself became load-bearing, and thus struts could be eliminated. Planes such as the Lockheed Vega of 1927 could thus carry the same load as earlier planes in a much smaller craft. Moreover, substantial increases in scale became possible. As well, the cumbersome water-cooled engines, such as the Liberty which had been used widely during the war, gave way in the mid-1920s to air-cooled engines such as the Pratt and Whitney Wasp.

Lindbergh's solo flight across the Atlantic in 1927 provided a great impetus to the nascent commercial aircraft industry. Only after Lindbergh was the financial and technical climate for the large scale development of aviation created (Gibbs-Smith, 1970: 190). Lindbergh himself, and other flyers, were surprised at the fame his flight achieved. Yet the stocks of airplane engine manufacturers increased in value ten times in the following eighteen months. An industry which had previously had difficulty obtaining credit now found itself with hundreds of millions of dollars to play with (Freudenthal, 1968: 76).

The 1920s can nevertheless only be viewed as a formative decade. Worldwide, passenger miles had expanded from one million in 1919 to 57 million in 1929. In the United States, the number of passengers rose from less than 6,000 in 1926 to 173,000 in 1929, and air mail expanded from 810,000 tons to 7.7 million. This was still only a drop in the bucket. There would be 4 million passengers by 1941. "By 1928 the volume of production of airplanes had not become sufficient to have a marked effect on business activity .... Nevertheless the industry was started during this period on a commercial basis" (Copeland, 1929: 323). The future looked bright in 1929, but had not yet arrived. Military-sponsored research had only just advanced by the end of that decade far enough for private commercially-oriented research to begin which could yield an economically practicable passenger aircraft (Phillips, 1971: 21). In particular much basic research on lift and drag was done in the 1920s, which would be utilized in the 1930s. As a result of aerodynamic testing, engines would be moved to the leading edge of the wing and new cowlings designed.[22] The latter alone so improved aerodynamics that the Curtiss pursuit plane could increase its top speed from 118 to 137 miles per hour. Improvements in aerodynamics were greatly aided by improvements in alloys of magnesium and aluminium and better steel (Gibbs-Smith, 1970: 194); metals research was in turn encouraged by aircraft development (Rae, 1968: 74). Wing flaps, which can be traced to World War One, were improved and re-introduced. Better and lighter engines emerged as well. The aluminium propeller replaced wood in 1925. From 1932, it operated at different speeds during take-off and cruising. New high octane fuels, based on chemical research done outside the industry, were adapted to aircraft. Various instruments were developed in the 1920s; the first "blind" flight occurred in 1929 and instrument flying would become commonplace in the 1930s. Retractable landing gear had been tried from almost the beginning of air flight; these were greatly improved as greater speeds required them. Private and government efforts combined in these achievements, but the government role was dominant at least in cowling design and fuel development.

It would take a decade (1925-35) for the industry to synthesize these various developments and produce a new generation of aircraft (Rae, 1968: 30). "In the 1930s, the accretion of technology from the previous decade began to produce handsome dividends" (Bilstein, 1984: 83). A steady stream of technological improvements does not necessarily yield steady advances in the airplane. The air-cooled engine was, for example, of limited importance until improved airframe design made higher speeds possible (Rae, 1968: 58). The variable-angle propeller also only became important as speed increased. Wing flaps allowed heavier more powerful planes to land safely. Sawers and Phillips have both emphasized the discontinuous nature of airplane development. Sawers speaks of three eras of rapid development, the pre-1914 period, the 1930s, and the 1950s. Phillips provides a similar view; "Successful major types of new aircraft appear discontinuously over time — not precisely at the times predicated on the carriers but rather as technological developments have permitted cost reduction or other performance improvement" (1971: 71).[23] While much of importance had occurred in the 1920s, the technological potential would only fully be realized in the next decade.

Phillips (1971: 50), has estimated operating costs per seat for the planes used by scheduled domestic air carriers 1928-1965. Before 1929, only three planes had a cost below 10¢/seat mile (1954 dollars) and the average was near 12¢. Of planes entering service 1930-1933, the lowest cost was the Fokker F-32 at 6.55¢. In 1934, the Lockheed L-10 achieved 4.7¢. Then, in 1936, the Douglas DC-3 weighed in with a cost per seat mile of 3.28¢. In a handful of years the cost of air passenger travel had been cut to barely a quarter of its previous level. The four-engined Boeing 727 introduced in 1954 would achieve a cost of 1.12¢/seat mile. "Measured in terms of operating costs, more progress was made from 1925 to 1935 from the trimotored Ford airplane to the DC-3, than was made during the following twenty years" (Klein, 1977: 17). The first profitable passenger-only air route was opened New York-Chicago in 1936. Along with decreased costs, the travelling public was rewarded with increased speed and comfort. In response to crashes by earlier planes (the death of Knute Rockne in a Fokker crash in 1931 had caused an extensive public inquiry) the DC-3 was also over-designed from the safety standpoint.[24]

The DC-3 is justifiably viewed as the most revolutionary breakthrough in aircraft design since the first flight. Within two years, it had captured 95% of U.S. air traffic. It would only be gradually displaced by four-engine planes in the postwar era. Even then, the essential features of such planes as the DC-8 were the same as the DC-3; refinements of DC-3 technology extended the speed, range, and size of piston-engined aircrafts. Rae hails the mid-1930s as the time of the birth of the "modern" airplane, and marvels at such an achievement by private companies during the middle of the Great Depression.[25]

Work on the next generation of aircraft had also proceeded far in the 1930s. Lockheed, Boeing, and Douglas all developed four-engine transport planes in the late 1930s (such planes were flown in the Soviet Union in the 1930s). The high-altitude potential of these planes led to the achievement of pressurized cabins in 1937. During World War Two these new larger planes were used extensively. That war not surprisingly saw enormous advances in engines and all types of equipment associated with aircraft. The first commercially successful four-engine plane was the Boeing 307, which was based to a large extent on that company's B-17 bomber. This new generation of postwar planes proved much more expensive than that which had gone before. The end result was that a whole new range of aircraft was technologically ready in the 1950s.

Research on the jet engine also began in earnest in the early 1930s (in Europe). While many important advances were made in that decade, the complexity of the jet required a large scale research effort which governments were only willing to finance with the coming of war. Many experts feel that the jet could have been developed a decade earlier, as the pressure and heat resistant titanium alloys and improved fuels which the jet required were available in 1930 (Constant, 1980: 186).[26] Still, even with the sustained research effort of the war, jet airplanes did not appear widely in the military till the 1950s and even later for civilian craft (in the United States, the key craft were the Boeing 707 of 1958 and Douglas DC-8 of 1959). The jet, then, was destined to have its expansive impact on output and employment long after the Depression.

A government commission in 1947 worried that one of the major problems of the industry was that over 80% of its output by value was sold to the military. Rae notes that this had largely been the case through the period of our study. Especially as development costs rose, technological advance in commercial aircraft must have been greatly slowed if not for military interest. This is undoubtedly the major area of technological spillovers from the military to civilian use. It highlights the positive role that government support of research can play. The government, through NACA and then NASA, also directly supported civilian development; close co-operation between government and private researchers ensured a flow of important discoveries from the former to the research labs of the latter (Mowery, 1989: 197).

Notes

1. "Man's commitment to the internal combustion engine is surely a central fact in the history of this [twentieth] century" (Bryant, 1978: 1024). He notes the lawnmower, the tank, and the chainsaw and bulldozer which revolutionized the timber industry, along with cars, planes, ships, and trains.

2. In England, the number of horses rose from 1.29 million to 3.28 million between 1851 and 1901, and the number used for transport from .5 million to 1.5 million. There were probably about 30 million horses in the United States by 1901, each of which required four to five acres of food. Barker (1982, 149) speaks of a looming energy crisis which was overcome by the automobile. We discuss the effects on agriculture elsewhere. The urban horse population fell from 3 million in 1910 to 1.7 million in 1920; the census thereafter lumped rural and urban horses together. Ville (1990) notes that the rapid expansion of the automobile is hard to explain by the more gradual changes in horsepower, and that the United States had the least to worry about.

3. At a time of small-scale production, and steady incremental improvement, survival was naturally difficult. Much of the early expansion of the industry was only possible through shifting much of the risk onto suppliers and distributors (Seltzer, 1928: 53-4). Further, most firms had most of their manufacturing done by parts suppliers; this meant that many of the smaller parts were standardized across carmakers.

4. One result was that motorcycles were emphasized in Britain; she was the leading consumer in the world of these before World War One. In Europe, only Britain, France, and Germany had a large enough national market to support a sizeable auto industry; Italian producers exported much of their output (Barker, 1982: 157-8).

5. Allen, for example, mentions the rise of suburbs, the decline of central cities, shopping centers, motels, industrial parks, parking problems, accidents, increased travel, labor mobility, and a feeling of freedom and self-confidence (1954: 118). Towns and villages found that much of their business was lost to urban centers. While agricultural distress was behind most rural bank failures, the access which the auto gave rural people to urban banks may have played a contributory role.

6. Transport planners could not foresee the tremendous future growth of the automobile. Many transit companies believed into the late 1920s that the automobile would fade away.

7. The 60,000 miles of concrete road built in the 1930s were basically responsible for the expansion of cement industry capacity from 150 million barrels in 1920 to 260 million in 1929 (Hogan, 1971: 1013).

8. The increased complexity of cars did counteract some of the labor-reducing effects of technological change. The addition of the starter, for example, actually caused the amount of labor per car to rise from 1924 (Abernathy, 1978: 157).

9. Kennedy (1941: 199) notes that luxury cars such as the Pierce-Arrow did the best in 1928-1929. This could be evidence of not just the role of income distribution, but of the stock market in supporting the economy in the late 1920s.

Internal Combustion 227

10. Abernathy feels that as mass production methods became firmly established in the auto industry they greatly inhibited product innovation, especially radical innovation (1978: 3-5). Certainly, there has been relatively little product innovation since 1925. In particular, Abernathy suggests that carmakers did not produce smaller cars for this reason.

11. Seltzer argues that already by 1924 GM had recognised that, while demand had exceeded supply in the past, in the future they should aim for stabilization and higher profits per unit. Then, sales took of, and Chevrolet output almost tripled over the next five years (1928: 214).

12. One reason was economies of scale. Chapin, of Hudson, maintained that if output fell just 10%, costs would rise too much to permit Hudson to maintain its existing prices (Seltzer, 1928: 49). The Depression would see the removal of most smaller firms, and pave the way for postwar domination by the big three.

13. Klein feels that it is the necessity of yearly style changes which squeezed the smaller carmakers out of the industry, and that it has also been responsible for the tightly controlled hierarchical nature of the modern car corporation (1977: 193, 196). He feels that there was very little real technological innovation after 1920. Bernstein argues that the modern industry learned its lesson and concentrated in the postwar period on improvements in appearance rather than durability (1987: 61). Note that technology facilitated stylistic differentiation; Duco lacquer made it possible to produce cars in colors other than black.

14. The independents had fared extremely poorly in the Depression. Those that did not succumb then failed early in the postwar era. Ford and GM had shown that companies with very large outputs could suffer substantial declines and still make money; Ford and GM each earned over one half of their 1929 profits in 1930 (Kennedy, 1941: 227).

15. Hogan (1971: 1310-4) describes many developments in the iron industry which facilitated improvements in the auto industry in the 1930s. Cold reduction was introduced in order to produce steel that was both ductile and smooth. Aluminum-killed steel was required for new fender designs. Over 70 new steel mills were built to embody the new technology.

16. Abernathy (1978: 3, 11). The postwar period does see unibody construction and steady improvements in mileage. Chrysler was perhaps more open to innovation due to its less vertically-integrated structure. It experimented with disc brakes in 1949, and introduced power steering in 1951 and the alternator in 1960. These innovations can generally be traced back to earlier efforts as well (Abernathy, 1978: 57-8).

17. Bernstein (1987) attributes much of this to problems in attracting capital to new product lines (though rubber manufacturers should have been able to provide internal finance). Since many of these new products were dependent on industries, including autos itself, which were faring poorly in the 1930s, other explanations are clearly possible. Again, the question naturally arises of why none of these new products had emerged earlier.

18. He likewise estimates that the pressed ware machine saved 80-92% and the blown ware machine 30-97%.

19. There was also room for entrepreneurship. National Steel, by concentrating on high quality products, adopting new technology more quickly than US Steel or Bethlehem, and pursuing an aggressive pricing strategy, was able to move from sixth place to first in profits 1929-1931 (Jaeger, 1972: 141). What is possible for one firm need not be possible for the industry as a whole.

20. Speaking of metal working in general, Williams remarks that "Perhaps the most obvious general changes would have been a steady increase in mechanical handling to reduce heavy physical toil; the replacement of overhead drives and belts by individual electric motors; and greater attention to the health and the comfort of the workers" (1987: 175).

21. Military orders, then as now, were a potential source of huge profits. Freudenthal (1968) decries the lack of competitive bidding, and, while recognizing the high cost of research, feels profits were extreme. The Federal Aviation Commission in 1936 remarked that "it has always to be remembered that this industry is peculiar in that it has essentially but a single customer" (Freudenthal, 100). Exports, mostly for military purposes, to Europe, China and Japan, became of increasing importance in the 1930s.

22. Bilstein notes that aeronautics was a relatively new science. Research had been supported by Guggenhein and others after the war, and many new engineers were trained in that period. "In short, the elaboration of a distinct technological infrastructure of aeronautics emerged in the post-World War I decade" (1984: 69-70). Aerodynamics in turn borrowed knowledge from the also fairly new field of fluid mechanics.

23. The Lindbergh-inspired financial expansion of the late 1920s had floundered in the early 1930s as private demand for airplanes faltered. Again, military demand held up much better. Still, companies felt it worthwhile to pursue commercial-oriented research efforts through the early 1930s. Before legislation forced their separation in 1934 airplane manufactures benefited from association with carriers who — due to mail routes — fared much better financially.

24. Barger looked at labor productivity in the airplane industry. On international flights, there was no clear trend over the 1930s (but productivity tripled 1939-1946). On domestic routes where technological advances had their clearest impact, output per worker more than doubled 1931-1935 and almost tripled 1931-1939. Average seating capacity had grown from 7 in 1932 to 18 in 1941 and average journey from 268 to 360 miles. There had also been an increase in the percentage of seats occupied (1951: 159-60).

25. Further evidence that the Depression need not stand in the way of new technology is provided by the small private plane market. Beech and Piper were both able to start successful new businesses in the 1930s (though a hoped-for truly mass market in private planes has never been realized). While government support aided large plane development, the development of smaller planes was more clearly privately determined (the Lindbergh example was important here). Even among larger craft, there were no government purchases 1932-1934.

26. Constant argues that each of the major constituent elements of the jet engine resulted from a complex interaction of scientific theory, prior technical prac-

tice, traditions of testability, and developments in competing and associated systems. The course of technological change appears to be far from pre-determined (1980: 99). The environment in which (and thus the timing of) innovative efforts are conceived determines the selection of intermediate goals, testing procedures etc., and these affect in unseen ways the final outcome.

10

Selected Other Industries

Energy

A cursory glance at modern industry might lead one to the mistaken impression that energy consumption per capita has been rising steadily for centuries. This is far from the case. As usual, different statistical series tell a somewhat different story, but all point to a decrease in energy consumption per person through the interwar period. Schurr (1960) has compiled statistics on energy use 1850-1955. Use of mineral fuels and hydroelectricity increases 180 times over the entire period. If, however, wood is included in the total, there is only a 17-fold increase, due to the much greater importance of wood as a fuel in the mid-nineteenth century (wood use naturally presents greater difficulties in estimation). Per capita increases over the period for the two series are 25 times or 2.5 times, respectively. Relative to GDP, the increase is 5 times or .5 times. If we include wood then, the ratio of energy use to output actually declined 50% over the century. Nor was this decline steady. The ratio had in fact increased to the 1910s and then fell. In per capita terms, energy use had risen from about 1885 to World War One, fell in the early 1920s, stagnated in the late 1920s, and fell again in the early 1930s, before rising back to the 1930 level in 1940. Decreased economic activity can provide a partial explanation for the early 1930s, but not for the 1920s. Energy consumption relative to GNP fell 15% in the 1930s (and a further 9% before 1955). Schurr points to four factors which drove the secular trends and especially interwar experience. There was a structural shift away from heavy toward light industry. Productivity advance (both labor and capital) naturally decreased the energy requirement per dollar of output (and thus has a labor-saving effect not captured in the original industry). Technological advance caused great increases in the thermal efficiency of energy utilization (see below). Electrification encouraged an improvement in the ratio of energy input to output (Schurr, 1960: 14-6, 145-56).

These four factors are all clearly tied to the elements of technological change and structural evolution we have discussed before. They combined to ensure that the energy sector would employ a declining proportion of the labor force. There was a further effect as coal was increasingly replaced by oil. Coal (after surpassing wood as an energy source in the 1880s) fell from 89% of the energy supply in 1899 to 78% in 1919, not quite half in 1945 and one quarter in 1958, while petroleum rose from 4.5% in 1899 to 12% in 1919 and 25% in 1929. Bituminous coal output fell from 13,325 million BTUs in 1920 to 13,079 in 1925, 11,921 in 1930, and 9,338 in 1935. Increased overall energy demands after the war would lead to some improvement, but the levels of the 1920s would not even be regained in the early 1950s (these were barely surpassed late in the war). Crude petroleum output, conversely, expanded from 2,634 million BTUs in 1920 to 4,156 in 1925, 5,652 in 1930, 5,499 in 1935, 7,487 in 1940, 12,706 in 1950, and 16,340 in 1955.

The shift from coal to oil can not, at least in the interwar period, be attributed to changing relative prices. While fuel oil prices fell 23% 1923-1929, the price of bituminous coal fell 44%. Still, in a longer term sense, relative price considerations were important. As Barger noted (1944: 44), "At present a simple man hour of labor in the mining industries will produce several times as many BTU in the form of oil as in the form of coal" (costs of refining, depletion, and transport must also be considered). Much of the shift was due to technical improvements in oil burning equipment, plus the marketing efforts of oil firms (see Copeland, 1929). More importantly, the shift from steam to internal combustion engines in industry naturally required a shift in energy sources. As well, with the rise of indirect energy use through electricity, oil gained ground by proving superior in electric power generation. Even when coal was used for electricity generation, there was still an efficiency gain over direct use in steam engines. (Barger, 1944: 44).

Oil

Oil production began in the United States in 1851 in Pennsylvania. A market for non-vegetable sources of illuminants (as literacy rates increased, in part) and lubricants (due to the increasing number of machines, steamships, and railroads) had emerged. The United States would dominate world production until well into the twentieth century. As the twentieth century began, the two major uses of petroleum were still for lighting[1] and lubrication. Gasoline, fated over the next decades to become the key petroleum fraction, was neglected. The contribution of petroleum to national energy consumption was almost insignificant. The next decades would see some important developments, but the rise of oil as an important energy source dates from World War One (Schurr, 1960: 38). This rise was naturally closely connected with the emergence of the automobile. Before this could happen, however, numerous chemical advances were necessary: "American oil was not a 'resource' until advanced chemical processing allowed its utilization in production" (Inkster, 1991: 175).

Exports were one-third of production 1860-1900. This reflected American leadership in the industry. Exports were dominated by refined products serving specific needs. While production grew faster than consumption before 1900, the reverse was true after and exports fell steadily. The United States would become a net importer after World War Two. During the interwar period, except for some years in the 1930s, she was a net importer of crude oil, but this was more than balanced by the export of refined lubricants and illuminants. Between one-third and two-fifths of the output of kerosene and lubricants was exported through the 1920s. This situation reflects the increased importance of gasoline in the United States and the emergence of foreign sources of supply.

Between the rise of electric lighting (which destroyed the market for illuminants) and the emergence of the automobile, oil output was kept from collapsing by its use in industry. Early use depended on the discovery of oil far from coal fields; only then was there a cost advantage (Schurr, 1960: 193). Railroads as well as industrial establishments proved willing to switch where coal was expensive. Fuels comprised half of crude production in 1909 (half of this being unrefined crude); there was naturally great regional variation. Oil only came to make serious inroads into the market for coal in the interwar period.

Still, it was the increased demand for gasoline which would have the greatest effect on oil output and technology. Gasoline was employed in stationary engines, but this was a small part of total consumption. Even in the 1950s, aircraft absorbed only 2% of the total and farm machinery 7.4% (Schurr, 1960: 118). We have discussed earlier the astonishing rise in automobile ownership in the United States in the 1920s (from 7.6 million in 1919 to 26.5 million in 1929). On top of this gasoline consumption per automobile rose, in large part as engine size increased, from 473 gallons in 1925 to 599 in 1930, 668 in 1935 and 733 in 1940 (versus 768 in 1950 and 790 in 1955).[2] Some scope existed for increasing "gasoline" output by expanding into the naptha fraction, and this was done early in this century. Further improvements in extraction of gasoline from crude oil would require chemical manipulation.

In the early decades of the industry, refining had involved merely the (imperfect) separation of fractions. No attempt was made to convert fractions. Those for which there was little demand, and this included gasoline through the nineteenth century, were often wasted. It had long been known from observation that at higher temperatures more gasoline could be obtained. The proportion of gasoline per barrel of crude oil rose from 13% in 1900 to 26% in 1918 (with the advent of cracking) to 42% in 1930. This required ever more sophisticated technology. Academic chemists developed the theory of petroleum cracking before advances in welding and pumping equipment in the 1910s made it possible to undertake large scale high-pressure reactions.

Cracking is the decomposition at high temperatures and pressures of heavy hydrocarbon molecules into smaller lighter molecules. As the technology evolved, it also became possible to transform lighter gaseous fractions into gasoline. The first commercially successful cracking process was unveiled in 1913, after four years of experimentation.[3] Enos (1962), in his pathbreaking work, identified 8 major and numerous minor cracking innovations which emerged in three waves, the early 1920s, around 1936, and in the 1940s. Each wave displaced the former. The timing of investment in the industry would have been quite different if the early 1930s had seen major advances in this technology.

Within a few years of 1920, four major and several minor continuous cracking methods were introduced. These were partly attempts to surmount Standard's 1913 patent; other companies had been spurred to innovative effort from 1915. They also represented the increased recognition across industry of the advantages of continuous processing. Independent innovators played an important role, but the development costs of such integrated technology could only be undertaken by large companies. The later 1920s were devoted not to developing new processes but to improving these;[4] this, however led to cost reduction as great as the original innovation (Enos, 1962).

Continuous processing not only reduced costs but increased yields and quality. By the end of the 1920s gasoline yields of 60% were attained, while the batch methods of the 1910s never surpassed 35%. Cracking, especially in the vapor phase, was also found to give gasoline good anti-knock (i.e. premature ignition) properties. Of course, the solution to the knock problem would come not from cracking but from the addition of tetraethyl lead. This was developed outside the oil industry through application of chemical theory (see Williamson, 1963: 409-14) in the mid-1920s and spread quickly once production difficulties were ironed out. The 1920s also saw numerous advances in the chemical treating of gasoline to remove impurities. Automobiles were able to travel farther per gallon due to improvements in gasoline.[5] Sulphuric acid and other chemicals were used to remove sulphur, oxygen, nitrogen and other odor-creating or corrosive elements.

The next major innovation would be catalytic cracking. This would yield by 1938 a 74% cost saving, but an even greater advantage was the higher octane rating possible.[6] Houdry had begun work on this in the 1920s. He had first searched for the proper catalyst, and settled on clay-type silica-alumina. Years of expensive effort followed, filled with battling the French government and private sources for funding, in order to maintain the effectiveness of the catalyst through processing and to achieve the best timing, temperature, and pressure. He first achieved experimental success in 1927. Almost another decade, with large-scale financing by American oil companies, would pass before equipment design problems could be overcome to allow commercial installation in 1936. Several years of institutionalized research would create Enos' third wave in the 1940s, where the Houdry process would be improved and rendered fully continuous.[7]

These innovations indirectly and directly conserved labor. Gasoline output rose 575% 1916-1927 but crude oil only 200%. Continuous processing naturally reduced labor input. Advances were made in the length of run possible between cleaning. Major savings in labor and fuel were obtained in the 1920s from the pipe still and the multiple fractioning tower which allowed more than one fraction to be taken off at a time (Hogan, 1971: 1044-5). Improved storage tanks reduced losses from evaporation. Increased scale of operation, often technically driven, also acted to reduce labor requirements through the interwar period. The result of all this was massive increases in labor productivity. While refinery output increased 270% in the 1920s (and gasoline output 450%), employment rose a mere 29%. One of the great growth sectors of the period thus had a marginal impact on employment.

There had been much investment 1926-1929 to take advantage of new technology. Combination units, which brought together various improvements of the 1920s, emerged in that decade but only became popular after 1932. The Depression and fears of technological obsolescence may have deterred smaller firms, but the big firms pushed ahead (Williamson, 1963: 437, 607). Refinery profits had been low through the 1920s due to overcapacity, at least at Standard; this stimulated research (Larson, 1971: 150), but could only discourage investment. The total number of refineries declined from 547 in 1925 to 427 in 1929 as smaller (generally obsolete) plants were squeezed out by large (continuous cracking technology required a scale many time larger than that of 1920).

The petroleum industry suffered less of a loss in output than most others. Total sales of oil products fell just 14% to 1932 and then rose. By the end of the decade, the industry would be producing half again its 1929 output (output growth averaged 9% 1909-1929 but only 4.5% 1935-1955). The decrease in the early 1930s was entirely due to decreased sales of fuel oil and lubricants. Gasoline use per automobile rose 1929-1931, and the number of cars registered 1929-1930. Use of gasoline by automobiles thus never fell below the 1929 level. Falling oil prices in the early 1930s accelerated the shift from coal to oil; kerosene and fuel oil sales thus grew from

1931. The decline in industry employment in the early 1930s was thus primarily due to continued productivity advance.

Oil Drilling and Exploration

Oil drilling activity fell much farther. It was at half the 1929 level in 1931. This might seem at first glance a response to the Depression. As usual, there is much else going on. Larson (1971: 59-60) argues that the world and especially the United States was characterized by overproduction from 1927. Inventories (of 600 million barrels) were far above normal requirements. The Depression only exacerbated an inevitable cutback in exploration and drilling. As for exploration, the feeling had emerged in 1928 that existing known reserves exceeded demand (Williamson, 1963: 336). Technological change had been reducing costs long before the 1930s.[8] Output per man hour in oil extraction increased 4.1 times 1902-1939; Barger notes that this was much faster than in coal mining and must have contributed to the shift from coal to oil (1944: 81). Alloy steels and then tungsten carbide greatly facilitated drilling; the former made the rotary drill, which was much superior to percussion devices, possible from 1910. Tall steel derricks replaced shorter wooden ones, and thus drill length could be extended (though well depth increased too). As for extraction, the 1920s saw successful research in, among other matters, how to use gas to recover more oil from a field. Even exploration benefited from seismography from 1923 and magnetic surveying in the late 1920s.[9]

Coal (Mining)

Schurr has analyzed long run trends in coal output. This had virtually doubled every decade through the late nineteenth century (the coal industry in the United States had been very small by European standards in 1850). Output increased less than a third in the 1910s, fell 18.4% in the 1920s, and 4.6% in the 1930s (falling and then rising later in the decade). After the wartime peak, output would decline 22.4% 1945-1955 (1960: 63). Schurr viewed this as a classic case of secular stagnation and decline. Peak production occurs just after World War One, a little over a decade after coal peaked as a proportion of the total energy supply. While movements were generally stronger in the United States, coal industries throughout the world were suffering from stagnation after World War One (world production fell 100 million tons 1913-1928 and again 1928-1938).

The rising importance of oil and natural gas in industry and home heating was far from the only reason for the declining fortunes of the coal industry. Railroad expansion had been a key to the rapid growth of coal output in the nineteenth century, absorbing 42% of output in 1885. Contraction in railroad traffic in the face of the oil-burning truck and car would provide a negative stimulus to coal production long before the diesel locomotive appeared on the scene. Moreover, thermal efficiency of coal use made great strides across a range of applications. The most noticeable advance was in electric power generation. It took 6.85 lbs of coal to generate one kilowatt hour of electricity in 1900, but only 3.39 in 1920, 1.62 in 1930, 1.39 in 1939, and .95 in 1955. Fuel consumption per unit of output also fell 32% in steam locomotives 1917-1936 (due to reduced heat loss, reduced friction, better load factor, and improved coking) and 16% in cement plants 1914-1935 (Gill: 1940). It took 22 cwt. of coke for each ton of pig iron in 1900 but only 13 in 1950. Williams estimates that overall the United States experienced a one-third increase in the efficiency of coal use 1909-1929 (1982: 30).

Employment in the coal industry fell from a peak of 622,000 in 1919 to 406,000 in 1932.[10] Start-up costs in coal-mining were high and mine owners thus tended to keep mines open even if they were not covering their variable (operating) costs. This helps explain the massive price declines of the 1920s. Much of the decrease in the early 1930s, then, must be seen as dishoarding against a backdrop of secular stagnation in a fragmented and competitive industry. Indeed, faced with intense downward pressure on wages, the leader of the United Mine Workers himself had taken the unusual step in 1923 of urging that 200,000 excess workers be laid off.

Still, the decline in employment in the 1920s greatly exceeds the decline in output. Beyond the impact of productivity improvements in coal-using sectors then, we must look to changes within coal mining itself. As similar developments are found in other types of mining[11] it is best to discuss these together. While total mining output increased 41% 1902-1939, output per man day rose 78% and per man hour 119% (decreases in the work week were concentrated in the 1910s and 1930s). Miners comprised 2.1% of the labor force in 1890, 3.2% in 1920, 2.5% in 1930 and 2.2% in 1940. The downward trend from 1920 is primarily due to technological advance (Barger, 1944: 78). Barger notes that the trends in output and employment are much more similar to those in agriculture than those in manufacturing.

All mining before 1850 in the United States had been surface mining. As pit mining became more common, an engineering approach which designed mines from an efficiency standpoint emerged in the late nineteenth century. Like agriculture, the mining sector imported technology from other sectors in the twentieth century, and proved quite open to the idea of mechanization (downward pressure in output prices likely encouraged this). While developments in railway tunnelling and steelmaking were important, by far the largest force was electrification, for it would have been infeasible in general to power mining equipment otherwise (Barger, 1944: 118).

Mechanical cutters driven by compressed air had been introduced in 1863 but were little used in the nineteenth century. Replacing air with electricity was a major advance, though pneumatic drills, themselves greatly improved in the early part of the century, remained important through the interwar period. While only 25% of American coal was cut mechanically in 1900, almost all was by 1950. The new cutting materials discussed in Chapter 8 were quickly adopted in mining. Cutting was also aided by improved explosives such as dynamite. Along with better mine design which allowed gravity to do much of the work, these changes caused the labor involved in ore cutting to fall from two-thirds of the total to less than one-third (Barger, 1944: 123).[12]

After cutting, the material must be loaded onto some sort of wagon. This task was rarely mechanized before the labor shortage of World War One. A variety of devices, including the shovel loader and conveyor belts, were introduced. Jerome estimated that mechanized loading saved 25-50% of the labor involved in bituminous coal mining; Daniel Weintraub estimated that 95,000 jobs were replaced 1920-1929 in coal mining (Jerome, 1934: 367, 382). In the 1930s, combined cutters/loaders were introduced.

Electric locomotives had been introduced in 1887, and became widespread on main tunnels in the first decades of this century. With the development of battery locomotives, use spread rapidly on side tunnels in coal and metal mines after World War One. The following years saw numerous advances in both engines and cars. There were also numerous improvements in hoisting machinery, which was generally electrified (Barger, 1944: 130-5). Ventilation, lighting, pumping water, and roof support (hydraulics) also improved, though it is not easy to estimate the resulting labor saving (Barger, 1944: 134). The whole could often be greater than the sum of the parts. While the addition of one machine might have little effect on the overall work tempo of the mine, widespread mechanization revolutionized the organization of mining.

A Note on Nuclear Energy

Environmental concerns have prevented nuclear power from achieving some of the prominence which enthusiastic boosters predicted for it in the 1950s. Nevertheless, it has clearly had important impacts on the price of energy, the labor demand in the energy sector, and the level of investment. The interwar period sees important steps toward the achievement of nuclear power but nobody at the time could reasonably have predicted that it would become a reality as soon as it did. The basics of atomic theory were just being worked out in the early decades of this century. Only in 1919 did Rutherford induce artificial separation of a nitrogen atom into hydrogen. In 1932 at Cambridge lithium was split into helium by electrically accelerating a proton projectile. This experiment provided the first verification of Einstein's 1905 formula. The possibility of inducing a chain reaction in uranium with slow neutrons was discovered in 1939 just before the war started. A handful of people in the 1930s began to consider the possibility of nuclear power, though Rutherford himself would die in 1937 convinced that it was impossible. Even for the optimists, "a clearer understanding of atomic processes was needed before power production came within sight of being a practical possibility" (Williams, 1982: 48). By the late 1920s such progress depended primarily on complex and expensive linear accelerators. While these were improved during the interwar period, it was thought that attaining the degree of sophistication necessary was decades off. Then came the war. Billions of dollars were devoted to research toward the atomic bomb. This appears to have accelerated the development of nuclear power by decades. It would still take years before the power of the atom could be harnessed to peacetime use.

Food Processing

While some forms of simple food processing have existed for millennia, the modern industry emerged over the nineteenth century in Europe. Mowery (1989: ch.3) attributes much of this development to transport improvements and urbanization in that century. The social prejudice against prepared food was only slowly overcome. The biggest breakthrough in large-scale food processing was the introduction in the 1830s of canning, the heat sterilization of goods within sealed containers (it was first made possible in 1809 using bottles, in an attempt to feed Napoleon's troops). Production of such products as cookies also expanded. Margarine was developed in 1870 in response to a prize offering by Napoleon III. Many nineteenth century advances were informed by science, especially after the discoveries of Pasteur.

Food processing has continued to expand in the twentieth century, and is now the largest industry in the United States. Part of this can be attributed to an increased desire for labor-saving in the home, though in fact the amount of time spent in housework has been reapportioned rather than reduced. Still female labor force participation rates have risen. There has also been a helpful trend toward diversity in food consumption. Transport improvements and urbanization still had a role to play, as foods were packaged to survive long trips. Perhaps more central was the innovation of self-service stores which encouraged the packaging of goods in cans, plastic, and paper. Not surprisingly, the increased output of the sector was closely tied to advances in the technology of food processing. Cellophane provides an interesting example of the latter two forces at work: Du Pont recognized that their product was most suited to the emerging supermarket; retailers viewed cellophane as the answer to high labor costs and long lines of shoppers at meat counters on weekends. While fresh meat was packaged in store, luncheon meats quickly came to be packaged centrally (Hampe, 1964: 177-83).

Food products output fell from 1928 but much less sharply than the economy as a whole. Bernstein concludes that there is some justification for viewing this sector as Depression-proof (1987: 53). People have to eat, of course, and food consumption itself barely fell in the 1930s. Some reacted to decreased incomes and increased free time by shifting some food processing back into the home. There were, though, inexpensive and nutritious canned foods which many families turned to in the 1930s. It might be thought that food processing was a sector which should have possessed some niches in which the unemployed could be put to work (the sector would expand significantly postwar). Changes in distribution methods and the introduction of home appliances should have created new product opportunities. There are some responses to these opportunities in the interwar period, notably margarine (long available, but previously uncommon) and breakfast cereals. Bakeries were one of the growth sectors of the 1920s as well. There was a marked increase in food product advertizing in the 1930s. Still, the most important areas of new product development, canning and frozen foods, were dependent on technological innovation. Much, though not all, of this new product innovation came "too late."

Canning

Canning had become quite common for milk and a variety of meat, fish, fruits, and vegetables by 1900. Entrepreneurs such as Heinz, Borden, and Campbell had already become household names. The trial and error efforts of the early part of the nineteenth century had created much knowledge of appropriate temperatures and pressures as well as the superiority of tin as a container. "But the real advances came after certain biologists pioneered the application of the bacteriological science of Louis Pasteur to the preservation of food" late in the century (McKie, 1959: 254). Scientific measurement of the exact conditions necessary to kill all bacteria occurred in the 1910s and 1920s (Hampe, 1964: 118).

Major nineteenth century mechanical innovations included the retort (1860s) which allowed high temperature cooking with pressure equalization. The 1890s and 1900s would see machines to peel corn, peas, and salmon. By the 1920s, continuous processing of the raw material was possible (Hampe, 1964: 117). There was rapid progress in speed, quality, variety, and cost of canned goods in the first decades of the twentieth century (Hampe, 1964: 118). The double seam open top can (which we still use today) was developed in the nineteenth century. In 1908, the sanitary can became possible with the development of a sealing compound which made a hermetic seal reliable; soldering was no longer necessary. Decades of research led by the early 1920s to enamel coatings which would not affect the taste of food but would prevent the bleaching out of the color of fruits and vegetables. This made it much easier to can meat; canned pork output doubled 1924-1925.

The 1920s, as usual, saw much process improvement. Continuous processing spread through the industry. The rotary cooker of 1919 solved for the first time the major problems in continuous cooking (Thorne, 1984: 156). Can making machines were steadily improved; output per minute jumped from 150 to 250. Peeling and grinding equipment was also significantly enhanced. New products were also introduced: tomato juice in 1924, luncheon meat in 1925, syrup in 1926, as well as baby food.

Can output had risen steadily for decades before World War One and doubled in the 1920s. After a slight decrease in the early 1930s, output took off from 1932, and had more than doubled the 1929 level by 1939. In both absolute and relative senses, the 1930s was the fastest decade of growth in canning in the last century. "The sales breakthrough of the 1930s was a tribute to the dedication of canners to quality products; testimony to the efficiency of sales, advertising, merchandizing, and public relation activities, and reward for tireless effort" (Hampe, 1964: 131). Some important new canned products, such as fruit cocktail, orange juice, peas, beer, and motor oil, only became available in the 1930s. For the most part, though, expansion of canned good sales in that decade largely represented increased recognition among consumers of both quality and cost attributes of products introduced in the 1920s and earlier. The technology for producing these goods was, naturally, subject to a series of minor improvements.[13] As canned goods caught on, changes in retailing and transport caused costs of distribution to fall (Bernstein, 1987: 621). The experience of canning, then, is testimony both to the fact that new products do not often achieve instant success, and that once recognized these could fare very well even in the depth of Depression.

Postwar expansion would hinge upon a range of innovations (plus wartime use of canned goods by the armed forces); quality and packaging of existing goods was enhanced (Hampe, 1964: 156-7). The corrosion and expansion problems of soft drinks were much more severe than for beer, and would only be solved in the 1950s. Aseptic canning, in which products were sterilized outside of the can, only became successful just before World War Two (it required that the entire system of canning equipment be sterilized). Suitable only for pastes and liquids which could be heated quickly, it was important for products like custard and ice cream mix which needed quick heating and cooling, as well as for packaging materials like milk cartons. The dominant aseptic canning process was developed by Dole in 1950 (Thorne, 1984: 161-6).[14]

Frozen food

While freezing does not kill all of the micro-organisms as canning does, it generally provides a more desirable product, and has thus provided stiff competition to canned goods in the postwar period (Thorne, 1984: 168). Complementary technologies only came together to make widespread consumption of frozen goods possible in the 1930s. Clarence Birdseye had begun experimenting after a trip to Labrador in 1915. He gradually discovered that if food was frozen quickly, smaller ice crystals were formed and thus much less damage was done to the food. Only in 1929 did he fully recognize the advantages of fast freezing and develop a viable process. General Foods purchased his patent and introduced a line of frozen foods in 1930. The market for these was, however, severely limited by the lack of freezers in both stores and homes. Not only was the electric refrigerator still a household novelty, but separate freezer compartments were only belatedly added to these as frozen foods entered the market. In 1933, there were only 516 stores nationwide which possessed freezers; there were some 15,000 by the end of the decade. Research over the 1930s led to both an improvement in product quality and price, and the introduction of display cabinets in stores. Frozen foods did induce considerable investment in production and distribution facilities, but only very late in the Depression. The impact could have been felt earlier but for the fact that electrical refrigerating was still a novelty in 1930.

Lumber

Coplon (1932) complained that the lumber and wood-using industry, though second only to agriculture in terms of employment,[15] was little studied by economists. That situation still largely holds true today. Nevertheless, the broad outlines of decline in this important sector can be described. Since 40% of wood output was employed directly in construction at the start of the 1930s, another 25% indirectly as sashes and doors, and 17% in auto production, hard times were assured for wood producers. Still, the downswing in the 1930s was greatly aided by secular forces. Output had been falling for decades. In the once important market for fuel wood, consumption declined 20-30% per decade in the nineteenth century and about 10% per decade through the early twentieth century (Schurr, 1960: 39, 47). More centrally, wood had been losing out as a building material to steel, cement, and brick; its share fell from 59% to 46% during the 1920s. This shift reflected improvements in competing materials, relative price changes, and tougher fire codes. Similar shifts occurred outside of construction. Steel largely superseded wood in shipbuilding, automaking, and railroad car manufacture. New materials invaded box, lath, and shingle markets (Brown, 1941: 30). Agricultural decline limited markets for repair and replacement in any case, but "competition of substitutes had been relentless" there as well (Jensen, 1945: 152). Coplon (1932) noted that a mere quarter century before there had been no viable substitutes for wood in most of its uses, but since then it had come to face competition in virtually every market. Jensen (1945: 20) argued that competition from substitutes had become increasingly severe from 1920 (and TNEC (1941: 108) noted that since wood was not malleable, was bulky, and could not be melted or poured into a desired shape, its replacement led to substantial labor-saving in use as well). World markets were glutted in the early 1930s. The Soviet Union was accused of dumping pulpwood. The domestic market, though, had always been the key to lumber industry prosperity. While the industry was responsible for half of world output, exports averaged only 3-7%.

"Depression began to afflict the lumber industry as early as 1925, when there were numerous requests and demands for curtailment of production to preserve prices" (Jensen, 1945: 151). Unemployment appeared there before it hit other industries, and was increasing rapidly by 1928. Still, the greatest collapse would await the early 1930s. By 1932, the industry would be operating at barely 20% of capacity. "The instability, unrest, and weakness in the lumber industry from the mid-1920s to the mid-1930s were symptomatic of the plight of an industry with unsolved chronic economic problems" (Jensen, 1945: 161). The lumber industry was ill equipped to deal with declining demand. The tax system encouraged chronic overcutting and overproduction (Jensen, 1945: 29). The small scale competitive nature of the industry (there were thousands of lumber companies and the largest four accounted for less than 5% of output) militated against production planning. Many small companies entered and exited the industry through the 1920s. Larger producers with high fixed costs in land and standing timber tried to keep cutting; their response to falling prices was to try to conserve on labor (Jensen, 1945: 25-7; Coplon, 1932: 162). The National Lumber Manufacturer's Association, in a presentation to the President in 1930, spoke of overproduction for the previous decade and a half (Coplon, 1932: 168).

The NLMA had no success in restricting output in the 1920s, but it would welcome government regulation in the 1930s. It had some success in encouraging standardization, grading, and other fair trade practices in the 1920s; these however could do little to stem the tide of competition from substitutes. The NLMA's most ambitious undertaking was a $1 million per year research program from 1927. There was some scope for new product development. Rayon, cellophane, glues, turpentine, flooring, and plywood did provide new outlets during the interwar period. Bernstein (1987: 87) feels that lumber firms were slow to invest in these new products.[16] However, the limited raw material requirements of these new uses must have limited their impact on lumber employment in any case.

Beyond its effect on output, technology also served to lower labor input. Bryant (1978: 1024) speaks of the revolutionary effect that the internal combustion engine had on the lumber industry through chain saws and bulldozers. On the larger estates, at least, transport of felled timber was also mechanized. Even greater changes occurred in sawmills. Forklifts, trucks, tractors, and electricity together yielded large gains in handling (Bernstein, 1987: 138). Jerome (1934: 109-11) surveyed a handful of sawmills. Horse drawn trucks were replaced in the 1920s by tractors, and then by overhead cranes or mobile carriers. Inside, the most common new machine was the sander, which allowed one worker to do the work of four. Carving machines replaced ten with one. Jerome also found bigger and more efficient edgers and planers.[17] The bank saw (its thinner blades reduced waste, and power use, while increasing speed and blade life) had been introduced in the nineteenth century but the first direct motor driven band saw appeared only in 1926; this eliminated belts and allowed constant speed and low cost maintenance (Brown, 1947: 75-6).

Textiles

"Textiles, like most nondurable goods, did better in output terms during the depression than major durable goods industries. But the economic performance of the sector was nevertheless quite poor" (Bernstein, 1987: 75). Textile output did fall and then rise with national income. While people were hesitant to cut back on food or housing in the Depression, significant cuts were made in expenditure on clothing (Wandersee, 1981: 41). However, the problems of the textile sector were obvious well before the Depression, and would continue long after. "The crash had compounded the depression that set in in the mid-twenties" (Bernstein, 1960: 255). Speaking of cotton in particular, Backman (1946: 135) could say, "Within less than a generation, this industry has found itself struggling for existence against a rising tide of liquidation, bankruptcies, reorganizations, and losses." By the start of the postwar period, textiles had ceased to occupy a central place in American manufacturing. Much of the market would be given over to inexpensive foreign competition,[18] with domestic producers holding on to fashion production, some cheap goods, and industrial products (Bernstein, 1987: 75).

The textile sector had begun the twentieth century in very good shape. High profits at the turn of the century had stimulated much investment.

Yet the cotton sector experienced a decline in earnings from 1907 to 1914. After the war, data from tax returns show that cotton profits trended downward through the 1920s. In 1926, the number of firms with deficits exceeded those with positive incomes; losses were 70% greater than net income. From 1927 to 1929, profits barely exceeded losses. From 1930 to 1932, there were large net losses for the industry as a whole, with over three quarters of firms experiencing losses (Backman, 1946: 135-9). The whole textile industry incurred net deficits in 1924, 1926, 1931-1933, 1935, and 1938 (Bernstein, 1987: 75). Even in 1929, only one half of all textile firms reported a positive profit. Share prices fell from 1923.

Within cotton, the number of spindles peaked in 1925, the number of broad looms in 1927, and the number of mills in 1922. Total net capital (assets minus inventories) was $7 billion in 1924, $4 billion in 1938, and $5.5 billion in 1942 (Backman, 1946: 12, 19). Despite the good times of the late 1920s, the textile sector actually employed fewer workers in 1929 than in 1923. Another 231,000 workers (net) lost their jobs in the next four years. While employment did recover later in the 1930s, the absolute level of employment in textiles has remained below the 1929 level throughout the postwar period. The apparel sector does achieve higher levels of employment in the postwar era, though of course there is a much higher population then as well.

What structural forces were at work in the textile industry? Wolman (1929: 55) recounted the general impression that expenditure on clothing had actually declined, at least relatively. Our data on consumer expenditure does show that expenditure on clothing rose barely 5% between 1923 and 1929, and fell thereafter. Expenditure in the later 1920s did not rise nearly as fast as population, much less national income. Price trends were favorable during this period, due to low raw cotton prices, the competition which results from excess capacity, improved technology and foreign competition. In real terms, cotton prices fell one third between 1922 and 1936 (Backman, 1946: 98). Demand appears to have been quite inelastic, and we must suspect some degree of market saturation. This might reflect the possibility that society had reached a plateau where individuals were relatively happy with their consumption of textile products.[19] More likely, the trends in income distribution we have discussed played a significant role in constraining aggregate consumption demand. There were some changes in taste which exacerbated the situation; shorter skirts and less extensive underwear both had a significant impact on the quantity of textiles demanded (Copeland, 1929). Whatever the case, textile producers could not, even by slashing prices in the cutthroat competition of the late 1920s, succeed in expanding output.

Copeland (1929: 417) argued that the lack of advertizing might have meant that effective demand was not fully mobilized.[20] Since shopping for clothing became dominated by people under thirty in the 1920s, rather than older women as before (Ewen, 1976: 222), this might be true. Given the lack of response to price changes, though, we might suspect that advertizing would only have affected the relative position of different firms, and had little effect on aggregate demand. Any net employment effect would only have been felt in the advertizing sector.

Structural problems involved much more than just market saturation in clothing. Some mills were able to take advantage of new opportunities outside of clothing in such products as tire cord, upholstery, and brake linings. More significant, though, were the number of inroads made into traditional textile markets by other goods. The paper industry gained in the production of towels, napkins, bags and bandages. The switch toward paper was due to the lower cost of paper good production (see Cohen, 1984), and could only have a negative impact on aggregate employment. Burlap, as well, took away some of the bag market.

The best known new competitor, though, was rayon. A new product technologically, it took 9% of total textile output by 1939. With the exception of minor downturns in 1932 and 1937, rayon output grew throughout the interwar period. Due to continued research effort, as well as mass production, the price per pound of rayon fell from $5 in 1919 to 54¢ in 1941. While employment merely tripled between 1925 and 1947 (to 60,000), output grew from 36,000 tons to one billion. Output per worker became much higher than in traditional textile sectors very early, and the increasing use of rayon for those products where it was a good (often superior) substitute, such as women's clothing, tire cord, or upholstery, would again lead to a decrease in total employment.[21]

The result of all this was chronic excess capacity. The number of idle spindles in the textile industry rose sharply from 1923 (Backman, 1946: 147). That employment did not fall at anywhere near a similar rate alerts us to the possibility that there was a significant degree of labor hoarding. Still, if textile employment had grown at the same rate as total employment after 1921 it would have employed 100,000 more people in 1929 (we should note first that not only is textiles a large industry, but that wages were a larger component of total costs than for any other industry). It thus plays a major role in the chronic unemployment of the 1920s. Beyond this, real hourly wages in the sector fell from 1921 to 1932. The multiplier effect on the rest of the economy should take into account not only those who were left unemployed but those who retained employment but saw their income fall.

Some producers naturally did fare better than others. Those who moved into the new markets mentioned above, or those producing style goods such as lace fared well (as income distribution worsened, perhaps). "The difficulty lay in the fact that the predominant share of the productive capacity of the industry was concentrated in product lines that showed little growth potential" (Bernstein, 1987: 77). There were, in fact, limited opportunities for expansion in any product lines. There may have been some potential in the garment sector, but this would have forced producers to deal with fickle consumer tastes. The four case studies Bernstein examines support this conclusion, for the entrepreneurs are found to be looking predominantly for new product lines outside of textiles (1987: 131).

There was one further structural problem. The textile industry had begun to move south in the 1920s. Plentiful labor, low taxes, and easy access to raw materials and water power were powerful incentives. The 1927 census reported that 67% of textile production by yard (56% by value) was located in the south. At first, we might anticipate a positive effect on both investment and employment. The northern mills would not die immediately. Over time, of course, this combination of forces must cause unemployment in the north. Structural unemployment of workers in northern textile towns would be severe throughout the 1930s. "It would be wrong, therefore, to suggest that all the problems which the industry experienced between 1929 and 1933 were the product of the depression, since the northern section had been in decline for some time" (Fearon, 1987: 98).

To the above must be added the effects of improved productivity. From the early 1920s, faced with weak markets, producers had adopted the methods of scientific management with two goals in mind. One was to improve quality. The second was to increase productivity, mainly by speeding up production processes or through automation. Workers were also spread over a greater number of machines. Bernstein (1960: 3) describes how mills strove to capture an increasing share of a decreasing market by forcing their workers to work harder. Output per worker in cotton rose 82.4% between 1919 and 1940 (Bernstein, 1987: 129). This occurred despite the fact that hours per week fell from over 60 in 1890 to under 45 in 1932 to 33.2 with the NRA in 1934, and 36.6 in 1939 (Backman, 1946: 81). Most of this productivity advance occurred between 1931 and 1936, due in large part to dishoarding.

The productivity advance was due primarily to a range of technological improvements. Soule (1947: ch.7) lists several labor-saving devices introduced in the 1920s. Backman (1946) speaks of the importance of long-draft spinning. Jerome (1934) surveyed a number of plants and found that the automatic loom required only one-third the labor, each warp-tying machine replaced ten to fifteen workers, each drawing-in machine saved three to five, and the high speed spooler-warper saved over half the labor involved in warping. Most important was the combined impact of electrification. As in many industries, electric motors replaced overhead belts in the 1920s (Williams, 1982: 151). By the mid-1920s the textile sector was using 1.75 million horsepower. Through the late 1920s and early 1930s, great improvements were made to electrical textile machinery. New metals, oils, bearings, cams, gears and shafts were developed. Among other effects, the speed of carding, spinning and weaving improved dramatically. As well, new instruments such as automatic humidity control were introduced (Hogan, 1971: 1335).

Apparel

Ready-to-wear clothing, such as jeans, had emerged in the nineteenth century to serve poor workers, sailors, and slaves who could not rely on home production. By the 1920s, clothing was increasingly bought by all classes. This happened first for men's clothing and a little later for women's. In 1894 the Sears catalogue had no women's clothing. By 1920, ninety pages were devoted to that product line (Cowan, 1983: 73-4).[22] Employment in apparel 1950-1970 is virtually double the figure for the interwar period (1.2-1.3 million, versus 580-680,000). To a large extent, this can be viewed as a change in consumption patterns associated with postwar prosperity. If this change could have occurred in the 1920s, hundreds of thousands of job opportunities would have been created, and the severity of that downturn significantly lessened. Once the Depression was underway, of course, those whose incomes declined were unlikely to consider buying products they were accustomed to making in their homes.

Technology did have a role to play in making factory production of clothing advantageous. Machines for sewing button-holes, and for sewing on buttons, emerged in the late nineteenth century. Jerome surveyed eighteen garment-making establishments in New York; seven had installed steam-pressing machines which achieved a 49% saving in labor in that operation, three had installed faster sewing machines and reported a saving of 15-20%, and two had introduced rotary cutting machines which reduced work crews from 22 to 8 (1934: 87). Large factories employing thousands emerged in the 1930s. Compared to those of 1900, by mid-century ready made clothes were of higher quality, fitted better, were available in many fabrics and styles, and were cheap in a world of expensive labor (Williams, 1982: 154).

Notes

1. Schurr attributes the fact that coal gas did not satisfy lighting requirements as in Britain to the fact that the coal industry had developed more slowly in the United States, and to the fact that a high proportion of the population, especially in rural areas, was a long distance from coal fields. A small amount of petroleum could satisfy lighting needs (1960: 98-9).

2. Low gasoline prices likely had little effect on car sales, but may have encouraged larger engine size (Williamson, 1963: 448). The price had risen before, and

fallen after the introduction of cracking in 1913. Prices rose moderately in the early 1920s, but this was arrested by continuous cracking. Prices fell in the late 1920s and early 1930s.

3. Many improvements were made in the later 1910s. The use of heating tubes in 1915 yielded a one-third efficiency gain. Perhaps of greatest importance for the future was improvements to pumping equipment so that this could withstand high pressure and temperature and thus continuous processing would be possible.

4. They were also devoted to massive patent litigation. A patent pool of four in 1923, extended to seven in 1931, escaped antitrust action and appears to have facilitated the spread of the new technologies (Williamson, 1963: 390-7). Technical improvements included heat-exchange, or the use of heated output for heating. Haynes, while critical of the oil industry for its lack of chemical analysis before 1930, notes that the 1920s witness a technical renaissance focussed on increasing yields, decreasing refining costs, and raising the quality of gasoline and lubricants (1948: 391).

5. Major advances in the quality of lubricants also occurred in the 1930s through better refining methods, the introduction of thousands of additives for improving performance, and the development of new laboratory methods for evaluation (Williamson, 1963: 635). The auto industry stimulated much of this research through concerns with corrosion, oxidation, and viscosity. Improvements in quality meant that lubricant use per automobile fell over the decade.

6. Williamson suggests that the Depression acted to restrain interest in investing in cracking technology, but the expansion in car engine compression ratios encouraged the achievement of higher octane ratings (1963, 606). Higher octane ratings would make possible better and smaller engines for both cars and airplanes (Hogan, 1971: 1322-3).

7. Problems in the cooling system arose and were solved by borrowing processing technology from the chemical industry. From 1940, synthetic catalysts were used (Williamson, 1963: 619); their development had been spurred by the Houdry process.

8. Discovery of large new fields in the 1930s which threatened to drown the market encouraged belated government regulation. This slowed extraction rates and increased well spacing (previously, oil could be pumped from under one's land by one's neighbor). The East Texas field in 1930 had possessed four times the necessary wells. Regulation enhanced productivity (Larson, 1971: 85), and likely prevented a surge in investment.

9. "The 1920s proved to be one of the most fertile periods of technological change in oil exploration, discovery, and production in the history of the industry" (Williamson, 1963: 313). He notes that the industry was slow to adopt some of these changes. Increased reliance on scientific expertise in exploration appears to have increased the rate of discovery; it may have increased employment marginally for a time.

10. The location of coal mines ensured that displaced workers would not find jobs easily. In southern Illinois coal towns in 1939, half of the unemployed had

been so for over a year. Another third were youth who had never had a job. Outmigration was almost impossible during a nationwide Depression (though some occurred), and had been unable to solve the surplus labor problem even in the 1920s (Brown and Webb, 1941).

11. Output in metal mining naturally suffered as iron and other sectors cut back on output. It was further hurt by technological developments there which saved on materials, and by increased use of scrap in iron, copper, lead, and zinc production. For lead and zinc, international commodity agreements fell apart in the 1920s, and the prices collapsed.

12. Mechanization tended to increase the amount of waste material included. However, improvements occurred in both cleaning of coal and separation of minerals. Flotation methods from 1912, which involved using chemical analysis to separate minerals in solution, allowed ores of limited metallic content to be economically mined. In 1923 selective flotation, where more than one mineral could be separated at a time, became possible. Improvements in the chemicals employed followed. This yielded savings at the refining stage as well, through increased purity of input.

13. It was discovered in the 1920s that corrosion depended not just on the thickness of tin coating but the type of steel base. Silicon, phosphorus, and sulphur were bad, while small quantities of titanium aided corrosion resistance. This led to new methods of tinplate production in the late 1930s which required much investment in new equipment (Hogan 1971:1377).

14. Outside of canning, there were also important product innovations in dehydration (soups, juices etc.) and freeze drying. The latter was first used for medical purposes, and later for foods, such as coffee, which could command a high price.

15. Brown (1947: 21) estimated that 8-10% of the population owed their employment to the forests. He noted that lumber provided a greater total tonnage to railroads than agriculture, and was exceeded only by coal and ore. Comprising 10% of traffic nationwide, lumber was the mainstay of railroads in regions such as the northwest.

16. Creosote, a wood tar derivative, greatly extended the life of railroad ties. It was applied to barely 20% in 1910 but over 75% in 1930. Tie sales fell from 7.3 million in 1929 to 1 million in 1935. The lumber industry was also hurt by the development of synthetic methanol and acetic acid.

17. Jerome also notes the existence of furniture carving and sanding machines. Output in that sector (less important than boxes) had suffered from the collapse of construction. It too was an industry of small establishments, and many firms were experiencing serious difficulties by 1929 (Fearon, 1987: 100).

18. While world production of raw cotton was fairly stable (5 million tons in the 1930s) the number of spindles in Japan expanded from 2 to 12 billion 1913-1939, and in India from 6 to 9 billion. Exports accounted for 3-4% of cotton production in the late 1920s. This figure dropped by 80% after 1929 due to competition either from the receiving country itself or from Japan (Backman, 1946: 72).

19. The Ewens (1976) describe how poor people who could not afford good housing or furniture would put a disproportionate amount of income into cloth-

Selected Other Industries 251

ing to impress others, as mass-produced clothing became possible. This process appears to be relatively complete by 1923. We may see a bunching of purchases similar to that which occurs with durable goods.

20. Copeland conjectures that the lack of advertizing might be due to the importance of patterns and design. Industrial organization may play a greater role. Not only were there numerous small firms, but there was little vertical integration between textile manufacturers and the converters who produced the final good. Bernstein has suggested that some of the hardship observed in the textile sector was due to the poor bargaining position of textile mill operators in a world of limited brand recognition (1987: 134).

21. For men's hosiery, inroads by both rayon and silk meant that cotton, which had possessed 71% of the market in 1927, was left with only 39% in 1939. Silk provided a different sort of competition, for it was not only more expensive, but harder to wash, and less durable.

22. The Ewens discuss the democratizing effect of mass production, while decrying the horrors of early textile sweatshops. After World War One the previous class distinctions in dress largely disappeared. Even the workers in sweatshops were able to dress like the rest of society. It was no longer possible to tell by looking at someone on the street what social class they belonged to (1982: 161).

11

Non-Manufacturing Sectors

Construction

While construction is always a volatile sector, its performance in the interwar period was extreme by any standards. Cornwall (1977) has attributed the economic stability of the 1920s primarily to the experience of the construction sector. Schumpeter (1939: 743) felt that if input sectors were properly accounted for, construction was the largest source of growth in that decade. Soule (1947: 170) concurred. Construction comprised over 8% of GNP each year 1923-1927. This is the highest ratio for an extended period seen in the twentieth century (Gordon, 1981), though higher ratios had been known in the nineteenth century. Construction expenditure, however, declined from 1926-27. Zevin (1982) and others have attributed much of the original downturn into Depression to the decline in construction; he argues that stock market speculation supported consumption and investment for the intervening two years. Certainly the 40% drop in employment 1928-1933 was worse than the decline in any of the other eight major divisions (Henderson, 1961: 45). New house construction fell from 937,000 in 1926 to 509,000 is 1929 to 93,000 in 1933. Expenditure in construction fell from $11.2 billion in 1929 to $3.7 billion in 1933. The Great Depression, like the two wars, would see net disinvestment in housing. Recovery would be painfully slow, lagging far behind all other sectors. The ratio of house construction to output would not regain half of its 1925 value even in 1940.

It was popular in the 1930s and 1940s to speak of building cycles with a duration of somewhat less than twenty years. Isard, Burns, and Hansen were among the authors of such works (while Schumpeter felt construction followed the longer Kondratieff cycle). Recessions in the declining phase of these building cycles were naturally severe. While the cycles were dominated by residential construction, other types participated. Hickman (1974) has dismissed the idea of endogenous regular cycles. His econometric work suggests that demographic changes provide the key to understanding long-term trends in construction (real income changes are also important). In the early post-World War One era, though, the link between population and housing was loosened, and much more construction occured than demographic variables would have predicted.

The rate of new household formation fell from the early 1920s through to 1933. Temin had attributed this to falling incomes, but Barber (1978: 444) notes that age specific headship rates hardly changed, and thus demographic variables are the more likely cause. The important variable for construction is the rate of change in the population growth rate; while population continued to rise it did so at a decreasing rate through the interwar period. Gordon (1981) calculated that the decline in immigration alone, all else unchanged, would have caused demand for housing to be 28% lower in 1933 and 40% lower in 1940 than in 1925. Simulation results show that housing starts would have been virtually 50% less in 1930 than 1925 even if income had continued to rise. With such negative demographic forces at work, the question is not one of why construction declined in the 1930s but why it did so well in the 1920s (Barber, 1978: 446). Potential investors should have been scared off by demographic projections.

Many authors have argued that secular decline was severely exacerbated by overbuilding in the 1920s. Gordon (1978) points out that, by 1925, construction outlays appear to have been much larger than warranted by prices or incomes. Why would construction expand faster than demand in the 1920s? Cornwall (1977) has argued that residential construction often gets crowded out by business investment. It has thus exhibited some tendency to be countercyclical in the postwar world.[1] An explanation for the 1920s, then, is that in the absence of many viable investment opportunities, and in the face of high savings rates, it was unusually easy to finance house building. Construction, then, could be viewed in a similar light to stock market speculation.[2] While quantitatively much less important than single family housing, there was much journalistic interest at the time in apartment buildings that never achieved high occupancy levels (apartment building activity would peak two years later than house building). Commercial construction was often speculatively financed (Soule, 1947: 170). At a time of falling agricultural land values, there were, more generally, speculative gains to be made from subdividing farmland on the edge of cities. Speculation in buildings must seem plausible, given the contemporaneous speculation in stocks and land (Bolch, 1970: 275).

Cornwall and others are correct in focussing on residential construction, for this plays a much larger role in the 1920s than in either previous or postwar construction booms (see Cornwall, 1977: 207). However, there is more than simply a deficiency of investment opportunities elsewhere driving the construction sector in the 1920s. In the first instance, there was a shortage of housing at the start of the decade due to the labor shortage during the war and a credit shortage in the immediate post-World War One years. This, though, was eradicated in a few years (Bolch, 1970) and could not on its own have led to an oversupply of housing (though it would still encourage a later decline in building activity).

The emergence of the automobile had a huge impact on both personal and business location decisions. Gordon (1961) and Rostow (1978) are among the authors who have noted the role of the car in stimulating construction in this period. Commuter railroads had previously allowed significant expansion in city size, but the automobile provided much of the workforce with unimagined mobility. Vast tracts of land on city fringes suddenly became viable sites for housing construction. "The suburb blossomed with expanded automobile ownership" (Bryant 1990: 134). "The extension of residential areas in and about cities made possible by the automobile and improved streets, and encouraged by the demand for outdoor recreation, has resulted in a remarkable suburban growth of detached houses" (Gries, 1929: 254). Gries noted that this had created a surplus of housing stock in inner city areas, especially given the decline in immigration. Developers of suburban housing likely underestimated the effect this surplus could have on suburban demand. Thus construction could peak even before automobile sales. The automobile and truck facilitated the movement of industrial and commercial establishments away from areas of high population density (Mieszkowski, 1993: 136).[3] Electrification also played a role. By powering pumps, stoves, and telephones, it made it possible to gain the advantages of the city in the suburbs (Platt, 1988: 273). The vast bulk of activity in the 1920s was in the suburbs. As with other new products, it should not be at all surprising if the small scale competitive construction industry of the 1920s[4] managed to saturate its market.

The rise and fall of construction has long been a centrepiece of Keynesian accounts of the Depression. This has meant both that there have been a number of studies of the extent of overbuilding, and that there has been considerable disputation of the results. In the 1960s, Muth had found that there had been too little investment in the 1920s, while Hickman had found too much. Bolch (1970) attempted to reconcile these two results. Muth had shown that there was too little investment per unit, while Hickman had found that there were too many houses. The latter result is the more important for our purposes. Bolch found that, while house construction had lagged behind household formation 1917-1921, it had thereafter been greater by too large a margin to be long sustained (1970: 261). While much debate is possible over more sophisticated econometric analyses, this difference appears incontestable. Further evidence comes from falling rents in the late 1920s and rising vacancy rates especially in 1929. Bolch concluded that the decline in construction activity was a major, though not the sole, cause of the economic downturn of 1929.

Bolch (1970) felt that increased interest rates and construction costs in 1929 could have contributed to the decline in this sector. Mercer (1972) feels that Bolch did not fully capture the degree of overbuilding in the late 1920s; in the dynamic sense of building occurring faster than demand was rising, the period 1926-1934 was characterized by saturation. In the more important static sense of actual stock exceeding desired stock, the period 1930-1936 was characterized by saturation. Moreover, it was not rising costs that choked off expansion but saturation of markets given the existing income distribution.

The stock of housing would rise rapidly after World War Two. Hickman (1974) attributes this to the lack of severe aggregate fluctuations (causation could, of course, run both ways), continuous population growth, and the gradual eradication of the postwar housing shortage. We could add the fact that the expansion of urban job opportunities facilitated a considerable movement of people from the country to the city. Unlike in the 1920s, when house building peaked in 1955 utility and non-residential construction were able to pick up the slack (Henderson, 1961: 42). Later, in the mid-1970s, a sudden decline in construction activity could be credited with encouraging economic decline (Mandel, 1980: 52-3).

Field (1992) has raised again the question of the quality of construction investment in the 1920s. Perhaps the mass of funds devoted to construction-related activity in the 1920s added relatively little to the real value of housing services. Thus, while investment figures show a 49% increase in the value of the housing stock 1922-1929, rental figures (including the imputed rental of owned homes) display only a 16.6% increase. While it might be thought that this simply reflects saturation of the housing market, it does not seem that the price of housing fell. Much of the differential must then reflect a very poor return on investment in housing. Field attributes this to the poor rules governing subdivision at the time. Urban planners were not yet used to the requirements of the automobile. Subdivisions were created with poor external access, but unnecessarily short blocks internally. Services were often unsuited to the needs of residents. Individual lots were generally disposed of and built upon in a haphazard fashion. The result by the end of the decade was that American cities were ringed by poorly designed and haphazardly built subdivisions. Grebler (1956: 6) recognized premature subdivision as an important cause of the wastage of funds during the interwar period. There may have been as many unused lots nationwide as families. Only in the postwar world with regulatory innovations in planning and zoning was it possible for builders to easily put together large plots of land so that they could realize economies of scale in construction. In the meantime, the institutional structure artificially deflated the return to home building (in the 1930s, it was often difficult to establish landownership due to failures among owners and banks). If true, this provides a further explanation of both the decline in investment in the late 1920s and the very sluggish recovery in the 1930s.

Eichler (1982) has discussed the reasons for the postwar expansion in house construction, and especially the rise of large scale house building companies. Beyond the regulatory innovations discussed by Field, he points to a great improvement in the mortgage market. Before World War Two, loans for over 60-70% of house value and with terms longer than five years were extremely rare. Second and third mortgages might be available, but these were expensive. Two thirds of families thus rented their accommodation. By the 1960s, two thirds of families owned their own home, and Eichler feels long-term low down payment mortgages were the key (he also credits road construction with an important role). In such a conducive environment, builders were able to significantly reduce costs by operating on a large scale; workers could specialize and be moved continuously from house to house, and standardization could be achieved in windows and doors (1982: 65).

It would appear that there was some scope for government action (at various levels) to improve the state of the construction industry in the interwar period. Eichler asserts that the Federal Housing Administration of 1934, while not important during the Depression, set the stage, along with Veterans Administration loans to servicemen, for postwar mortgage innovation. In the depth of Depression, more than regulatory tinkering may have been required. Government construction expenditure (20% of the total in the 1920s) held up through 1930 and 1931, then collapsed severely to 1933. Even the efforts of the WPA from 1936 did not restore the original level of expenditure. The New Deal also financed some housing projects. Such government efforts only served to transform a pathetic performance into merely a poor one (Hogan, 1971: 1324). In Britain, government subsidies helped the construction industry to a much greater degree, and this supported economic recovery (Rostow, 1978: 224).

The interwar period does not see productivity improvement in construction of the magnitude which we see in the postwar period. Given the severity of the collapse in demand, there was limited scope for further labor displacement in any case (if Field is right, improved productivity might well have led to expanded output). Nevertheless, even in this unique sector, the forces we have described elsewhere still had a role to play. Technological change was most notable in larger buildings. Empirical research led to the elimination of impurities from concrete and steel and a much greater understanding of the properties of these materials. Steel skeletons had become common in the 1890s. The development of light gauge steel advanced quickly in the interwar period. Steel flooring with space for electric wiring was introduced in the 1930s. The electric elevator emerged in the 1890s and the (conveyor belt induced) escalator a decade later. Twentieth century innovations included excavators, mixers, paint sprayers (and better paints) and riveting machines. Advances in welding were stimulated by pipeline construction in the 1920s. Welding not only required less labor than rivetting, but allowed design to be more fluid.

Other factors than technology may have been more important in stimulating such cost decrease as occurred in residential construction in the interwar period. Grebler (1956: 13) points to migration to warmer climes, smaller house sizes, lighter materials, and the increasing relative importance of the rural non-farm population, versus the addition of garages, closets, and heating systems, and rising incomes. He calculates that the real cost per dwelling fell 40% between the 1890s and 1940s.

The decline in construction activity naturally crippled a number of other sectors. Construction was responsible for one-sixth of railroad ton miles, and took most of the country's brick, stone, cement, and lumber production. Glass and paint producers were also seriously affected. Construction was the third largest consumer, after automobiles and railroads, of steel. While residential construction peaked in 1926, the building of large structures which employed steel in large amounts peaked in 1929. (Some projects, like the Empire State Building, were continued in the 1930s). Construction of bridges, buildings, and factories would decline sharply thereafter, and this would be a major factor in declining steel output (Hogan, 1971: 1323).

Process innovation in materials sectors exacerbated the employment decrease there. The pug mill of the 1920s saved labor and allowed the production of a uniform clay free of air products (Williams, 1982: 189-90). Katon (1939) describes how technology had depressed employment in the crushed stone sector (important for concrete, railroad ballast, and flux for steel production) from World War One (employment peaked at 67,000 in 1913). After 1925, "the rise in output per man was accelerated, so that employment fell gradually from 1925 to 1929, despite expanding markets." Output per man hour tripled 1915-1929, due to better explosives, trucks and tractors, conveyor belts, power tools, and other innovations.

Transport

Transport services as a whole grew much faster 1889-1920 than 1920-1946. If the growth rate of the earlier years had continued, freight transport employment would have been twice as large in 1946 as it was and passenger transport employment three times. For passenger transport, the switch to private cars is likely the predominant cause of growth retardation. Private trucking, though, can only bear part of the blame on the freight side. Over the whole period 1889-1946, productivity grew faster in transport than in either agriculture or manufacturing (Barger, 1951: 49-57). Moreover, the revolutionary role of the railroad in tying together isolated markets had been completed, and freight traffic at best could only expand with output.

Railroads

We would expect transport services in aggregate to suffer as goods producers suffered. As mines and quarries provided more than half of rail traffic over the entire period 1889-1939 (Barger, 1951: 85), we would not be surprised if the sector was hurt severely. Yet, railroads were in many ways the sector which was hardest hit in the 1930s (Jaeger, 1972). Moreover, railroads were unhealthy throughout the 1920s.

The task of geographical expansion in the United States had been completed by World War One. The length of railroad actually fell during the 1920s. Due to double-tracking or yard extension, there was a slight 5% increase in railroad track mileage during that decade. Before 1916, trackage had been expanding by 2% per year. Between 1929 and 1939, over 11,000 miles of track were abandoned. Railroad construction was thus employing far fewer workers throughout the interwar period than it had before the war. Those who supplied inputs to railroad construction suffered as well. Four times during the 1930s rail production fell below one million tons; the last time this had happened was 1878 (and rails had become heavier over time; Hogan, 1971: 1299).

The railroads were also facing increased competition from road and to a much lesser extent air transport during the 1920s and 1930s.[5] The diesel truck had appeared about 1930, and soon became common (Bryant, 1978: 1008). Still, in 1940 over half of transport employees were with railroads and only one fifth were in trucking. In the post World War Two era the shift from capital-intensive railroads to the more labor-intensive truck would have a positive employment effect within the transport sector itself. If we take into account the effect on the capital goods sector, the net impact is less clear-cut, and likely negative.[6] For railroads themselves, freight ton-mileage, which had expanded mightily for decades, grew less than 10% in the 1920s, much less than the growth rate of industrial output. Ton-miles remained below the 1929 level through the 1930s. Revenue passenger miles fell steadily through the 1920s from 47.4 billion in 1920 to 31.2 billion in 1929 and 26.9 billion in 1930. The fall continued through the 1930s. Most of these passengers were lost to the automobile. People were thus transporting themselves rather than employing others to do it for them.[7] Not surprisingly, producers of railroad cars and locomotives suffered. Railroad car construction fell, though unsteadily, from 1923 through the 1930s. There were 2209 cars produced in 1929 but only 7 in 1932. Locomotive production also fluctuated in the 1920s, but fell overall through the period (Hogan, 1971: 995-8, 1300-1). Despite the fact that there was a shift from wooden to iron cars, railroads dropped from consuming 25.4% of finished steel in 1923 to 10% in 1932 and 6.1% in 1938 (Hogan, 1971: 1297).

When the railroads were returned to private hands after World War One, there was talk of consolidation, for it was recognized that many railroads were too small. The next decades would see line closures and mergers. The government decided that instead of forcing consolidation it would support railroad profits through rate regulation. The Interstate Commerce Commission was given the task of trying to maintain railroad returns. In downturns, when demand slackened, rates were not allowed to fall and indeed were often raised. Railroad rates were raised 30% in 1922, and then stayed constant while costs fell through the rest of the 1920s (the same happened to public utilities). Jacob Viner was amazed that nominal freight rates had actually increased in the early 1930s, while price levels overall fell dramatically (O'Brien, 1989: 84). Not only did freight rate regulation decrease railroad employment, but it had repercussions on various other sectors which were unable to obtain raw materials or deliver output as inexpensively as they should have been able to. O'Brien (1989) has suggested that as much as 8% of the drop in employment between 1929 and 1931 could be attributed to railroad rate regulation.

There was also significant technological advance within the railroad sector. While we might think of the interwar period as the dawn of the diesel, the most important advance for our purposes was the increased power of steam locomotives. The first decade of the twentieth century had seen the advent of superheated steam, which had cut fuel consumption by one quarter. The average tractive power of locomotives was 36,935 lbs. in 1921, 45,225 in 1930, and 54,000 in 1946. Thus, railroads were able to haul much longer trains than before. Despite an increase in ton-miles, the number of locomotives fell over the 1920s from 69,122 to 60,189. These locomotive figures likely roughly approximate the decreased need for train crews, for the size of crew is generally invariant to the length of the train (locomotive up-time may have increased). Barger calculated that average train length increased 11% 1929-1939, but attributed this to the concentration on long hauls which resulted from truck competition (1951: 89-90). There was a 41% drop in employment in the maintenance of equipment and stores 1929-1939, which Barger largely ascribed to such technological improvements as metallurgical advances, cast steel frames integral with the cylinders, better lubrication, roller bearings, rust resistant steel cars, modern machine tools, and assembly line repairs (1951: 102-8). As many of these predated the 1930s, there may well have been a degree of labor dishoarding. Heavier rails, chemically treated ties, and improved ballast and tie plates put down in the 1920s also seriously decreased employment in the 1930s. Treated ties lasted three times as long. Track-laying was also mechanized. Employment in track maintenance fell 53% 1923-1939, though much of this may be due to deliberate under-maintenance (Barger, 1951: 100-8). As well, Jerome (1936) estimated that almost 40,000 job opportunities were lost 1924-1929 due to the introduction of automatic signals at grade crossings. Even in the office, employment of service and administrative staff fell 40% 1912-1939, due to business machines and centralization (Barger, 1951: 100-1).

Overall, Weintraub estimated that 345,000 workers were replaced by increased technical and managerial efficiency in the 1920s. Output per worker grew 20% 1920-1929, 31% 1929-1939, and a further 36% 1939-1946 (Barger, 1951: 98; he is amazed that productivity growth does not slacken as the industry ages). In sum, due largely to changes in output, technology, the cessation of construction, and railroad consolidation, railroad employment fell from well over two million in 1920 to about one million in 1939.

Diesels only emerged in 1934, and were only used for yard-switching, with a few exceptions, such as areas where water or fuel for steam was difficult to obtain, until the war. During and after the war, they would be found not only to be more fuel-efficient, and quicker, but to be less wearing on the rails and equipment, and to be available 90% of the time compared to the 50% for steam locomotives. One of the big problems with diesel was the transmission of power to the wheels (which was easy with steam); this was done electrically. Diesels were based on the gas-electric passenger cars (somewhat like modern Budd-cars) developed by Hamilton for use in suburban passenger service in 1924. These were incapable of pulling many cars. In 1930, Hamilton was bought out by General Motors and after intensive research the first mainline diesel engine was introduced in 1934. This might well be viewed as a Menschian innovation, for the bulk of the research was performed in the Depression. However, the diesel engine itself had been around for decades, and had undergone years of development which had largely solved the problems of compression and timing. Not only had the diesel truck preceded the diesel locomotive, but the first diesel ship had been introduced in 1911, and half of new ships were diesel by 1927. It is worth noting that, while Hamilton's car had been around for years, the development of the diesel occurred under the auspices of a company which had no previous interest in either railroads or locomotive production. Both railroads and locomotive manufacturers remained hostile for some time after its introduction (Hogan, 1971: 1301-3).

Urban Railroads

Urban electric railroads peaked in terms of passengers, revenue, and employment shortly after World War One. The cause of their downfall was the car and the bus. They employed about 250,000 workers at their peak. We have figures on all electric railroads. These include a few interurban services established in the first two decades of the century and quickly abandoned in the face of the automobile (employment in interurban services fell 48% 1922-1929 and 70% 1929-1939). The number of passengers on all electric railroads fell 8% 1920-1929 and 38% 1929-1939. Employment fell 18% 1920-1929 and 46% 1929-39 (Barger, 1951: 112-4).

Pipelines

Pipelines were one transport sector which showed output growth in the interwar period. Total ton-miles quadrupled 1921-1929 and increased 38% 1929-1939. Employment, however, did not quite double in the first period and fell 13% in the second. While this was in part due to fluctuations in construction activity, it mostly reflected larger pipes, better pumping machinery (diesel replaced steam from 1935; remote control electronic pumping stations emerged in the 1930s), and perhaps also longer hauls (Barger, 1951: 124-6). Pipelines not only displaced labor elsewhere — railroad oil freight cost three times as much per ton mile (Williamson, 1963: 675), and fell through the early 1930s — but failed to contribute to employment growth themselves.

Oil had been transported at first by barge and railroad. The first wooden pipeline appeared in 1862, and a five-mile wrought iron pipe followed in 1865. In 1878, a 110-mile line was laid across the Appalachians, and the development of inland oil fields encouraged further construction. By 1910 there were 45,000 miles of pipe, by 1921 67,000, by 1930 110,000, and 115,000 by 1936. The growth of the oil industry and its westward expansion were naturally key factors in stimulating pipeline construction. Decreased costs of producing and laying pipe were also important. The 1920s were likely the period of most rapid technological change before World War Two; "The outstanding single feature common to all the innovations of the 1920-1930 period was the marked substitution of capital equipment for manual labor" (Williamson, 1963: 362, 371). Better ditchdigging equipment was developed. Welding replaced screwing, and then electric welding was introduced. Pipe was produced in longer sections. Seamless pipe appeared at the end of the 1920s. With carbon steel, pipes became lighter and stronger, and pumps could be placed farther apart.

Pipeline expansion in the 1920s was largely due to the development of new fields like West Texas far from markets, and the invasion of midwest markets by southern producers (Williamson, 1963: 347). One third of construction in the 1930s occurred 1930-31, and was a continuation of this late 1920s surge. Specialized gasoline pipelines, in large part a response to high railroad freight rates, were also important. Slow growth from 1932 can only in small part be blamed on the Depression; the oil industry, after all, suffered much less than others. "The great expansion of 1927-1931 left the crude oil pipeline system with more than enough capacity for the next several years" (Williamson, 1963: 571). Without the discovery of new fields, and with the onset of government regulation which limited the output from new discoveries, there was little reason for new construction. As well, improvements in pumping equipment increased the capacity of existing lines. Expansion later in the 1930s would in large part be due to the extension of the West Texas field and development of the Wyoming field.

Shipping and Shipbuilding

Less than 10% of transport employees in the United States worked in internal shipping. The output in ton-miles of the sector fell 4% in the 1920s and 13% in the 1930s.[8] Employment, however, fell 41% in the 1920s and 16% in the 1930s. Labor productivity had been growing by almost 3.5% per year since 1889, much faster than in railroads. This reflects the introduction of bulk cargos, especially in oil and iron ore (Barger, 1951: 148-50). It must also reflect the new generation of ships produced during the war, which were faster and larger than those before. The switch from coal to oil burning significantly decreased the labor required on board; fuel oil use in shipping expanded from 41 million barrels in 1919 to 106 million in 1929. The figures do not include port personnel; their numbers declined also due to mechanized loading and unloading (Barger, 1951: 150).

Total American deadweight tonnage increased by ten times 1913-1920. At the end of World War One, the U.S. government had a huge surplus of ships which it slowly sold. In 1925, almost one third of total tonnage was laid up. The American shipbuilding industry, with higher costs than its competitors, had little access to foreign markets. There was thus little scope for production. The Merchant Marine Act of 1928 did provide some boost to domestic production. Shipbuilding declined with foreign trade in the early 1930s. Military procurement picked up late in that decade. Production of merchant ships recovered only from 1935. In part this was due to the obsolescence of World War One era ships. Still, this recovery fell far short of bringing prosperity to the industry. That would come only with the war (Hogan, 1971: 1049-54, 1360-5).

Agriculture

It would be a mistake to assert that the 1920s were a bad decade for farmers. It was a decade in which the world-wide overproduction which has plagued that sector throughout most of the century became a severe problem. However, in large part due to government support, farmer's incomes rose through most of the 1920s (after a severe drop 1919-1921). Per-capita farm income was $314 in 1920, $191 in 1921, $243 in 1923, $287 in 1925, $279-$281 1926-1928, and $295 in 1929. The immediate post-World War One reconstruction period kept foreign demand for food unusually high until 1920. Farm incomes had doubled during the war. Thus, if we compare the later 1920s to the 1920 figure, it appears that farmers are suffering. However, if we choose the pre-war period as our reference point, we see that farmers incomes are higher in the 1920s (farm labor fared less well; wages were the same in 1929 as in 1914 and had been stagnant since 1923).

They were not, though, rising as fast as national income, especially after 1925 when farm incomes stagnated. Moreover, the land speculation boom before 1921 meant that many farmers were carrying very high mortgages in the 1920s (farm values fell through the 1920s and 1930s). The debt load rose 25% 1920-1925. Walton and Robertson assert that it was the debt load which was the source of farmer discontent in the 1920s, especially in the midwest (1983: 496). (Schumpeter, 1939: 732-6 argued that productivity advance was a more serious problem.) There were numerous farm bankruptcies throughout the decade, and many farmers were in a very weak financial situation when the Depression hit. Almost one million farms were repossessed 1930-1934.

Many countries had increased output considerably during the war and reconstruction period. When the countries which had been devastated by war finally returned to normal levels of production, there was a huge glut on the market. Between July and December 1920, the price of cotton fell from 37¢/lb to 14¢/lb and the price of wheat from $2.94/bu. to 97¢/bu. Throughout the 1920s, the price trends of most agricultural products remained below the price trends of industrial goods (including inputs to agriculture). Farmers reacted to falling prices by increasing output, which simply depressed prices further. For most products, output continued to rise through the early 1930s (until government policies were introduced to curb output). The First World War exacerbated a secular trend. Demand for food is very inelastic and 92% of crop acreage was devoted to food. As population growth rates tapered off, aggregate food demand ceased to expand as it had before. However, technological change led to steady increases in output per acre.[9] In the United States alone, output had grown less fast than population 1899-1939, but the loss of export markets meant that supply grew faster than demand. Many European countries decided to forego inexpensive supplies of foreign grain in favor of protecting local producers. Even more severe import restrictions were imposed during the crisis of the 1930s (see Tracey, 1972: 95-8). Countries, like Canada and the United States, which relied heavily on export markets, were hardest hit. Those, like the United States, Soviet Union and Argentina, which lay outside the charmed circle of preferential trading arrangements, were almost shut off from export markets. Imports (generally of products not grown in the United States) exceeded exports for the first time in 1922, and did so every year 1925-1940. Decades later, we have still not figured out (or at least applied) the appropriate public policy to deal with a world in which farmers are able to produce so much more than we (in the developed world) want to eat. Income distribution was important here; Stricker (1983: 29) estimates that 40% of the population was not getting the minimum dietary requirement, and Schlebecker (1975: 232) asserts that the interwar problem was not overproduction but the low incomes of most consumers. The rise of synthetics limited whatever scope there might have been for increased industrial outlets for agricultural products.

Various governments responded to these adverse trends by stockpiling agricultural output during the 1920s. Thus, the incomes of farmers could be protected for a period. During the Depression, governments became less willing to hold stocks than they had been during the 1920s. This, along with the fall in aggregate income, meant that prices fell much faster in the 1930s than the 1920s (Temin, 1976). In the United States, the Federal Farm Board, without power to control output, found it impossible to long sustain price supports in the 1930s (Walton, 1983: 533).[10] Agricultural prices fell by about half between 1929 and 1933, much faster than industrial prices. The ratio of output to input prices fell one third 1929-1933 (Fearon, 1987: 102).

Keynes felt that changes in relative prices should have no effect on aggregate demand; purchasing power would simply be transferred from some agents to others. Kindleberger (1973) has argued that this large shift in relative prices could have had an important impact on aggregate demand for people whose income falls (must) react much faster than those whose income rises. In terms of our discussion of income distribution (in Chapter 3), it is noteworthy that it was not the average worker who appears to have gained at the farmer's expense but rather those with higher incomes and lower propensities to consume. Since the agricultural sector comprised only 9% of the population, we could not possibly attribute the entire fall in consumption demand to this (Temin, 1976: 150). However, this sudden shock to farmer real income must have had some impact on industrial output and employment.

In the American case, there were further influences acting to depress prices. Mitchell (1929) spoke of three of these. First, Americans were eating less food per capita, perhaps as they were performing much less strenuous labor on average (and as the ideal of thinness, especially for women, gained currency). Second, the American diet was changing. Some farmers, those who produced dairy products, eggs, fruit and vegetables, benefitted, but most farmers suffered. Third, as the textile producers lost ground to paper and rayon, they naturally decreased their demand for raw cotton and wool. The export situation was no better; "The decline in cotton exports from the late 1920s to 1937-8 meant a decrease in labor requirements equivalent to approximately 300,000 man years" (Hopkins, 1973: 166). Thus cotton fared as poorly as food crops.

To many Americans and Canadians who lived through the Depression, that experience is inextricably intertwined in their memories with drought. The years of below-average rainfall were regionally concentrated, naturally. Canada's wheat sector saw output per acre fall from 23.5 bushels in 1928-29 to 10.8 in 1933-34. Aggregate statistics on acreage do not show large (or any) decreases in the 1930s. The dominant source of low farm income in the 1930s is to be found in price rather than output (the drought-stricken areas were a small part of total world supply). Those farmers who were hit by both low prices and low output were, of course, in severe difficulty.

From looking at farm income figures alone, we might have expected employment to have been rising in the 1920s and falling in the 1930s. Much the opposite was true, however. Henderson (1961) has analyzed agricultural employment statistics. He finds that agriculture and mining were the only sectors where the absolute level of employment fell between 1919 and 1959. Agricultural employment fell in every decade. However, employment proved relatively unaffected over this period by cyclical fluctuations. The possibility of subsistence farming counterbalanced the decreased demand in a downturn. Thus, while employment does fall over the 1930s, this is part of a secular trend, and is severely curtailed by movement of the unemployed into subsistence farming.

Productivity rose in agriculture in the 1920s, but not as fast as elsewhere (Gordon, 1961). Given an inelastic demand curve which was certainly not moving outward, any increase in labor productivity must, however, result in decreased employment. Output per worker grew by 26% between 1920 and 1930 (Soule, 1947: 231). Much of this was realized in increased output, which ended up in government stockpiles. The Bureau of Labor Statistics estimated that the 1919 output could have been produced in 1927 with 2.5 million fewer workers; increased output meant that only a third of these were displaced (Jerome, 1934: 381). There was a movement of well over a million workers out of agriculture into urban employment in the 1920s. Nor did this outflow remove all underemployed labor from the agricultural sector. Gill (1940) suggests that through much of the 1920s, close to two million workers per year left the country for the city, but about one and a half million moved back (the latter tended to be much older than the former). Industry was simply unable to absorb this increase. While these numbers are rough guesses, it is clear that the employment figures give an underestimate of the degree of technological unemployment in agriculture.

Likewise, in the 1930s, people would have been expected to leave the land at a similar rate as in the 1920s, for productivity expanded at a similar rate. The outflow ceased with the onset of the Depression, and reversed after 1934. This reflects the even greater difficulty of finding urban employment, and the advantage of being close to food supply. "That portion of the population which would have migrated to the cities in more normal years was backed up on farms awaiting opportunities for employment elsewhere" (Hopkins, 1973: 11). The number of paid farm workers actually declined by a half million 1929-1933 as hired hands were replaced by family members returning from the city (Chandler, 1970: 40). Gill in 1940 could still speak of a huge and growing reserve of unemployed buried in the agricultural sector. Many of these workers ended up on a plot of land too small to adequately support a family. The number of farms rose in the 1930s as empty farmhouses on abandoned land, especially in New England, were reinhabited (Fearon, 1987: 104).[11] The experience of World War Two highlights the extent of underemployment in the 1930s. Despite a loss of 1.8 million men (half replaced by women) the agricultural sector was able to expand output dramatically. More combines and tractors were added during the war than throughout the 1930s.

The major source of productivity gain was technological change.[12] The interwar period can be considered the glory period of American farm mechanization. The most notable example was the tractor. The gasoline tractor had appeared in 1892. In 1918 (boosted by the wartime labor shortage) there were 80,000, in 1920 246,000, in 1925 549,000, in 1930 920,000 and in 1940 1,545,000. The basic innovation was the all-purpose tricycle tractor by International Harvester in 1924. Before that time, the tractor could not be used for cultivation and was thus largely limited to grain crops. Improvements included enclosed transmission, replaceable parts, and high grade steel in the 1920s, rubber tires from the late 1920s (only 14% of tractors had then in 1935, but 85% by 1940; these yielded a 10-20% fuel saving plus greater comfort and allowed travel on highways), power takeoffs and power lifts for implements in the 1930s, the first diesel in 1931, smaller tractors suited to farms of less than 200 acres in the 1930s, and high-compression engines (which relied on improved gasoline) from 1935. It is probably impossible to overstress the role of the tractor in American agriculture (Hogan, 1971: 1399). Not only did it have a direct impact on labor input (and an indirect effect by inducing farmers to work longer hours), but it also served to replace farm animals which had previously been required for motive power. (It was aided in both these tasks by the increased use by farmers of cars and trucks; it was estimated that each truck replaced one or two horses). The number of farm animals fell by 19.5% between 1920 and 1930. The United States possessed 27 million horses in 1916, 22.3 million in 1925, and 15.4 million in 1938. All of the labor (and enough land to feed 16 million people) previously devoted to looking after these animals and raising the crops to feed them, was released. Both the employment and grain surplus problems were aggravated.

The Great Depression did not prevent this technology from spreading. The number of tractors did hardly change 1932-1934, but Hopkins attributes this primarily to low fodder prices (1941: 47). Its advantages were so great that even in an environment of inexpensive labor, limited markets, and financially strapped farmers, it was rapidly adopted. The most notable periods of tractor adoption were World War One, the late 1920s, and 1934-1937 (Schlebecker, 1975: 249). The even more rapid adoption during the Second World War indicates that adoption would have been even faster in the 1930s if the economy had been healthy.

The arrival of the tractor encouraged the development of many farm implements. "Advances in harvesting methods between 1914 and 1945 consisted mostly of applying tractor power and making larger machines" (Schlebecker, 1975: 245). Simply widening implements could yield large labor savings. Not only were tractors faster and stronger than animals, but power which did not depend on forward motion could be supplied to implements and hitching was greatly simplified. There were over a million wheat binders by 1938; these were in turn superseded by combines. There were only 61,000 combines in 1930 but 190,000 in 1940, with an average labor saving of 80% or 3.5 manhours per acre. The army of migratory harvest hands in the central states disappeared in the 1930s (Hopkins, 1973: 73). There were only 10,00 corn-picking machines in 1920 but 50,000 in 1930 and 110,000 in 1940.

The 1930s were also the decade of rural electrification. While often celebrated primarily for its effect on the rural lifestyle, this also affected productivity. Electric pumps, milling machines, and refrigerators were only a few of the possibilities once farmers gained access to low cost electricity. There were a number of biological improvements as well. The work of Mendel on genetics had been rediscovered at the turn of the century. Hybrid corn (the first in 1926 doubled or trebled output per acre), rust-resistant wheat, and improved cotton increased yield per acre, improved the quality of output, and decreased the variance in output. By so doing, they naturally further decreased the number of workers needed to feed the nation. The same is true of sulfa drugs from 1935 (and later penicillin) which greatly reduced livestock disease, as well as improvements in animal breeding (e.g. milk per cow rose from 380 lbs. 1909-1913 to 452 lbs. 1927-1931), and feeding.

We have in the above dealt with agriculture as a unit. We thus might be thought guilty of the sin of over-aggregation which we have accused others of. There certainly were exceptions to the picture we have painted. Some farmers were able to specialize in crops finding increased favor in the American diet, such as citrus fruits (which also gained export markets). Others were able to take advantage of trucks to supply fresh vegetables to urban centres. Still others successfully moved into new products like soybeans. On the other hand, per capita wheat consumption fell one third 1900-1930. To the extent the "new" products required more labor than the old (many proved difficult to mechanize) some limited scope was provided for employment increase. Nevertheless, there appears to be much less need for disaggregation here than with either industry or services. The experience of the major products, wheat, corn and cotton was quite similar. The vast majority of farmers were affected in the ways we have described.

Services

The service sector[13] had come to employ over one third of the American labor force by 1930. Moreover, since the 1920s at least, the proportion of consumer expenditures devoted to services has expanded in every decade. Any sectoral analysis of the interwar period must come to grips with this enormous sector. Some services will be seen to contribute in an important way to the onset of Depression. Even if this were not true, we must understand why the resources released by other sectors could not find a productive home there. In this context, the lack of attention paid to the service sector in the literature on the Depression is nothing less than shocking.

In less developed countries, a common response to a shortage of employment in the formal sector is to find (under)employment in the service sector, performing shoeshines or selling pencils. Within the developed countries, the percentage of the labor force in services has risen above the trend rate in postwar recessions.[14] In the case of the 1930s, there were many counted as employed who only worked minimal hours in retail trade. At the very least, then, we might wonder why the service sector was not able to make the unemployment statistics look better in the Depression.

We would expect services to play the role of employers of last resort because they are generally less capital intensive. Laundries and cleaning and dyeing establishments which do involve a considerable capital expenditure have occasionally been included in the census of manufactures for that reason (Stigler, 1956: 158). Investment, of course, is only part of the entrepreneur's problem. If demand conditions or technological change is not conducive to increased hiring, we will see little difference between the service and other sectors.

Service sector firms tend to be very small relative to manufacturing firms. This varies a bit across service categories (see Stigler, 1956: 56-7), but on average is true to this day. Before World War Two, the corporate form was much rarer in services than in manufacturing. With relative ease of entry, we might have expected service firms to have arisen to employ the vast supply of inexpensive labor in the 1930s. Again, of course, the restraining factor must be the lack of potential markets.

The absolute level of employment did rise for the majority of service occupations in the 1930s (Henderson, 1959: 62). As a whole, the sector continued to absorb a greater percentage of the total labor force through the Depression, as it had since 1870. From 27% in 1910 it rose to 27.6% in 1920, 36.4% in 1930, 40.3% in 1940, and 43.7% in 1950. It is notable that the 1920s recorded the largest rate of increase of all time (this may in part make up for the small change in the 1910s). As we will see below, the overall performance in the 1930s masked the experience of a number of sectors which lost ground.

The secular rise of services is due to many factors. As incomes rise, some service sectors such as insurance, health, travel, and entertainment come naturally to take a larger portion of expenditures. Family budget studies in both the United States and England support this hypothesis. There also appears to be a naturally positive relationship between higher average incomes and the relative importance of government and marketing. Specialization occurs along with economic growth, and one result is that service firms arise which perform tasks previously handled by goods producers themselves.[15] The service component in goods also seems to have increased. The research, accounting, marketing, and financial-legal expenditures are much larger relative to materials and labor in the modern car industry than in the early car industry (Grubel, 1989: 192-3). Urbanization has meant that previously home-produced services can become market-supplied. Demographic variables have a less clear effect. Families with many children may demand more educational services, but be less likely to patronize restaurants. The gradual aging of the population appears to have a net positive effect on services (Stigler, 1956: 159-65).

Many writers, such as Baumol and Fuchs in the late 1960s, have argued that the increased relative importance of services may be due to the fact that productivity has grown more slowly in services. This effect would be counteracted somewhat by relative price changes which would shift effective demand toward goods. There is empirical evidence that postwar, and especially recent, shifts may be due to this productivity growth differential (Costrell, 1988: 102). The basic problem is that, due to the intangibility of the output, it is difficult to accurately estimate output and productivity in services.[16] While Gershuny (1983: 38) has argued that the rising relative price of services is evidence of slower productivity growth, this could merely reflect unmeasured quality improvement. In general, it is likely that we underestimate productivity growth in services, and overestimate it in manufacturing (Costrell, 1988: 103). Productivity growth in services might be accounted for elsewhere; Kuznets in 1980 noticed that productivity gains in agriculture and manufacturing were often due to gains in transport, communications, and finance (Stanback, 1981: 4-5).[17] Properly accounting for quality changes could conceivably eliminate any differential.

Julius Klein, writing in 1928, noted that agriculture and industry had each lost about a million jobs since World War One. The displaced had largely moved into services (though not quite all of them). The recipient service occupations were chauffeurs and bus drivers, gas station attendants, electric and oil servicemen, various movie-related occupations, construction managers, insurance agents, teachers and hotel and restaurant workers. This list can be easily divided into two groups. The first few occupations were ancillary to developments in the goods sector. We might expect that the growth in employment for bus drivers closely paralleled the growth in bus production. However, buses need only be built once, while a driver is needed throughout the productive life of the bus. Since over half of consumer services were associated with food, clothing, automobiles, and construction, we might expect service employment in these to fare poorly. The last few occupations were not strongly connected with goods sector developments, but must reflect the secular forces discussed above. Services that were closely tied to goods production might be expected to decline or at least stagnate as their respective goods-sector industries declined. Other services would have been expected to continue the secular trend in the absence of changes in aggregate variables.

Government

Whereas government employed only 2.23% of the labor force in 1920, this figure was 2.39% in 1930, 3.38% in 1940, and 5.62% in 1950, and of course has reached much higher levels since then. Government, indeed, could be considered one of the leading sectors of the postwar period (and thus more rapid interwar expansion could have significantly reduced the severity of the Depression). Direct government employment is only the tip of the iceberg. Gramlich (1985) estimated that one third of the services produced and consumed in the United States were sold to the government. This proportion had increased to the mid-1950s, but has been relatively stable since then.[18] A number of authors have attempted to explain the seemingly natural tendency, observed in all developed countries over the last century, for the size of government to grow faster than the economy. Positively, it can be seen as an increased interest in public goods, or in creating equality of opportunity, as income rises (or as reflecting an increased desire for regulation, if it is felt monopoly power has increased). Negatively, it can be viewed as the natural tendency of bureaucracies to expand their domain.[19]

Historical events shape public opinion of the role of government; bureaucrats may seize the opportunity to serve their own interests (Higgs, 1987: 15-8). Hughes notes that particular areas of governmental control have emerged as responses to particular historical crises; these have rarely been abandoned when the crisis passed (1977: 238). It was inevitable that the Depression would eventually trigger further government involvement (but this could not happen instantaneously). The longstanding suspicion by the American people of the untrammelled market was given new force. Beyond the emerging belief that government was responsible for full employment, the rest of the New Deal built upon previous traditions of government intervention (the big break with tradition had come with the progressive movement from the 1870s). World War Two led to a massive increase in government involvement, and the Cold War served to perpetuate this. Korea, like every other war since the Civil War, served to permanently ratchet up government expenditure. People became accustomed to higher taxes than New Dealers could have dreamed of (Hughes, 1977). Total state, local, and federal government expenditures had been 6-7% of GNP before World War One, rose from 8 to 14% during the New Deal, fell from a wartime peak of 46% to 11-15%, rose to 22% during the Korean War, and remained above 20% thereafter (Higgs, 1987: 21). We should note that Hughes does not provide a total indictment of government. He recognizes in particular the value of New Deal relief efforts, and suggests we might well consider similar policies again. It is the haphazard nature of growth, not government per se, which is criticized.

The major components of government spending diverge in important ways. Military spending took up a larger proportion of GNP in the 1930s than in the 1920s (2.84% in 1932 vs. 1.86% in 1922). The more startling comparison is with the postwar world, for military spending absorbed 17.12% in 1942 and 7.79% in 1948, and would not fall below the latter figure through the 1960s. Scarcely anyone would advocate military spending solely for its employment characteristics, but it must be noted that this component of government spending is (within our sectoral framework) the biggest single reason for unemployment being so much lower in the postwar period. We can hardly doubt that a greater perception of the threat to national security in the 1930s would have decreased the unemployment rolls.

Highway construction was one component with a favorable employment trend in the 1930s. While it had absorbed 1.75% of GNP in 1922 and 1.89% in 1927, these figures were 3.04% in 1932 and 2.54% in 1938, and were between 1.12% and 1.88% through the early postwar decades. Education actually lost ground in terms of the percentage of the total labor force in the 1930s (2.84% in 1920, 3.45% in 1930, 3.36% in 1940, and 3.57% in 1950). Some of this could be attributed to the demographic forces discussed above; the rest can be attributed to misplaced government budget-cutting efforts.

Professionals

Despite the decline in the educational sector as a whole, the number of college professors did grow faster than the population, as it has through most of the twentieth century (research labs still found it easier to hire topflight researchers due to lesser academic competition; Hounshell, 1988: 316). Interestingly, while the professions as a whole were relatively Depression-proof, doctors and lawyers fared less well than college professors in terms of employment. The number of lawyers grew less fast than population, and the number of doctors per person fell 17% (Stigler, 1956). The latter was largely due to increased specialization, improved technology, and the effects of higher training standards on supply (Stigler, 1956: 158). The growth rate in professors was much higher through the whole first half of the century. While the numbers of lawyers and doctors (and clergymen) each expanded by 70% over the half century, less fast than population, the number of college teachers multiplied eight times. In the postwar period, all of these professions have expanded rapidly. While much of this may be tied to higher incomes, some may be construed as increased investment in human capital through health and education[20]; such investment might have been enhanced earlier.

Business Services

The economics profession had paid relatively little attention to producer services until the 1980s, even though this has been the largest and fastest growing subsector for the last three decades (Grubel, 1989: 151, 188). It had instead lavished its attention on consumer services, the smallest of the service divisions. This concentration created the impression that the United States was turning its back on goods production. Instead, most service growth is intimately tied to goods production (the other subsectors of significant growth being government, health, and education). With product differentiation (itself a response to advertising, urbanization, and the development of numerically controlled machine tools), there has been an increasing tendency for goods and services to be coupled: repair services, market research, instructions. More complex manufacturing technology has increased the need for engineers and planners. Increased firm size has increased the need for personnel and consulting, and is associated with an increased need for advertizing and consumer finance. Government regulation has also encouraged the growth of some producer services. "The logic of this process highlights one of the fundamental roles of producer services — that of integrating or binding together the increasingly differentiated, specialized parts and functions of the market system" (Stanback, 1981: 51).

Producer services employed roughly 6% of the labor force through the interwar period and into the late 1940s. Since then, the proportion has increased by roughly 2% per decade (Stanback, 1981: 14). Stagnation in the early postwar period indicates that the secular forces described above had only become powerful at that point. Stanback, surveying the last century, concludes that service employment first expanded in distribution, then in government, and only then in producer services (1981: 18-9). Given the decline across goods-producing sectors in the 1930s, then, it is to be expected that employment in associated services would decline to the same degree, except insofar as different labor practices prevailed.

Finance and real estate experienced a decline in the 1930s (2.84% of the labor force in 1920, 3.45% in 1930, 3.36% in 1940 and 3.5% in 1950).[21] Henderson (1961) has shown that these decadal figures give a misleading picture. Employment in this sector rose considerably during the financial booms of the late 1920s and late 1950s. Allowing for that, there is little secular trend between the late 1920s and late 1950s. It is possible that some of the growth which can be observed in this sector after 1948 (and which gives it the lowest unemployment rate of any sector) is due to unnaturally slow growth during the Depression and World War Two, but the magnitude of such an effect would be exceedingly small. The decrease in employment within this sector in the early 1930s re-established the level of employment of the mid-1920s. These displaced workers should be seen in the same light as those released by the automobile or construction sectors. It is worth noting as well that the bulk of increased employment during the late 1920s was devoted to the investment of funds. Since the degree of mechanization in processing information increased over the 1920s, there was also some degree of technological unemployment.

Clerical workers are counted among the employees of all sectors. Thus, while they perform a service function, they are often counted as manufacturing employees. These workers, who had only comprised 1% of the non-agricultural labor force as recently as 1870, amounted to 10% in 1930. Growth in this category over the preceding decades was largely driven by technology. The typewriter, from the 1870s, had been the innovation which spurred this growth the most. Carbon paper, address machines, calculating machines, cash registers, mimeographs, and dictaphones were also important (Rotella, 1981: 69). Employment growth and office equipment production were closely correlated through the period. The new equipment lowered the cost of clerical work and thereby increased the quantity demanded; demand was apparently elastic. Changes in the organization of production, government regulation, and the growth of clerical-intensive sectors like finance and government also played a significant role in stimulating demand for clerical work. Supply of labor increased at the same time since more women entered the labor force generally and clerical work particularly.

The decadal growth rates for the clerical sector ranged from 6.4% to 9.73% 1870-1920, while the labor force grew by somewhat over 3% on average. In the 1920s, clerical growth slumped to 2.8%, only slightly greater than the non-agricultural labor force growth rate of 2.16% (Rotella, 1981: 64). It appears that the introduction of new machines had begun to work against continued employment growth at that point. Barger, in his survey of the transport sector, speaks of clerks being replaced in the 1920s by business machines, "a trend characteristic of all business activity" (1951: 100-1). Over $33 million of adding machines were produced in 1925 and 1927, and over $25 million in 1929. Over $10 million worth of both calculating and addressing and mailing machines were produced in the latter year as well. Labor-saving technology can either increase or decrease employment depending on the elasticity of demand. In the 1920s, the trend was certainly turning away from employment creation. In the 1930s, the process of replacing workers with machines would continue.

Consumer Services

Recreation, amusement, and travel perform well in the 1920s and in the postwar period, but relatively poorly in the 1930s. Due to the relatively small size of these sectors, we can ignore the structural and technological forces at work here, and attribute the experience of these sectors to the changes in aggregate income during the 1930s.

Non-professional personal services grew in relative importance throughout the period of interest, from 3.95% in 1920 to 5.29% in 1930, 6.2% in 1940 and 6.38% in 1950. This category includes barber and beauty shops, laundries, hotels, shoe repairs, and funeral parlors. We need not deal with myriad minor personal service categories in detail. There appear to be two conflicting forces at work here in the 1930s. One was the naturally negative impact of declining incomes on demand.[22] In many cases, labor-saving technology (the washing machine replacing laundry services)[23] further reduced the demand for labor. The other was the impact of declining wage demands on the supply of personal services. The fact that employment rose through the 1930s does indicate that there was some ability in this sector to absorb excess labor. This is where the shoe-shines and haircuts could be performed by otherwise unemployed workers.

Gershuny (1983: 84-5) has argued that there need be no secular trend toward increased use of consumer services (and they have remained a relatively constant proportion of GNP since the 1960s). He notes that the provision of the service function has four parts: equipment (cars, TVs), collective infrastructure (roads, networks), non-material activities (TV programs, maintenance), and unpaid labor (driving, watching). Society chooses the low-cost method of provision. Technological change or infrastructure provision can thus unleash a sudden transfer of resources between goods and services. The introduction of the automobile, and the resultant displacement of delivery men, would be a good example of goods replacing services.

Stigler (1956) has analyzed the barber/beauty parlor experience in detail. There appears to be a downward trend in the ratio of barbers to men from 1910 due to changing tastes which allowed men to be less compulsive about their hair care, and the introduction of the safety razor. This trend appears to have reversed in the early 1930s (only decadal figures are available). The number of hairdressers per woman increased through the 1940s and then declined. The permanent hot wave machine of 1905 encouraged the early increase, and improved home perm kits in the 1940s the later decrease. Taking the two subsectors together, total employment rose 160,000 1920-1930 and 40,000 1930-1940. With limited entry barriers (capital costs for hairdressing were not trivial, but licensing requirements which required training only emerged for either subsector postwar), this would appear to be all the sector was capable of absorbing.

The same is broadly true of other personal services. Varied taste and technological forces would cause some to contract while other expanded. In all, though, the whole sector could never reasonably be expected to absorb more than a few per cent of the labor force. Even those it did absorb tended to earn low incomes (hence the successful efforts to install entry barriers, which did succeed in raising barber/hairdresser incomes after World War Two). There might have been greater scope for the introduction of new services. The postwar world sees such basically new services as health clubs and diet centers. Such developments require the emergence of new technology and new effective consumer wants. Such things take time. While new services *may* be easier to introduce than new industries, they are still far from a simple or automatic response to hard times.

We might expect that domestic service would behave much as the personal services discussed above. While comprising 7.4% of the labor force in 1870, 4.12% in 1920, and 5.4% in 1930, domestic service only accounted for 5.22% in 1940 and 2.96% in 1950. The postwar period has seen a further decline. If we compare the number of servants instead to the number of families, a more extreme picture emerges. While there were 94.1 servants for every thousand families in 1910, there were only 61.7 in 1920, 67.7 in 1930, 60.2 in 1940 and 34.4 in 1950.

This long-term decline occurs despite the fact that there is virtually no change in the ratio of average household incomes to servant wages. Some of the long-term result may be due to changing income distribution, since only those at the top of the distribution hire servants. The worsening of the income distribution during the interwar period may explain part of the slowing and reversal of this longstanding secular downturn (this being one of the few cases in which an uneven distribution has a positive employment effect). Much of the postwar experience may be due to the creation of other employment opportunities for black women, and decreased immigration in the immediate postwar period. These two groups have tended to comprise the majority of household servants.

All of the above are only partial explanations, however. The answer would appear to lie in the development of labor-saving household technology. Stigler (1956: 158) is sceptical of this explanation but the work of Cowan (1983) argues forcefully in its favor. The thrust of her work is that despite well over a century of dramatic change in household technology, the workweek of the average housewife has remained surprisingly constant. While the labor of men, children and servants has been largely displaced, the labor of women has only been redistributed. While it is not the main purpose of her book, she does conclude that the decrease in the number of servants per family in the first half of this century was basically due to the advent of a variety of new appliances (1983: 99). Indeed, she notes that many appliance purchases during this period were expressly motivated by the desire to replace servants.

We can too easily forget how labor-intensive simple household chores were before the advent of washing machines or vacuum cleaners. It is important to recognize that some of the simpler, earlier appliances had a major impact on household labor time. Throughout the 1920s and 1930s, then, the introduction of new appliances steadily decreased the demand for servants. In the 1930s, the fact that people were willing to work for much less than before did result in the hiring of another 400,000 servants. This was limited both by the decreased incomes of the time (though again the uneven distribution helps) and the new technology which made servants less desirable. If not for this new technology, hundreds of thousands more would have been able to find employment as domestic servants through the 1930s.

Distribution

Wallis and North (1986: 112) have estimated the resources used in trade as a percentage of GNP on a decadal basis for the century after 1869. As we might expect from an increasingly complex and market-oriented world, the trend is upward. From 16% in 1869, the proportion rises to 19% in 1899, stagnates for a decade, reaches 19.6% in 1919, and then falls to 18.7% in 1929. From 20.5% in 1939, the proportion rises to almost 22% in 1948 before falling to 18.25% in 1972. In the interwar period, and again postwar, there are factors at work which decrease the relative importance of trade. Organizational change (often technology driven) emerges as the most likely candidate.

All too often, economic historians concentrate on production, and neglect the equally important process of distributing goods. There have, of course, been exceptions such as Chandler's pathbreaking work. For the interwar years, however, the mainstream literature virtually ignores distribution, despite the obvious fact that significant changes occured both in organization and employment. We will focus on retailing.[24] Employment statistics are available only from 1929. There is data for previous decades on the number of dealers. These show that the number of grocery stores increased faster than population from 1870 until after World War Two. The number of restaurants grew even more rapidly. These two trends can be attributed to rising income and urbanization. Stigler has estimated that by far the dominant influence was that of urbanization. Farm families purchased about two thirds of the amount of retail goods of city dwellers. Smaller family sizes and increased female participation also contributed to the increased importance of restaurants. Labor-saving technology such as cash registers and calculating scales tempered somewhat the increase in retail employment. So also did the increased importance of durable goods; sales per worker in durables tend to be about 50% higher than for non-durables. While the long term trend was upward, there was a sharp decline in retail employment after 1929. Some of the poor showing in the 1930s can naturally be attributed to the reversal of the previous trends in income growth and rural-urban migration. As well, particular declines, such as those suffered by retailers of cars, furniture, and building materials, can be seen as results of the poor experience of those goods during the Depression. However there is much more than this going on. Indeed, the period following World War One is often referred to as the postwar marketing revolution (Clewett, 1951: 766).

Unemployment rates in trade-related services have been slightly above the economy-wide average through the postwar period (Sinclair, 1981: 26). Still, we might have suspected that many unemployed would at least try to support themselves through self-employment in retail trade in the 1930s.

In some areas, this indeed occurred, as one experienced marketer noted before a government inquiry in 1939; "A large number of people, unable to find employment in their accustomed occupations, sought a livelihood by opening retail stores. The tendency was particularly pronounced in those retail trades where the investment required is relatively low and where, in the general belief, the problems involved are not complex. Restaurants and filling stations meet these requirements particularly well, and the number of these two types of retail outlets increased greatly" (Williamson, 1963: 681). Restaurants have always been a line of business with a high rate of entry and exit. Beyond new market niches opened by the automobile, and roadside diners likely drew customers from other locations, we should not be surprised that there was not unlimited scope for net additions to the restaurant business in the 1930s. The increase we do see reflects the spreading of business over more entrepreneurs. Highway construction and urban street network changes encouraged new filling station construction in the 1920s and 1930s. Old stations were not abandoned as long as variable (operating) costs were covered. Daily sales per station fell from 327 gallons in 1929 to 225 in 1935 (while total gasoline sales rose 15%). This figure had fallen in the early 1920s as well, and profits were already low. Price wars, giveaways, and various attempts to improve station quality (including technical improvements to pumps and repair facilities) ensued.[25]

While this sort of forced worksharing, at least at the management level, was occurring in a couple of the distributive trades, employment across the sector was falling. To understand why, we must again look at forces which had been underway for some time. Chain stores in the food and variety fields were relatively unimportant until after World War One. A&P had opened its first store in 1859, but by 1900 there were only about 300 food chain stores. It was still a world full of specialized butchers and bakers, and streets clogged with pushcart peddlers; these employment opportunities would largely disappear before the 1930s. By 1910, chains had captured perhaps 20% of the grocery market. In 1912 A&P introduced an economy store with no credit or delivery, limited fixtures, and a small markup, and was rewarded with a big turnover. The sales of chains swept upwards at a breathtaking pace until 1929. From 4% of retail sales in 1919 they reach 22% in 1929 (Clewett, 1951: 766). There were 21 food chains in 1900, 62 in 1910, 180 in 1920, and 315 in 1928. By the end of the 1920s, they had captured 32% of the food market. That share would remain between 31% and 41% for the next three decades (Hampe, 1964: 277-8). In 1925 A&P added fifty stores per week.

Independent store owners responded with co-operative wholesaling and self-service stores (which lowered their labor requirements in turn). The first co-operative wholesaling setup was Red and White from 1921.

IGA followed in 1926. Co-operation and improved record keeping would lower the wholesale margin from 11% in the 1930s to 3-4% in the 1960s (Hampe, 1964: 297-307). Only after 1935 did the ratio of chain sales to total retail sales decline slightly.[26]

The chains were able to achieve much greater levels of productivity than independent retailers. For groceries, workers were twice as productive. For liquor stores they were three times as productive (after the repeal of Prohibition, of course). Overall there was an average 30% differential, measured in sales per employee. As chains had come to comprise one quarter of retail sales by 1929, employment was lowered by 7.5% as a result. Since chains were able to undersell their competitors, using dollar sales as our measure significantly underestimates the true employment effect of this organizational innovation. The major reason for the advantage of chains appears to be the larger store size they were able to sustain (Stigler, 1956: 67-72). The number of stores fell through the 1930s as volume per store rose. Advertizing was important here. The dominant facilitating factor, though, was the automobile, which allowed consumers to go beyond walking distance to purchase bulky goods.[27]

In the 1930s a new phenomenon, the supermarket, emerged. The supermarket borrowed various ideas already extant: no credit, combination stores, self-service (the first store designed for self-service was Piggly-Wiggly in 1916; by 1928 there were 2200 Piggly-Wigglys), and volume sales with low markups. Weekly sales were ten times the average for grocery stores. There were 94 in 1934, 300 in 1935, 1200 in 1936, and 3066 in 1937 (over 10,000 in 1946 and 33,000 in 1960); from the mid-1930s chains began opening supermarkets (the Depression itself may have encouraged an increased consumer preference for low price over service). Refinements such as shopping carts and shelf islands were introduced. Thus, through the 1930s, the number of retail grocery outlets decreased as one large store displaced a handful of smaller centres. A&P alone had 10,000 fewer stores in 1945 than 1929. As with chain stores, the increase in size led to savings on labor, and was largely due to the rise of the automobile. Another factor was important here, though. The essence of the supermarket was that a large amount of refrigerated and frozen goods could be made accessible to the consumer. This in turn depended on improvements in refrigeration technology. The consumer, to fully appreciate the advantages of supermarkets, needed access to trustworthy home refrigeration as well. The postwar era would see a further expansion. While there was one grocery store for every 78 households in 1940, there was only one for every 481 in 1980, despite a 38% increase in real food purchases per household (Oi, 1988: 351).

With the advent of the automobile, the custom of delivering groceries and a wide range of other goods almost ceased. It had been uncommon for women to drive horse-pulled vehicles, but it was common from the beginning for women to drive automobiles. As American families acquired automobiles in the mid-1920s, the American housewife replaced the corner store delivery boy as the final conduit of distribution (Cowan, 1983: 79-85). Along with the in-store gains in productivity wrought by chains and supermarkets, then, there were further losses in employment felt across almost all retailing categories as a result of shifting the task of delivery to the consumer. All of these influences were concentrated in the late 1920s and early 1930s.

There were other changes of lesser magnitude. Department stores (along with large grocery and drug stores) squeezed numerous small cigar, jewellery, and dry goods stores out of business. The department store revolution had begun with Filene's in Boston in 1912. Specialty stores had remained dominant till the 1930s. Department stores differed only slightly in labor productivity, though, and thus only had a very slight negative impact on employment. Only in the postwar era, with access to electronic data processing and print and electronic media advertising would chain department stores establish a clear advantage over their rivals. The mail order business, which had a 50% productivity advantage over department stores in the goods they competed in, boomed 1900-1920 and again in the 1930s, but never amounted to much more than 1% of retail sales.

Increased sales per hour can reflect a reduction in quality of service rather than an increase in productivity. Schwartzmann (1985) has tried to estimate the productivity effects of the slight increases in capital intensity and capacity utilization over the whole period 1929-1963, as well as the degree of economies of scale, and finds that all three are swamped by a decrease in the quality of employees (younger and less educated).[28] While estimating scale economies [especially] is always difficult, and his estimates of labor quality change are at variance with authors such as Denison, his results do suggest that a decline in quality of service has occurred. While consumers often prefer self-service, he argues that stores seem only to have moved to it as labor costs rose. If correct, his results would fit well with those who argue that productivity has grown more slowly in services; an increasing relative price of quality would induce consumers to demand less of it.

The fact that self-service is generally preferred by consumers, and that those with cars find it easier to "deliver" their own purchases, indicates that the decrease in service, if such exists, is not a simple reaction to rising prices. Rather, the workers displaced should be viewed in much the same light as those displaced elsewhere by technological change. Nor were all changes in the interwar period service-reducing. Free home trials of consumer durables, the extension of return privileges, free repairs, air conditioning, and concealed lighting have to be weighted against self-service, less delivery, less credit (for groceries) and packaging by manufacturers. Moreover, Schwartzmann calculates that 45% of the increase in sales per manhour was due to an increase in average transaction size; this alone generated an increase in sales per manhour of .78% per year. While various factors, including rising incomes, contributed to this in the long run, it would seem likely that the carrying capacity of the automobile (along with home refrigeration) must have been the prime source of the increase during the interwar period. As well, increased advertizing meant that retailers no longer had to provide as much advice to customers, nor was their reliability as essential (Oi, 1988: 330-1). Advertising has become a substitute for sales personnel.

Notes

1. Cornwall calculates that if residential construction had behaved in the early 1930s as it did in early postwar downturns, expenditure would have increased by $1.5 billion rather than decreasing by $4 billion. With multiplier effects, this alone could explain much of a $20 billion decline in output (1977: 218)

2. "Similar speculative developments [to the stock market] involving the inflation of capital values occurred in the real estate field. Homes, apartment houses, office buildings, and hotels were built with almost reckless abandon under the spur of promoter's profits and the ease with which securities could be sold to finance the cost of construction" (Gordon, 1961: 417).

3. High profits and new technology stimulated office building construction. Mieszkowski (1993) surveys the literature on suburbanization; while flight from urban blight has played a contributory role postwar, transport improvements were the predominant cause earlier in the century.

4. While much less organized than they would become in the postwar period, developers played a much larger role in the 1920s than ever before (Soule, 1947: 170). Self-employment always accounted for at least 30% of the work force before World War Two (Henderson, 1961: 39).

5. Agricultural products which required speedy delivery were among the first to shift modes. Livestock haulage by road became common in the 1920s; by 1932 more than half were sent that way. Milk was also often carried by truck (Schlebecker, 1975: 229). Cunningham (1929) felt that trucking had only taken away the worst business from railroads and allowed them to specialize in long-haul bulk transport. At least in terms of profitability, he felt that government regulation was a more serious problem.

6. The expense of capital goods will after all mostly comprise the cost of the labor it takes, perhaps at various stages of production, to produce them. If trucks prove less costly, this could well be due to a decrease in the net labor input required. The temporal distribution of employment could be affected significantly. While there were only 1.1 million trucks in 1920, there were 3.5 million in 1930, and these of larger capacity. Three quarters of truck haulage is estimated to have

been done by producers (Hogan, 1971: 1014-8). It is thus difficult to pinpoint the ton-mileage carried by trucks. It is estimated they handled only 4% of intercity ton miles in 1929.

7. "Rail passenger travel reached peak levels in 1919 and 1920 — levels which were not surpassed until World War II. The movement over the five decades 1889-1939 suggests a rather clear case of retardation in growth". The major loss was in short-distance travel. Especially with the settlement of California, long-distance rail travel held up (Barger, 1951: 69). Larger buses and better roads encouraged the formation of the Greyhound bus network in 1929, but buses only served 4% of intercity travel in 1929; innovations in packaged bus tours served to almost double this figure in the 1930s (Walsh, 1990).

8. The end of the war had brought overcapacity and a fall in demand which caused freight rates to fall. The government stepped in with subsidies. It also devoted efforts to river dredging etc. after 1922.

9. There had been fears at the turn of the century of an impending food shortage as supplies of natural nitrate fertilizer dwindled. Methods of isolating nitrogen from the air overcame the danger (Haber, 1971: 32). In the late nineteenth century, other fertilizers such as ammonia sulphate had become readily available as a byproduct of gas works and steel furnaces. The largest single chemical sector in 1914 was fertilizer (Haber, 1971: 175).

10. One of the main goals of the New Deal was to increase farm incomes by restricting output. Before the Supreme Court ruled the Agricultural Adjustment Act unconstitutional in 1936, farmer's incomes had been increased by nearly 50%. Later efforts fared poorly in comparison (Rasmussen, 1982: 122). Other countries also pursued output control policies after 1933 (Garraty, 1986: ch. 3).

11. The movement back to the farm is observed also in France, Canada and Colombia. Some governments even encouraged it through subsidies (Garraty, 1986: 123). Fearon (1987) notes that there were 2.5 million farms of less than 50 acres in 1925. While some were profitable truck farms, most were very poor.

12. It was not changes in the capital/labor ratio but in the composition of capital which increased productivity. Efforts toward soil conservation and improved rotations which spread work temporally also helped. Changes in farm organization, including diversification, had only a small effect on productivity and none on employment. On the other hand, diminishing returns, expansion to poorer land, soil exhaustion, and new pests all acted to decrease productivity (Hopkins, 1941: 7, 51-2, 112).

13. The difficulty in defining services stems from the intangibility of output. Different authors define it differently. We will include government in our discussion here, but discuss construction, utilities, and transport separately.

14. "Service employment is far less sensitive to the business cycle than goods employment because compensation tends to be more flexible in service industries. Thus, the service share of employment rises especially rapidly in recession. Similar influences arise when major structural changes depress the output of goods-producing industries for longer duration" (Leveson, 1985: 94).

15. The U.S. government in 1980 estimated that 25% of GNP was services used as inputs by goods-producing industries. This was more than manufacturing itself. Not all were counted as services. 30% of those classified as industrial workers actually provide services. Contracting out can occur because monitoring costs increase with firm size, specialist firms are better able to keep up with changes in knowhow, improved communications make it easier, or to escape union control (Grubel, 1989: 195-6).

16. Kendrick has argued that there is no inherent reason why productivity growth should be harder to measure for services. The quality of goods changes over time as well. He estimates about a 1% per year differential in total factor productivity growth (1.3% services, 2.2% goods) 1948-1981. Others would argue that, at the least, the practical problems of productivity measurement in services are still unsolved.

17. The service sector has in many ways aided productivity growth in manufacturing. Engineering, scientific, computing, and advertizing skills have obvious effects. The development of a more reliable and speedy payments mechanism has also aided industrial productivity. "The producer service industries are one of the main ways in which human and knowledge capital accumulation, increased specialization, and roundaboutness manifest themselves in a growing economy" (Grubel, 1989: 244).

18. Stanback estimates public sector employment as 9% of the total in 1929, 17% in 1939, 14% in 1948, and levelling off around 20% in the 1960s (1981: 14). Increased spending, by these figures, appears to be far from an inconsequential part of the employment recovery in the late 1930s.

19. Wolff (1987: 165) argues that the "unproductive" portion of state and local government expenditure, that concerned with allocation rather than production, was the fastest growing (6% per year) through the postwar period. While only 25% of government employment in 1947 was unproductive, 38% was by 1976. He feels a similar process is visible in producer services, and blames these trends on oligopolization.

20. Stanback's non-profit sector largely comprises health and educational services. Roughly 2% of employment in the interwar period, it grew steadily to 6.34% in 1977. Neither the institutional structure nor the trained personnel could be put in place overnight. Stanback attributes the rise in non-profit services to society's increased desire for social equality, and producers' increased demand for healthier better-educated workers (1981: 14, 20).

21. Wallis and North (1986: 113) look at resource use by sector and estimate that finance, insurance, and real estate absorbed 8.28% of GNP in 1920 (up from 4.19% in 1869 and 7.96 in 1899), 12.61% in 1930, and 9.88% in 1940. The postwar period sees a slight rise from 10.45% in 1950 to 10.61% in 1958 to 12.15% in 1972.

22. Modern estimates of the elasticity of demand for services vary, in part due to different definitions of the sector. Grubel finds an elasticity of virtually one in cross-section but only .6 in time series. Green and MacKinnon (1990: 367) suggest that, since those who lost their jobs were mainly low income earners, the demand for [many] services was only slightly affected.

23. Commercial laundries boomed in the 1920s, then declined in the 1930s. They would be almost destroyed in the postwar period by the electric washing machine. The rise in the 1920s was likely in large part due to the automobile.

24. We only have data on wholesaling from 1929. Moreover, much wholesaling is done either by producers or retailers, and thus it is not at all clear what the numbers tell us. It appears that independent wholesalers lost ground to both in the decades after 1929, due in part to the rise of chain stores, and increased advertising. For our purposes it is safest to assume that wholesaling largely follows the experience of manufacturing in the interwar period (Stigler, 1956: 139-48).

25. One potential labor-saving innovation, the self-serve station, was experimented with in the 1930s, but this ran afoul of fire regulations. With the rise of the specialized filling station in the 1920s, the large refiners had moved into distribution; for various reasons including capital costs, monitoring, incentives and government regulation, they had come to prefer franchising to direct ownership.

26. There were also state laws passed which penalized chains. Huge battles erupted in Congress over the same issue. This arguably was due not just to pressure from independents but a recognition of the employment effect.

27. Copeland (1929) attributed the rise of chains, the eclipse of village stores, and the decline in mail order sales in the 1920s largely to the automobile. Hopkins (1941: 65) estimated that between one-quarter and one-third of farmers changed their principal trading centre to one farther away after acquiring their first automobile.

28. If we assume that retail services are proportional to sales, we can derive the result that productivity grew more slowly than in manufacturing. For the 1959-1979 period, Oi (1988: 339) finds that retailing productivity grew .47% per year while manufacturing productivity grew 2.07% per year. Barger, noting the many changes in retail practice over time, doubts that sales, or sales weighted by margin, are reliable measures of the output of retailers (1955: 29).

12

Summary Of Sectoral Studies

We have seen considerable evidence that market saturation characterized not just the automobile market in the late 1920s, but also a number of domestic appliances, radio, and construction. Decline in these sectors would not only be felt in other goods sectors which supplied their inputs, such as iron, wood, rubber, and glass, but also in the important area of business services. If we accept that limited investment opportunities led to excess savings in the late 1920s, then the rise and decline of the financial services sector can be attributed to the same forces. The existence of inelastic demand can be discerned in sectors other than those listed above; textile producers for example, were unable to expand output by slashing prices.

Market saturation depends on income distribution. Widespread undernourishment could not be translated into effective demand for agriculture. Farmers themselves were limited in their ability to purchase from others by slumping incomes. A more even income distribution must at least have limited the scope of the Depression.

The decline in investment can be understood in two ways. First there were reasons why both building and machinery purchases were concentrated in the 1920s. The rise of the automobile had radically changed locational decision making, and thus induced a surge of residential, commercial and industrial construction. We have also seen that such diverse areas as railroad construction, pipelines, and oil well drilling were destined to see decreased investment levels. The coming together of various technical strands created a radically new generation of machine tools in the early 1920s; improvements over the next decades could be incorporated into these machines. In the absence of new product or process innovation, then, there was little scope for investment in the 1930s. We have surveyed the products which would evince growth potential in the late 1930s and the postwar period, and seen that only a couple of these (e.g. the refrigerator) were technologically ready in 1930. The steel industry provides an example of the investment-inducing potential of some process technology. While investment slumped in most areas of iron manufacture, it surged in the area of rolling mills because of radically new process technology. The oil industry also saw much process-induced investment, but the timing of innovation ensured that this largely fell outside the early 1930s. Most process innovation of the 1930s could be readily incorporated into the existing capital stock with a small investment outlay.

Once market saturation in some sectors and declining investment across the economy had started the downturn, the stage was set for massive dishoarding of labor. The productivity advances of the 1920s were felt quite unevenly across industry. Industries in which both output and productivity were expanding failed to provide much job creation in the 1920s. Many other sectors, in which output was relatively stagnant in the 1920s, did not release all of their surplus labor. Agriculture was one of these, but there the institutional setup prevented a massive bloodletting in the 1930s (though hired hands were replaced with unemployed family members). Many textile manufacturers, especially in the north, stumbled through the 1920s, and then collapsed in the 1930s. Mine owners, faced with huge startup costs, kept their mines operating in the 1920s, but often closed them in the 1930s.

Movements in labor productivity might simply reflect changes in output or employment, rather than inducing these. We have striven, therefore, to describe the technological roots of both the very high levels of productivity advance in the 1920s, and continued advance in the face of surplus labor in the 1930s. Electrification, assembly lines, and continuous processing were behind most, though not all, of these advances. It is not just in manufacturing that technology had its effects; farmers, miners, railroaders, clerical workers, delivery men, and domestic servants, among others, found themselves replaced by machines. We have also seen indirect sources of labor-saving that still had a technological base. Thus, paper towels replaced cloth towels, and cars replaced railroads. As well, productivity advances across the economy had a further effect by decreasing per capita energy consumption.

Structural unemployment was often induced by innovation. The most obvious cases were miners and textile workers, who suddenly found themselves unemployed in one-industry towns. The phenomenon affected a much more diverse group, though, including vaudeville performers and musicians.

Workers displaced in the early 1930s, even if not facing structural difficulties in job seeking, had virtually no place to go. There was almost no new product growth in manufacturing. Mining was contracting. The market for construction had been saturated in the mid-1920s. The agricultural sector was faced with stagnant demand and increasing productivity; some small scope existed for a return to subsistence farming by the desperate.

The less capital intensive service sector might seem the obvious home for displaced workers. Business services naturally suffered with the industries they served. Government could only expand appreciably as public attitudes were transformed. Servants, clerical workers, and delivery men were not the only service workers to be replaced by machines. Employment of barbers was limited by both tastes and the introduction of the safety razor. The retail sector, especially restaurants and gas stations, provided some scope for self-employment, but at the price of severely depressing incomes. The rise of chains and increasing store size severely limited employment opportunities in the retail sector. The specialty stores and pushcart vendors whose numbers might have swelled with the unemployed were instead squeezed out of business.

Du Pont could claim in 1937 that 40% of its sales came from products unknown in 1929. Airplanes and food processing were other areas in which new products emerged as the 1930s progressed. Why was the timing of innovation such as it was? Virtually every individual innovation in our period, whether product or process, depended on previous developments in electronics, chemistry, and/or internal combustion. All three strands encouraged process innovation in the 1920s, and all three engendered new products before and after but hardly at all during the late 1920s and early 1930s. With internal combustion, we speak simply of the assembly line, the automobile, and the airplane. With electricity, the 1920s saw widespread adoption of electrical machinery. The preceding decades had seen the rise and fall of numerous appliances as well as radio. Later decades would witness television, computers, and a range of goods made possible by the transistor.

Despite the diversity of chemical application in the modern world, the same time path was evinced within that stream of innovation. Continuous processing, which originated in chemicals, spread to other industries, notably oil and food processing. Cost savings also occured as fixation reduced the cost of nitrogen, and electrolysis the cost of alkalis and chlorine. Early new products played their greatest role in supporting developments elsewhere. Both automobiles and radios were heavily dependant on plastics for components. The most notable new product of the 1920s was cellophane. The fastest growing chemical sector of the period was rayon, which replaced more jobs elsewhere than it created. Major new product innovations would await the 1930s (in part due to scientific advance in the study of the atom, temperature, and pressure). Nylon, neoprene, much improved plastics, teflon, vitamins, and a range of new pharmaceuticals emerged in that decade.

While the timing of a particular innovation can never be perfectly predicted beforehand, there is a certain logic to the process. Thus, a generation of simple electrical appliances and machines naturally soon followed the solution of the problems of electricity generation and transmission. More complex electrical devices came later as more sophisticated electronic components were introduced. The development of modern pharmaceuticals followed a logical path through experiments with dyes to success with sulphanomides, to success with antibiotics. Likewise, vitamins could not appear until careful experiment had first identified the physiological effects of trace elements. Simultaneous discoveries, as occurred with the vacuum tube and almost with the transistor, indicate that some points in time (for both technical and social reasons perhaps) are conducive to particular research agendas.

Nevertheless, the emergence of the industrial research laboratory must be seen as having an important impact on the direction of research, at least for some firms and industries. Bell focussed on pressing telephone problems until 1925 when Bell Labs was formed and given a broader mandate. Du Pont had emphasized development and process technology, but shifted toward basic research and new product development in the 1930s. Pharmaceutical companies focussed on testing early on; the rash of new laboratory construction in the 1930s reflected a commitment to basic research. More contact with universities in the 1930s encouraged a greater focus on basic research throughout industry. The experience of the jet engine or the computer indicates that many major product innovations would have been possible years earlier if the research had been adequately funded. The interwar success with synthetics in Germany provides more certain proof of this.

World War One had greatly stimulated technical developments in a couple of areas, notably aeronautics and chemicals. Military funding of research would continue to play an important role for aircraft in the next decades. Still, the effect on civilian technology was much less than that which would result from the Second World War. The existing body of technology naturally limited the scope of research in each war. Social attitudes conditioned the degree to which government interest would be sustained when war ended.

The long-term nature of innovation has been clear throughout. This alone should create scepticism of Schumpeterian or Menschian explanations. Decades of gradual development along many lines yielded the DC-3, the transistor, the movie, and the automobile. Research may be hastened or slowed at certain times, but no major innovation is entirely a response to the influences of any short (less than a decade) time period. Generally, major innovations are the combination of quite different lines of research, often focussed on different ends. Thus, the DC-3 required advances in engines, fuels, materials, instruments and designs. The automobile had depended on improved gasoline, lacquer, alloys, lubricants, antifreeze, and plastic. Alternatively, electrochemistry produced better graphite which improved the efficiency of electrical apparatus, better and cheaper fertilizer, acetylene for use in plastics, and aluminium which helped the isolation of such metals as tungsten which were to have farreaching applications. The automobile industry and the machinery industry mutually reinforced each other's development.

Establishing Plausibility: An Exercise in Addition

When more and more people are thrown out of work, unemployment results

Calvin Coolidge

It would be ludicrous to pretend that the employment losses by industry which resulted from technological change and market saturation could be estimated with precision. We can, however, arrive at ballpark estimates. Our main purpose here is to show that we are plausibly able to explain much of Depression-era unemployment experience.

We apply the most powerful mathematical technique of all, simple addition. Those who might question the approach must attack our numbers, not our technique. Only sector by sector can the quantitative importance of this line of argument be destroyed. Again, our present purpose is to establish plausibility. This entire book will have succeeded in its purpose if the likelihood that much of the Depression was caused by the forces we have outlined is accepted, and debate moves on to the more detailed quantification of the effects. Since traditional methodological approaches have been unable to explain the Depression to the satisfaction of the profession, a novel methodology such as this would seem to be warranted. Industry-level analysis which is based on qualitative as well as quantitative information, and which relies on the interplay of scholarly argument, should provide a more compelling approach to understanding the interwar period.

Automobile industry employment more than doubled from 212,000 in 1921 to 447,000 in 1929. As output rose much faster, we can already see the effect of productivity improvements. Employment tumbled to 285,000 in 1931 and 243,000 in 1933. It is obvious that the auto industry fell both faster and farther than the rest of the economy. Faced with market saturation, and continued process improvement, it is likely that most of the 200,000 difference in employment 1929-1933 would have happened anyway. The decline to 1931, especially, can largely be seen as the result of saturation, while continued low employment to 1933 must be viewed as largely a result of low aggregate demand. The loss of some 150,000 jobs in the auto industry itself is only a small part of the story though. The auto industry indirectly employed one eighth of workers in manufacturing, or about 1.3 million. If similar proportions held in input industries, then some 450,000 workers in total were destined to lose their jobs in the early 1930s, even without the Depression.[1]

Electrical machinery employment (which includes appliances) fell from 328,000 in 1929 to 214,000 in 1931 before climbing back to 305,000 in 1933 and 432,000 in 1935. The downturn can largely be ascribed to saturation in industrial machinery, radio, and other appliances, and the early recovery attributed to new products such as the refrigerator. The upward trend from 1931 clearly indicates that expansion was possible if new products were available. Attributing just half the 1929-1931 decline to technological forces yields over 50,000 jobs, and the effects on input sectors at least as many more. Employment in utilities fell by over 80,000 1929-1933, and most of this, some 50,000 jobs, was due to productivity increase.

The decline of 830,000 in construction employment 1929-1933 (from 2,319,000) continued a trend begun two to three years earlier; 90,000 had already lost jobs by 1929. We could conservatively attribute half of the decline to secular forces; other authors have made much stronger claims. The greater than proportional decline in input sectors such as wood and stone and clay may well reflect a lag in adjustment to demand from the construction sector. As these sectors were characterized by small producers and a tendency toward cutthroat competition this result is likely. Employment in stone and clay, which had been about 280,000 through the mid-1920s, fell to 260,000 in 1929, 172,000 in 1931, and 125,000 in 1933. Recovery to 166,000 in 1935 could be largely attributed to expanded road building and WPA expenditure. In the crushed stone sector, we saw that productivity advance was responsible for tens of thousands of job losses as well. As many as 100,000, even in this small sector, lost their jobs in the early 1930s because of process improvement and the decline in construction. The experience of lumber is even more sobering. Wood had through the 1920s faced increased competition from other materials. Lumber employment had already tumbled from 932,000 in 1923 to 867,000 in 1929, but then collapsed to 509,000 in 1931 and 454,000 in 1933. Wood, already replaced to a large extent in car production, suffered from that source as well, though to a much lesser extent than steel, glass, or rubber manufacturers. It also suffered as furniture production collapsed (employment there fell from 193,000 in 1929 to 127,000 in 1931 and 105,000 in 1933); this collapse was closely connected with the decline in house construction. First degree unemployment in lumber and furniture which can be traced to autonomous declines in construction, autos etc. must amount to at least 100,000, and second degree unemployment another 200,000.

We would not want to double count. The declines in employment in iron and steel (225,000 of 880,000), rubber (43,000 of 149,000), leather (36,000 of 318,000), and glass (18,000 of 67,000) are in line with what we would expect from the experience of the sectors they served. All, though, concentrated their decline in the early years of the downturn; glass employment stabilized 1931-1933, and the other three sectors rose. This likely reflected both labor dishoarding and the search for other outlets. Continued process innovation inflated considerably the job losses which resulted from declines in auto and other production. The steady fall in non-ferrous employment from 313,000 in 1929 to 209,000 in 1931 and 155,000 in 1933 provides a less extreme example of the leading role played by these sectors in the Depression.

It is not double-counting to add the effects on the transport sector. Transport and public utilities laid off 1,242,000 of 4,071,000 workers 1929-1931. The bulk of freight tonnage came from sectors which contracted severely in the 1920s and 1930s: minerals, coal, lumber, stone. Mines and quarries alone provided over half of railroad traffic through the period. On this basis alone, we could safely attribute at least half of the employment decline in transport to the forces outlined elsewhere. As well, employment decline can be traced to the completion of the national rail network, the time path of pipeline construction, self-transport of passengers in automobiles, and widespread labor-saving technology (much of the rest could be due to government regulation). On top of contributing to chronic unemployment in the 1920s, this sector, on its own and through reacting to autonomous declines elsewhere, easily added another 600,000 to the unemployment rolls. Further, employment in railroad equipment production had fluctuated between 38,000 and 80,000 in the 1920s but fell to 18,000 in 1931 and 14,000 in 1933. This decline was aided by the rise of trucking competition (recovery later in the decade is largely attributable to the diesel). Employment in shipbuilding fell from 55,000 in 1929 to 30,000 in 1933.

A negative secular trend in energy consumption was severely exacerbated by the shift from coal to oil in the interwar period. Oil output expanded through the period, but employment barely increased in the 1920s and contracted in the 1930s. The coal sector contracted sharply. As elsewhere in mining, considerable productivity improvement exacerbated the effect of output decline. All told, a hundred thousand would be laid off in the late 1920s, and another 340,000 in the early 1930s. The vast bulk of these job losses must be viewed as a cause rather than result of the Depression.

The best known new product of the late 1920s, the talking motion picture, likely had a negative employment impact. Hiring of projectionists more than compensated for the loss of jobs in theatrical orchestras (ignoring structural unemployment difficulties). The dominant effect, though, was the demise of vaudeville. Centralized production of entertainment decreased total labor input. Advances in still photography were too late to encourage expansion in camera or film production.

Employment in textiles, after rising from 1.5 million in 1921 to 1.7 million in 1929, fell to 1.42 million in 1931 before rising to 1.47 million in 1933 and 1.65 million in 1935. The sharp fall to 1931, and recovery thereafter, indicate that some dishoarding of labor had occurred. Real consumer expenditure on textiles fell as much 1931-1933 as it had 1929-1931. The shift from New England to the South, overproduction, the rise of rayon, and a variety of other technological improvements point to there having been considerable excess labor in 1929. Well over 100,000 workers, at the very least, must have been let go much earlier than would otherwise have been the case. In the early postwar period textile employment also fell but this was more than counteracted by rising employment in apparel. Even with lower incomes, there was perhaps some scope for employment expansion there in the interwar period.

Agriculture had released a million workers through the 1920s, and would have released hundreds of thousands more if there were urban job opportunities. Employment in agriculture fell 360,000 in the early 1930s; much of this, though, involved the replacement of hired hands by family members. Very little of this can be attributed to the Depression itself, at least directly, for demand for food held up very well. To the extent that the Depression in its early days made the government less willing to stockpile surplus as it had in the 1920s, an indirect case might be made.

In agriculture, unlike other sectors, the Depression likely improved the employment figures, as those who lost jobs elsewhere moved back to the farm. This phenomenon became increasingly noticeable as the 1930s progressed. A glance at employment totals after World War Two shows that another million workers could well have been shed in the 1930s if there had been any place for them to go. While the agricultural sector thus served to make unemployment totals look better than otherwise, this was done by further depressing average farm incomes. Kindleberger has argued that hundreds of thousands of jobs were lost elsewhere due to declining farm incomes over the interwar period. The employment effects of agricultural decline are much harder to pin down than is the case for other sectors. In the interest of always biasing our results downward, we might attribute a mere 100,000 to this source.

There are a number of sectors which show dramatic postwar employment growth but are relatively stagnant through the interwar period. Paper products employed about 200,000 through the 1920s and 1930s but 5-600,000 postwar. Printing and publishing employed about 300,000 through the 1920s and 1930s, but from 700,000 to well over a million postwar. Food processing employed 6-700,000 interwar, but roughly twice that in the postwar period. Pharmaceuticals employed 25,000 interwar, but this rose from 100,000 to 170,000 postwar. Oil refining fluctuated about 70,000 interwar, but was double that in the 1950s. Aircraft manufacture employed hundreds of thousands postwar, but under 15,000 through the 1920s and 1930s. These sectors are indirectly responsible for the Depression; it is truly said that if you are not part of the solution you are part of the problem.

The decline of 212,000 (from 1,591,000) in financial services is almost entirely a fallout from the stock market crash. This provides the most noticeable way in which speculation served to artificially support an ailing economy in the late 1920s, and concentrate job losses in the early 1930s. Much of the 529,000 jobs lost in other services reflected the natural decline in business services associated with declining industries.

Employment in wholesale and retail trade fell from 8,376,000 in 1929 to 6,958,000 in 1933. Much of this can be attributed directly to market saturation in certain product lines. Automobile sales alone comprised several per cent of total retail sales. Productivity increases (chains led to a 30% saving in sales per employee) had caused job losses in the 1920s, and likely considerable dishoarding as well in the early 1930s. Surely, 100,000 jobs at least were lost to each of these forces, and likely many times more. The rest of the job losses in distribution resulted from the natural working of the multiplier mechanism.

Other elements of the service sector would prove able to absorb workers, but this absorption would take time, and only in the late 1930s would the sector become a net job creator. We need not rehash here the reasons why every sector had limited absorptive capabilities in the early 1930s. The simple fact is that the workers laid off had nowhere to go.

For unemployment of the first degree we have, then, over 2.5 million job losses accounted for (autos 150,000, inputs 300,000, electric machinery and appliances 50,000, inputs 50,000, utilities 50,000, construction 400,000, stone and clay 100,000, lumber and furniture 100,000, transport 600,000, transport equipment 30,000, mining 300,000, agriculture 100,000, and services 400,000). We can account for at least another 400,000 unemployed of the second degree just in lumber, textiles, and services. Even without exhaustively focussing on every subsector of the American economy, we can account for virtually 3 million job losses. The fact that employment fell much further than output in the downturn implies that we have seriously underestimated the degree of labor dishoarding. Most of the sectors we have focussed on had experienced the bulk of their decline by 1931. This indicates that these are correctly viewed as the leading sectors of the downturn. In the absence of sectors prepared to substantially expand employment in 1929, we have isolated the recipe for disaster. A multiplier effect of well under four would suffice to generate the standard unemployment figure of some 13 million at the trough of the Depression (and remember that millions were already unemployed in 1929).

Recovery

Economists have tended to focus their attention on the initial downturn. When they do look at the recovery, they tend to wonder why it was so slow. Perhaps, though, we can ask why it happened at all. An economy that could suffer as it did 1929-1933 seems at first glance capable of even greater horror. Certainly, aggregate expectations were no longer so salubrious by 1933. Never having seen a Depression so severe, prognosticators were not at all sure where it might end.

We have not argued that equilibrating mechanisms are nonexistent, just that they are much weaker than mainstream theory (theory unable to come to grips with the Depression downturn) predicts. By 1933, the forces of decline had largely run their course. Net investment was negative in 1932. Consumer durables purchases approached zero (with the exception of new products such as refrigerators and radios). The multiplier effect had had four years in which to operate. In 1933 then, even weak regenerative forces could dominate aggregate movements. Moreover, both capital goods (even if unused in the interim) and consumer durables increasingly deserved replacement. Automobiles employment rose from 243,614 in 1933 to 387,801 in 1935. Electrical machinery and appliance employment rose from 305,040 to 432,896, (214,734 in 1931) and was one of only two sectors with higher employment in 1935 than 1929. Of industries with over 100,000 workers, these two displayed the greatest relative growth. Gross private domestic investment grew steadily from .9 billion in 1932 to 11.9 billion in 1937.[2]

We shall not enter into the debate over the role of the New Deal in the recovery. It is certainly likely that some policies, such as agricultural price supports (though these largely transferred income from workers), and some elements of the NIRA, improved the income distribution. It is noteworthy that farm equipment was the other sector in which employment was higher in 1935 than 1929 (41,663 in 1929, 19,264 in 1933, 52,866 in 1935). Employment relief programs, at least late in the decade, should have had some expansionary effect. Rural electrification and dam building could have had major stimulative effects in some regions. By modern standards, though, fiscal policy was hardly expansionary, especially in 1933. Moreover, some New Deal regulations may well have slowed recovery. The general verdict appears to be that the New Deal, while providing important direct employment relief, was at best a minor factor in recovery.

Discussion of recovery can not neglect the service sector. While much of the 1.7 million increase in employment in trade 1933-1939, the 150,000 increase in finance, insurance, and real estate and the nearly 700,000 in selected services were results of increases elsewhere, it is clear that services grew much faster than the rest of the economy. Including the 700,000 new jobs in government, services was responsible for well over half the job creation 1933-1939. This sector did prove to have substantial absorptive capacity, but could not display this overnight.

Notes

1. We could, of course, trace these effects through other sectors by the use of input-output tables. Von Tunzelman (1982), studying the effect of new industries on total output in interwar Britain found that his results were very similar whether he employed input-output tables or simply looked at their share of final demand. We follow the latter method here.

2. Bernstein (1987: 51) has calculated net investment in 1937 as a proportion of 1929. Chemicals (369%), stone, glass, and clay (850%) petroleum and coal (132%) tobacco (130%) and food (178%) displayed increases while figures for apparel (33.9%), rubber and plastic (22%) transport equipment (35%) and leather (19%) remained very low.

13

International and Regional Comparisons

In Chapter 12 we summarized the industry level evidence in favor of our view of the Depression. There are other ways in which economic activity can be disaggregated, however. One of these is geographically. The main thrust of this chapter is to show that the geographic experience of the Depression, both internationally and internally, is poorly explained by existing theories, and that our approach is able to cast much light on the question of why some areas suffered more than others. At the end of the chapter, we turn our attention briefly to gender and racial differences in employment experience, and again conclude that these can only be understood within a sectoral approach such as ours.

The Depression Abroad

I have elsewhere (1991) argued that comparative research is undertaken too little in economic history. The mass of technological and institutional detail which has had to be dealt with has prevented such an approach here. This work would soon have become unmanageably large if we had not concentrated on the United States. International comparisons have from time to time been included. The ability to handle the differential national experience of the interwar period must be the final test of any attempt to explain the Great Depression. It is possible, of course, that different forces were behind the Depression in different countries, but this would be an incredible coincidence. A cursory view of the literature on other developed countries shows that the approach taken here can be applied outside of the United States.

For some nations, the explanation is relatively straightforward. Those who relied to a large extent on exporting raw materials to the developed world were destined to suffer as the United States and others cut back severely on imports and investment funds. While these trade and financial flows might be small relative to the developed country's GNP they could be huge with respect to the less developed country's economy.

Stripped of the foreign exchange with which they had previously purchased manufactured goods, and operating in a world with little opportunity for international borrowing, these nations were forced in upon themselves. They lost not only the gains from trade but suffered the adjustment costs of shifting resources toward import-substitution industrialization. Many Latin American countries at least accelerated their trend toward inward-looking policies in response to the shock of the 1930s (Thorp, 1984: 2). Industrialization was pushed for its own sake, a fear of urban unemployment, and a recognition of the dangers of relying on primary product exports (Drummond, 1972: 232). While the terms-of-trade explanation of long waves hypothesizes that inelastic supply conditions in the developing world would force industrial economies into decline, the chain of causation appears to have worked in the opposite direction in the 1930s. Economic crisis in the developed countries drove them to reduce their imports from LDCs.

The effects of decreased demand were severely exacerbated by institutional changes.[1] World trade in foodstuffs fell 25% 1929-1937; this must be seen as primarily a result of trade restrictions (encouraged by world-wide productively improvements). Livestock, grain, and sugar were the commodities most affected. Countries outside regions of imperial preference suffered the most. Even within the Empire, though, "When prices and output collapsed in the metropolitan areas, the effect in the dependent areas of the Empire was immediate, varying in intensity from painful to catastrophic" (Drummond, 1972: 212). Canada, like the United States, first tried stockpiling, and then resorted to subsidies. Argentina and Australia could not afford to aid their farmers as much; Australia only introduced subsidies in 1938. The late 1930s would see some success in reciprocal trade barrier removal, but this would largely aid developed countries who both imported and exported.

Interestingly, most of the Third World began its recovery from the Depression long before the developed world was able to re-establish previous levels of import demand (the terms of trade fell by 30% in the early 1930s and recovered slowly) or foreign investment (Thorp, 1984: 12). Government spending, devaluation, and shifts in income distribution may have contributed to expansion (Thorp, 1984: 2-3). Yet if we accept that the Depression was a structural crisis of a particular stage of development, it is no longer a surprise that countries which were relatively backward fared comparatively well in the mid-1930s (and aided recovery elsewhere slightly by increasing their demand for developed country exports). The fact that the Third World recovered quickly despite the Depression makes it much more difficult to attribute the Depression in other developed countries entirely to the United States.

Why do other developed countries suffer in the 1930s? Given the limited trade links (trade was only 5% of United States GNP in the 1920s, versus 10% in the 1980s), attention has naturally focused on financial markets. American foreign investment fell in the late 1920s. This has been alleged to have induced bank failures and economic contraction in Europe. Temin, looking at the case of Germany, found weak links between the balance of payments and the credit market, and between credit contraction and GNP. He concluded, "The European Depression and the currency crisis that was a part of it therefore was due largely to events outside the United States" (1976: 159). More recently (1993), he has suggested that (following Eichengreen, 1992) demand shocks were transmitted through a rigid system of exchange rates (governments chose deflation over devaluation), and that financial panics could be transmitted internationally. As with the Keynesian/monetary debate, there is still much room for independent causes.

One potential transmission mechanism is of particular interest to us. Kindleberger (1973) notes that in the nineteenth century when Britain was the leading industrial nation it had expanded its foreign investment at the same time that it had restricted imports. That is, when the domestic economy was booming, investment was focussed there, but outlets were sought abroad during domestic recessions. This helped cushion other countries from British economic cycles. The United States in the interwar period cut back on foreign investment and imports simultaneously. This, Kindleberger feels, was enough to cause a severe depression on its own (1973: 306). He suggests that investment in the United States must be supply-driven rather than demand-driven, and attributes this to the country's innocence as an international lender. We can see this, again, as evidence of there having been a shortage of good domestic investment opportunities in the prosperous 1920s. Technological forces not only crippled the American economy, but caused it to decrease both foreign investment and import demand at the same time, and thus impose a demand shock on other countries as well.

If it is impossible to find a suitably powerful transmission mechanism, then it is likely that the same causes were at work in other developed countries. Some writers have noted that the European economies may have been inherently weak, and thus relatively small external shocks could have major effects. Aldcroft (1977) has highlighted the role of World War One and the treaties which followed in making the European economy unstable. For the new countries of Eastern Europe,[2] and perhaps also Germany, this is certainly creditable, but it would seem an unduly harsh judgment for, say, France in the 1930s. Government officials in France, Britain, and Germany in the late 1920s felt that they had overcome war-related problems and were on the way to prosperity (see Garraty, 1986: 29-31). Temin

TABLE 13.1 Average Unemployment Rates, 1921-29, 1930-38, 1971-79, 1980-85

	Galenson & Zellner		Maddison		OECD	
	1921-9	1930-8	1921-9	1930-8	1971-9	1980-5
Australia	8.1	17.8	-	-	4.1	7.6
Belgium	2.4	14.0	1.5	8.7	5.0	12.2
Canada	5.5	18.5	3.5	13.3	6.8	9.9
Denmark	18.7	21.9	4.5	6.6	-	-
France	3.8	10.2	-	-	3.9	8.3
Germany	9.2	21.8	4.0	8.8	2.4	6.4
Netherlands	8.3	24.3	2.4	8.7	3.9	11.1
Norway	16.8	26.6	-	-	1.7	2.5
Sweden	14.2	16.8	3.4	5.6	2.1	2.8
UK	12.0	15.4	6.8	9.8	4.5	10.9
US	7.7	26.1	4.9	18.2	6.2	7.7

Source: Eichengreen and Hatton, 1988, 9. Reprinted by permission of Kluwer Academic Publishers.

has argued that there is no evidence that the interwar economy was unstable; only the inability to uncover a large shock has caused scholars to consider it so (1990: 6). It is worth noting that some European countries had entered a recession before the United States. Unemployment rates exceeding 10% were seen in Germany and Scandinavia in the late 1920s. While arguments can be made that this was partly due to the earlier withdrawal of American capital, and that minor recession was turned into imported Depression in 1929, it is still evident that there were some independent recessionary tendencies in Europe (Aldcroft, 1977: 201). We would expect these to be similar to the forces operating in the United States.

A number of countries were close behind the United States technologically, and were thus in a similar situation, though generally at a somewhat lower income level (and thus we would expect that consumption would fall faster in the United States than elsewhere; this is what Romer (1993) finds). We would expect that those with the greatest similarity would suffer the most. Canada suffered two ways: as a country whose industrial structure looks much like that of the United States, and as a country which exported a variety of raw materials to other developed countries. Germany, the second most technologically advanced nation, also suffered greatly. France, the largest automobile producer in Europe (Citroen having quickly copied American mass production techniques) fared poorly. Britain, which had lagged in the 1920s, and where new industries had not boomed spectacularly and thus saturated the market for consumer durables, did relatively well (relative to previous experience and the experience of others) in the 1930s. Automobile production continued to expand in the 1920s. Across western Europe those countries which had boomed less in the 1920s suffered milder slumps and better recoveries (Jaeger, 1972).

Other developed countries would be expected to suffer along with the United States from the lack of new product innovation. Yet they had clearly not achieved as much market penetration in terms of the existing product mix. The simple fact that 85% of automobiles were produced in the United States in the 1920s is ample evidence of that. In the nineteenth century, followers such as Germany continued to experience rapid growth while the leader, England, stumbled in the 1870s. It might seem, then, that the 1930s provided an opportunity for European nations to close some of the gap between themselves and the United States (as Japan would do). This only occurred to a limited extent; while American industrial production fell more than elsewhere, it recovered at the same rate on average (Bernstein, 1987: 29-30). Market saturation, however, is not an absolute concept, but depends on the relationship between production and effective demand.

Svennilson noted that European nations had not proceeded as far as the United States in terms of the secular change in consumption patterns. They were not ready to absorb large quantities of some of the new goods. Thus, on the eve of World War One there was a large technological backlog in Europe. "Indeed the unused stock of new techniques and innovation on which post-war progress could be built was very large" (1954: 46). This involved both product and process technology. Electrification, the use of oil, and the use of scrap were adopted quickly. This hurt the large coal and steel industries. As well, textiles were suffering from foreign competition and shipbuilding from overcapacity. New products such as cars, radios, rayon, and cement did expand, but not fast enough to pull workers away from the stagnating industries. Sluggish real income growth meant that cars and appliances spread much less rapidly than in the United States. Svennilson attributed sluggish structural transformation to such factors as the effects on firms' financial position of inflation and deflation, collusion, unions, and government intervention (1954: 34-5). To a large extent, though, it must simply reflect the limited market opportunity for the new products. In any case, he recognized that with many stagnating industries, and few growing industries, there was a tendency for the former to hoard labor.[3]

The experience of the 1920s thus bears some similarity to that in the United States. Process technology was quickly adopted. New products were not able to absorb labor rendered redundant in other sectors. The fact that autos and other new products did not take off as across the Atlantic meant that income growth was slower in Europe (causation ran both ways). In some countries, such as Britain, unemployment levels were much higher than the American counterpart as well. The 1930s could be correspondingly less severe, as auto production continued to expand.

There were other common characteristics. House construction surged in the 1920s to replace wartime destruction; this led to a natural decline in activity in the 1930s (Svennilson, 1954: 51). The only exception was the United Kingdom where government support of public housing caused increased activity in the 1930s. The interwar period also witnessed an aging of the population such that a larger proportion were of working age. As in the United States, this decreased dependency ratio may have limited consumption (Svennilson notes that the Netherlands and Scandinavia had both the highest population growth and income growth). Along with the outflow from agriculture, it helped pave the way for high unemployment levels.

We would not want to create the impression that European counties had similar interwar experiences. Even in America, different regions fared quite differently. Differences in the level of development of European nations guaranteed that there would be wide differences in growth rates. We will focus on Britain, which admittedly was the European country most likely to share the American experience. There is, in any case, a very limited literature on most other countries.[4] Toniolo, for example, has noted that there are less than ten works on the Italian Depression (Eichengreen and Hatton, 1988: 221). The range of industry-level studies on which we have been able to base our study of the United States is also lacking in most other countries.

Britain

Writers on interwar Britain have long recognized that structural transformation was an important component of interwar experience. Keynes himself noted that not all of the unemployment was due to general demand deficiency, and that microeconomic measures as well as macroeconomic were needed to restore full employment. The Bank of England also emphasized the importance and rapidity of structural change, but did not attempt to explain this (Weir, 1989: 376). Heim (1984), as we saw earlier, has pointed to the importance of structural unemployment in Britain. Saint-Etienne (1984), convinced that government policy in Britain was correct, attributed interwar unemployment to demographic and structural problems. Von Tunzelman (1982) has shown that much of the new industry/old industry debate was misguided in thinking there were large productivity differentials between the two. However, he leaves intact the possibility that the trajectory of declining and growing industries may have had an important direct effect on unemployment.

The fact that there was a shortage of new product technology in the 1930s is masked in the British case by the sluggish adoption of the new products of the 1920s.[5] This would cause the British interwar period to look at first worse and then better than in North America. "The American experience is instructive. The industries which led the boom in the United States, construction, services, transport, and the new consumer durables trades, were much less buoyant in Britain. The weak boom helped to soften the impact of the subsequent depression in Britain and left something in reserve for recovery" (Aldcroft, 1977: 203-4).

The growth of output 1924-1937 is similar to the growth rate of previous periods, despite the high levels of unemployment. The explanation, of course, must lie in more rapid productivity growth. Without new sources of output growth, this could only result in unemployment.

The attention paid to industrial analysis in Britain is striking. The chronic unemployment of the entire interwar period found there may have struck scholars as being more amenable to structural explanation than the sharp American downturn. Yet the American boom and bust can be easily ascribed to the experience of a few sectors. The role of technology is obscured in Britain, for the 1930s saw the adoption of products which had earlier saturated Americans markets. The underlying problem is the same, though; deficient product innovation against a background of dramatic productivity improvement.

The reasons for the earlier development of the auto industry in the United States are a larger market, higher incomes, more equal distribution, and previous exposure to mass-produced typewriters and sewing machines; entrepreneurs thus could more easily envision a large market (Church, 1981). While the United States possessed a mature market in which most buyers were replacing existing cars by the late 1920s, this did not happen in Britain until the late 1930s. As in the United States, demand was inelastic and people did not quickly replace cars. Miller and Church characterized demand in the 1920s as coming from intramarginal users becoming acquainted with cars. Only in the 1930s were a new group of purchasers introduced; the growth in sales of low horsepower cars is indicative of the widening of the market. Auto output does not decline but simply stagnates 1925-1933, before accelerating (1979: 188-91).

British experience with the new consumer durables was similar to that with automobiles. Per capita electricity consumption in 1929 was low not only relative to the United States and Canada, but France and Germany as well. The 1930s would see catching up largely due to government rationalization efforts which replaced small inefficient power generation (Catterall, 1979: 242-4; he argues that interwar growth was largely dependent on public electricity supply). Catterall recognizes that consumer durables were one of the most important sources of economic growth and employment creation in the 1930s.

Canada

Canada suffered two ways. As a country which exported a variety of raw materials to the United States, it naturally suffered as that economy declined. As a country with similar income levels and technological attainment, it was the most likely to suffer the same structural problems as its southern neighbor. It is therefore no wonder that it experiences one of the worst Depressions in the world.

Exports had come to comprise 25% of GNP by 1930. These were not only focussed on the American market, but often on the sickest parts of that economy. Pulp and paper production was Canada's largest industry, and exports to the United States its major outlet. Competition from radio had decreased American demand even before 1929. Nevertheless the industry had expanded massively in the late 1920s (doubling capacity 1925-1929). It was also hurt by Swedish competition in the 1930s. By 1936, almost half the industry had failed. Canada was also dependent to a large extent on foreign investment; 40% of non-farm capital came from abroad. The sectors which relied most on foreign capital saw the largest declines in investment.

American and Canadian GNP follow an extremely similar pattern through the 1930s, which lends credence to the idea that the latter imported the Depression from the former. Safarian (1959) noted that this similar aggregate performance masked important differences; the government played an even smaller role in Canada,[6] agriculture was more important (and suffered the same difficulties as its American counterpart, including the severe drought which hit the western plains; the output decline from the drought rendered self-sufficiency impossible for thousands), and durable investment recovered more slowly.[7] The Canadian downturn slightly preceded its American counterpart. Green and MacKinnon (1987: 354) have noted the existence of overcapacity in the boom industries of the 1920s. Some of these, such as newsprint, were export oriented. Others, like automobiles, were not. Canada was the only country with a similar car per person ratio to the United States in 1929. It had absorbed other new consumer durables at similar rates in the 1920s as well. It thus was destined to suffer in these sectors in any case. Canada's economy had perhaps been the most dynamic in the world in the 1920s and construction, machinery, electrical products, chemicals, and motors were its most dynamic sectors (Aldcroft, 1977: 212-3; the Soviet Union may have grown faster). These sectors were destined to lead the economy into decline in the 1930s.

7. Canada in the 1930s saw a similar divergence between low and high risk interest rates to that in the United States. This occurred despite the fact that Canada's branch banking system was immune to the runs and panics we see in the United States. This provides further evidence that it was demand-side factors behind the fall in investment.

8. Unemployment rates tend to be higher in the postwar period in agricultural areas and in regions farthest from the economic centre (Sinclair, 1987: 27). Regional differences were much greater in the 1930s.

9. The mere fact that industries can not relocate overnight allows us to pass over the determinants of initial locational choice. The fate of regional economies in the 1930s was largely, though not entirely, determined by the nature of industries which had previously chosen to locate there. Such analyses are available. Warren (1973) discusses the reasons for the slow spread of steel manufacture in the south and west. Rae (1968: 11) deals with why aircraft manufacturers chose the west coast.

14

Conclusion

Due to the time path by which the technological potential of the Second Industrial Revolution of the late nineteenth century was realized, in concert with the fact that early industrial research laboratories emphasized process over product innovation, there was an abundance of labor-saving process innovation and a lack of new product innovation in the late 1920s and early 1930s. Thus, the average propensity to consume fell. Likewise, there were limited opportunities for investment. These contractionary tendencies were exacerbated by adverse trends in population growth and income distribution. Employment fell in consumer durables due to market saturation. As well, many industries which had hoarded labor in the 1920s as productivity rose dishoarded these workers in the early 1930s. Workers who lost their jobs had no place to go due to the lack of new product technology. Entrepreneurs in such an environment could not profitably hire the unemployed at any wage above subsistence.

The evidence for this chain of argument can only be found sectorally. In part two, I traced the course of technological change, market penetration, investment behavior and opportunities, and employment sector by sector. The job losses which reasonably could be attributed to changes in technology and other secular forces were large enough that, with a reasonable multiplier, they could account for the entire unemployment experience of the 1930s. I encourage scholars to further refine my estimates sector by sector. In Chapter 13, I discussed the degree to which this line of argument can be applied to countries other than the United States. As well, I showed that regional, gender, and racial variations in employment experience can more readily be understood within this book's approach than within more aggregative theories of the Depression.

It is likely that the argument that there were too few growing industries will be more easily accepted than the conjecture that industries in either sudden or protracted decline were instrumental in causing the downturn. The fact that the vast bulk of postwar employment generation has depended on technology developed in the 1930s or after is too striking to be casually dismissed. The question, "What were the unemployed to produce in 1930?" has no simple answer. In the absence of new product technology, it may well have taken decades for them to all be re-absorbed into the workforce. Alternatively, can anyone really doubt that the Depression would have been much less severe if, say, television, the DC-3 and sulpha drugs were ready to go in 1929?

However, to accept the second conjecture must at least weaken resistance to the first. Industrial decline unmatched by industrial expansion can only mean economic decline. The question becomes a matter of degree, and one which can only be answered sector by sector. This has been the guiding principle of this work. If it has succeeded in diverting a bit more attention away from highly aggregative research, it has achieved much.

"A technological innovation is like a river — its growth and development depending on its tributaries and on the conditions it encounters on its way"(Braun, 1982: 1). We have been able to trace most of interwar technological history back to its "roots" in the three big technological revolutions of the nineteenth century. It appears that the relative timing of process and product innovation could well be the "natural" outcome of the order in which the ramifications of these three revolutions were worked out. Yet research does not operate in a vacuum, and various social forces determine which questions innovators choose to tackle at a given point in time. The question remains as to what degree institutional changes, and the rise of the industrial research laboratory must be the most important of these, may have accentuated the search for process technology at one point in time and product technology at another. We have seen that the earliest labs indeed focussed on process improvement, as well as minor product development to protect patent positions. Only after a handful of new product successes were labs likely to devote considerable effort to this longer-term and riskier task. The modern American lab devotes two-thirds of its effort to product research, though much of this is minor product development (the Japanese devote two-thirds to process technology). This seems a far cry from the earliest labs. The transition, if such it can be called, has hardly been smooth. Even after its breakthrough in radios, General Electric's lab still devoted much of its research effort to short-term projects suggested by the manufacturing arm. More recently, unable to produce another product quite as revolutionary as nylon, Du Pont has had to rethink its research strategy.

Even today, we must recognize that development is by far the most important part of R&D lab work. While the growth in size of research establishments may have enhanced the potential for major breakthroughs in one sense, it has been accompanied by increased bureaucratization of the research process, which likely encourages routine problem-solving and may discourage new ideas. Philips (1971), studying the airline industry, argued that industrial R&D occurred as a response to exogenous (in this case government financed) technological change. There has, through this century, continued to be an important role for small firms in making basic innovations. Mowery and Rosenberg (1987) have argued that scientific inquiry itself has often been driven by technological developments which raised important questions. We must thus expect that, while social forces deserve greater investigation, the particular forms of technology being developed will remain as important determining factors in the economic history of the twentieth century.

A New View of the Economic Process

"The [neoclassical] theory conceals either in aggregation or in the abstract generality of multisector models, all the drama of events — the rise and fall of products, technology and industries, and the accompanying transformation of the spatial and occupational distribution of the population" (Tobin, 1975). Our purpose has been to show that we can only hope to understand the movement of aggregate variables by comprehending the dramatic events going on at the industry level.

Faced with the fact that economic growth occurs at widely different rates at some times and places than at others, the average taxpayer might be forgiven for thinking that much of the brainpower of academic economists is devoted to understanding why these differentials occur. They might be shocked to find that most economists focus on much narrower, though still valuable, concerns. Growth theory, with its mechanistic focus on savings rates and capital-output ratios, has provided little insight. Economic historians and development economists, lately marginalized within the profession, are the only ones who regularly concern themselves with the long-run determinants of growth. Even they, faced with the intractability of the big question, and an overwhelming temptation to mimic "mainstream" economists, tend to focus on narrow issues.[1]

There tends of course to be a tradeoff between the importance of the question and the precision of the answer. Yet humanity does not abandon the study of philosophy because it is not possible to define perfectly the meaning of life. Nor should economics turn its back on the big picture simply because it is less amenable to the precise tools of the mathematician.

Economic historians, when surveying world development up to the close of the nineteenth century, tend naturally to a sectoral focus. They often pursue this to extremes: to understand the English Industrial Revolution is not just to understand cotton and iron; Germany a century later is more than electrical apparatus and chemicals. These are, however, key pieces of the puzzle, and the expansion of these sectors is crucial to the expansion of the larger economy. Sheer economic complexity has perhaps dissuaded scholars from more readily extending the same form of analysis into the post-World War One world.

There are equilibrating forces in the economy, but these are sluggish. If technological change were to suddenly grind to a halt, the economy would likely go through an extended decline before reaching a new full employment equilibrium (perhaps the steady state long dreamed of). At any point in time, the state of the macroeconomy depends on the course of technological change in the various sectors of the economy. We know without doubt that it is the combination of product and process innovation which is primarily responsible for our world being so different from that of the eighteenth century. It is not a great logical leap therefore to suspect that the course of innovation largely determined both cross-country differentials in growth (and therefore render the comprehension of the forces behind successful technological transmission so important), and temporal differences as well.

We should not overstate our case. Industries interact within an aggregate economy, in which they both compete for resources and supply each other with markets. The rate of growth of an industry may be hindered by resource competition from others, or aided by increased demand from other sectors. These various effects may be difficult to disentangle. But we must recognize that much of what we think of as aggregate forces are simply, as the word implies, the sum of sectoral forces.

There are exceptions. Government policies, interest rates, and some types of expectations only operate in the aggregate. The danger lies in over-emphasizing their importance. Our study of investment is instructive here; aggregate expectations appear to have been high and nominal interest rates were low, but investment decisions are made at the firm level and depend primarily on the prospects of particular product lines.

In addition to its overly aggregative nature, macroeconomic analysis is guilty of being ahistorical. Economic variables change only slowly. Growth, decline, and structural change do not happen overnight. Even the 5-10% growth rates recorded in some postwar economies could only significantly change the ordering of nations in terms of per capita income over a period of decades. Economic historians must generally concern themselves with even smaller margins. Yet a difference of just a tenth of a per cent in annual growth rates can over centuries have a massive impact on a country's well-being. Growth and structural change happen slowly, but the forces behind these often change discretely. There truly are turning points in history, though one might find little evidence of these in much of the New Economic History. The Russian Revolution may have hardly changed

Conclusion

GNP in the 1910s, but would transform the entire structure of the economy in the next decades. The automobile can be viewed in the same light.

This can not be emphasized too much; it is a huge mistake to think of economic development as being driven by gradually evolving forces simply because the result is gradual change. We can never comprehend growth if we make that initial error. As with people, so with societies: momentous events occur beneath a calm exterior. Only rarely, as in the Great Depression, does this chaos clearly break the surface. Even then, disaster can not happen overnight, but takes years.

Usually, there are counterbalancing forces at work which prevent a disaster such as the Great Depression. It is these, not the natural equilibrating mechanism of the economy, which keep us in general a bit closer to full employment. The equilibrating forces are slow and weak. They placed some limit on the severity of the Great Depression, but we would do well not to depend on them.

The Role of Economic History

The use of the word "History" to define the field of Economic History has undoubtedly contributed to the peripheral role its practitioners play in the average economics department. It would perhaps be better if both economists and economic historians saw the latter's focus as the study of economic growth. Introductory textbooks often discuss the major foci of economic analysis: efficiency, equity, growth. They then proceed to dismiss the latter and focus on the first two. The same is largely true of the discipline as a whole. Growth theory never has come to grips with enough of the basic complexity of the growth process to be more than a subject of curiosity. The tendency of economic historians at times to eschew the methodology of the mainstream might be taken as due to the nature of their subject matter rather than their preferences.

Unless society's rate of time preference is unduly high, the question of why growth occurs faster in some times and places than others must be reckoned the most important question facing economists. (At least some of the astonishing rise in Japanese studies or Pacific Rim studies in recent years reflects this fact). The only contender for primacy would be the cause of business cycles. We have seen, though, that fluctuations can not be comprehended fully unless informed by an understanding of the causes of growth.[2]

Growth depends on technology, and technology is complex and unpredictable. Elegant modelling exercises will never dominate a truly scientific exploration of economic growth. Economic historians must have a greater willingness to pursue interdisciplinary research. The limited contacts between them and historians of technology (Staudenmaier, 1989) is incredible. The danger exists that the study of issues of central importance to economic understanding will be completely divorced from the discipline of economics. Economic historians must have a wider perspective than mainstream economists. They must aggressively use this wider perspective to show that economic history can provide essential knowledge to practitioners in other fields. They must move beyond the application of mainstream theory to develop a new theory of economic growth.

Will It Happen Again?

L.V. Chandler, writing in 1970, could be extremely optimistic that Keynesian economic theory had advanced to a state where another severe Depression was impossible. Since that time, of course, the economy has been rocked by recessions of much greater severity than those of the 1950s and 1960s. Keynesian theory took a beating along with the economy, and many economists came to advocate a hands-off approach to economic management. It is unlikely that public opinion would allow a government to pursue Hoover-like policies in a downturn like that of 1929, but nobody writing since the 1970s can be quite as complacent about economic salvation from the wisdom of the economics community (especially as debt burdens limit governments' room to manoeuvre). There has been some convergence of views in macroeconomics. While there are still hardliners, most would see some truth in both Keynesian and non-Keynesian theory. While recognizing that the natural functioning of the economy can keep us away from full employment for long periods of time, they also recognize that governments, with the existing policy tools, can not work miracles. Lags between inception and effect of policies, difficulties in forecasting, and negative effects of government spending on private sector expectations, are among the major reasons for this.

During the 1930s, and since, we have built a number of automatic stabilizers, such as unemployment insurance, into the economic system, and provided deposit insurance and other safeguards to protect the banking sector. This, plus changing public sensibility about the role of government, does make it likely that a downturn as severe and lengthy as the Great Depression will not happen again. Still, this provides only very limited consolation.[3]

Further reassurance comes from reconsidering the technological roots of the Great Depression. A handful of industries played the dominant role in fomenting that Depression. As the economy has become more diversified, it has become more difficult for the same proportion of the economy to collapse at once. In particular, we are now well into the consumer durables revolution, whereas that sector in the 1920s was dominated by just a couple of fairly new products. Perhaps, also, firms have become some-

Conclusion

what more farsighted and thus less likely to saturate their own markets. The effects of the forces we have discussed can still be observed in the postwar economy. They cause nowhere near as great a calamity (though we should note that in Britain, unemployment rates were actually higher in the 1980s than 1930s). The interwar period, witnessing a major structural break in consumption patterns, was likely a more unstable economy than today's.

Casual inspection indicates that the forces we have discussed continue to dominate the postwar economy. On the face of it, a couple of decades of rapid growth, mild recessions, and low unemployment, followed by a couple of decades of sluggish growth, instability, and high rates of unemployment is more amenable to description in terms of long-term forces than with the standard theoretical constructs of the business cycle theorist.

While new technology was adopted during the 1930s, business conditions naturally slowed this process. Then the Second World War both encouraged technical development and clamped down on private consumption. It is not an exaggeration to say, therefore, that there was a technological backlog available at war's end. Along with a range of new products, it was also true that a number of the products of established industries were in short supply; there had been no automobiles produced for four years, for example. A range of industries were therefore ready to greatly expand production. Freeman (1982: 67) speaks of the coincidence of four or five bandwagons.[4]

Some authors (e.g. Gordon, 1961) credit a considerable role to government support of returning servicemen (the GI bill) in fomenting growth. Transfer payments will, of course, have their greatest impact when there are a range of goods and services which consumers are anxious to purchase (otherwise consumption will fall far short of income, with disastrous consequences). Another factor which contributed to growth over the next decades was the baby boom. Though little recognized in the literature, any parent can attest to the effect on their propensity to consume of having children.

The automobile industry may have learned its lesson but was not totally immune to saturation (auto output fell 25-36% in all major industrial countries in 1974-75). Appliance sales also collapsed in the mid-1970s after decades of expansion. Pocket calculators and television had saturated their markets (Mandel, 1982: 59). Transistor and semiconductor producers operated at 50% of capacity in mid-1975. Facing declining demand from the automobile, shipbuilding, and construction sectors, steel orders fell by one third 1974-1975. While these sectors contracted, others, such a pharmaceuticals, remained strong through the 1970s.

It would take another work as long as this one to trace the course of technical change through the postwar era. It appears, however, that the rate of product innovation observed in the 1950s was not sustained.[5] The fact that R&D expenditures, which rose through the 1950s and 1960s, fell in the 1970s is both symptom and cause. Process innovation continued and perhaps accelerated. Katsoulacas cites empirical evidence that the product innovation of the 1950s and 1960s had larger employment effects than later process innovation. Employment fell much further than GNP in the mid-1970s slump. Some of this may reflect dishoarding, but much was due to the steady advance of automation. Even after recovery in the late 1970s unemployment remained high.[6] Overcapacity in existing product lines kept investment levels low; firms were unwilling to borrow the funds which banks were willing to lend (Mandel, 1982: 94).

The above is merely indicative of the line of argument which could be followed. Again, it seems likely to yield greater insight than standard macroeconomic studies have to date. Bernstein, again, tries to combine the two approaches. The OPEC shock acts like the 1929 Stock Market Crash (but unlike that of 1987) in shaking confidence and initiating economic decline. This environment makes it more difficult for the structural changes of the time — towards services, chemicals, fabricated metals, and instruments — to proceed. He also credits a role to the baby boom (1987: 207-14).

The greatest damage to the reputation of the economics profession has been done by those who purport publicly to be able to predict the future. It would be the sheerest folly to try to forecast the future course of technological change in any detail.[7] Without that facility, other sorts of prediction are likely to be grievously in error. We can deal with a more general question: What is likely to happen to the overall rate of innovation? It would seem that the next century should see considerable product and process innovation.

History can be interpreted in two ways. The rate of innovation would seem to have been growing for centuries. We might then posit that the future will see an even more dramatic rate of change than the recent past. However, until recently world population exhibited a similar trend; those who predicted a population disaster have witnessed the slowing down of population growth rates across much of the world. We may have peaked; we may have exhausted much of the available technological potential. Future generations may look back on the last couple of centuries in awe.

Stanback in the 1980s worried that the consumer durable revolution was largely a spent force. Microwaves, VCRs, and video cameras have shown that it still has some life. Can those who have suggested that we have already exhausted much of our potential be rejected as casually as those who speak of imminent resource constraints? Are they as mistaken as the general manager of the National Machine Tool Business Associa-

Conclusion

tion, who worried in the mid-1920s that the age of invention was drawing to a close (in Wagoner, 1968: 130)? Even in the present, it is not easy to speak of the rate of innovation with any precision. Nell has argued that we are actually in an era of unprecedented innovation, but most of this is labor-replacing or market-destroying (1988: 167). If so, we have seen that such a situation need not last forever.

It could be argued that the rising importance of the service sector bodes ill for the future; lower levels of productivity growth there will eventually cause it to absorb so much of the labor force that product innovation must ground to a halt. We have already seen many reasons to doubt the basic premise; in particular, business services greatly aid productivity growth elsewhere. Alternatively, the fear has emerged that information technology may create widespread unemployment by replacing service workers; previous complacency about the insatiable ability of services to absorb labor has disappeared (Gershuny, 1983: 1). Among possibilities in the future, we must consider the effect on labor of ordering merchandise by two-way cable television.

The service sector also appears capable of substantial product innovation. Gershuny, indeed, feels that modern computers and communication technology provide the base for a new wave of innovation in entertainment, information, education, and medicine; he compares this to the wave of innovation in the 1950s and 1960s (1983: 2). However, Grubel points out that since service consumption takes time, there is a limit to how many services a consumer can demand.[8]

We would seem, though, to be a long way from that point. Nor should true consumer satiation be viewed as dangerous, but rather as the goal towards which the system is striving. We discussed in Chapter 3 the ordering of human wants posited by some psychologists. We may have satisfied our basic needs for food and shelter (for most in the developed world at least) and be well on our way to satisfying convenience wants (though we could all, I am sure, list wants yet unsatisfied) but we are clearly a long way from fully realizing human potential.

It may be that the technology to meet higher-level wants, or to further supply convenience so as to free up consumer time, is more difficult to conceive than the technology that has taken us this far. This is not obvious, and, even if true, is merely an argument for greater support of research. As long as we recognize how far we are from the satisfaction of all human wants, we should not be too pessimistic about the potential for technological innovation.

Gomulka (1990) has argued that the innovation rates of the last decades may be unsustainable for the simple reason that they have been the result of research expenditure continually rising both absolutely and as a proportion of national product. This trend can not continue forever; at some point in the not too distant future we might anticipate a levelling off of research expenditure (subject to public policy changes, of course). Studies of many sectors have shown that output rises faster when inputs are rising. Even if we put aside the question of whether research can be expected to behave like other activities, we could still suspect that it is the size of the research effort, rather than its rate of growth, which primarily determines the rate of innovation. There is certainly no reason at this point to foresee a drastic decrease in the resources devoted to research. Moreover, Gomulka himself points to a factor which should increase global research efforts. Research is heavily concentrated in a small number of developed countries. As more and more less developed countries achieve rapid economic growth, the human resources of vast parts of the world can also be turned to the betterment of mankind. Some day, we can hope, the whole world will be developed. In the process, we could well expect continuing increases in the rate of innovation.

Public Research Policy

> "I should conclude, therefore, from this study of the origins of the radio industry that a flourishing program of fundamental research is vital to our future industrial progress" (MacLaurin, 1949: 243).

We have described decades of economic transformation in terms of the gradual working out of practical implications of scientific/technological breakthroughs. While we have good reason to suspect that the future time path of innovation will not be such as to cause another depression, there is much scope for government research policies to limit future bouts of unemployment. We are not alone in suggesting that continued technological innovation is the key to sustained economic growth in the future.[9] The Panel on Technology and Employment has "concluded that the rapid generation and adoption of new technologies are essential to maintaining and expanding U.S. employment and wages" (Cyert and Mowery, 1988: xxxii). Stanback concludes his study of the service sector by noting that without government support for innovation and diffusion, the economy may be subject to a lengthy period of decay or decline (1981: 130-1, 8). Katsoulacas cities empirical evidence that countries which are technological leaders gain an employment benefit (1986: xxi). Freeman has long advocated the support of basic research because this will open up new areas for practical innovation. He emphasizes product innovation and market-widening process innovation, which accords well with the approach taken here. Most case studies to date have found a social rate of return on research and development of 20-50% (even the private return is found to be higher than the return on physical capital; Gomulka 1990: 38).

The short term negative effects of labor-saving technology should not blind us to the beneficial long-run impact. We would have gone far astray if we had become modern-day Luddites (though these were not the simple machine-smashers they are often portrayed as). If other policies are in place to ensure that growing industries can absorb displaced labor, then process technology becomes an unmixed blessing. Likewise we need not fret too much about the possibility that a new product may have deleterious effects on existing lines of production. In the long run, after all, product and process innovation are complementary; a vast array of new products is only possible if labor is released from the satisfaction of preceding wants. In the shorter term, economic well-being may hinge on domestic producers adopting process innovation at least as fast as foreign competition.

The government does not therefore need to decide, at least on those grounds, which sort of innovation to support.[10] If would appear that the private sector may be better equipped to pursue process innovation. Thus, some bias toward product innovation may be desirable. The key distinction, though, should be between basic and incremental innovation. It is the former whose cost and uncertainty may frighten off private enterprise. It is also the type of research where it is impossible for the original innovator to capture a large part of the benefits of their discovery. Decades of development in unforeseen directions will benefit society to a much greater degree than any royalties the original creators receive. Positive externalities of this sort provide an economic rationale for government support: private enterprise will do too little basic research.

There may be other rationales for government intervention. Noble (1977) has argued that the modern research lab reflects the takeover by corporate capitalism of the course of technological change. However, the counterfactual is not articulated: what would have happened otherwise? Much, though not all, of modern technology was beyond the scope of the independent tinkerer. The research lab, even if perverting the direction of change, has likely accelerated it. Moreover, the role for small firms, especially in radical innovation, has not at all been eliminated. Biases, beyond those toward incremental process innovation which we have discussed, must be clearly elucidated before countervailing government policies can be imagined.

While the argument so far, as well as the very uncertainty of research itself, points to a broadly-based research strategy, particular preferences may be desirable. Among the many policy options available to deal with environmental issues, it might be felt (for political if not economic reasons) that supporting environmentally friendly technology was worth emphasizing. Society might wish to choose in other ways the direction in which to focus fundamental research, though it should be aware of the dangers of doing so. Since we have seen that we cannot objectively decide which research path is superior at the start, we should be careful of strongly biasing research activities.

We have seen in numerous cases the necessity of close personal interaction among scientists, engineers, production people, and marketers. This is a major reason why research is generally performed in-house rather than contracted out. It is also one reason why large firms have an advantage in some sorts of research. Mowery and Rosenberg (1989) have emphasized the need for cooperation between innovators and producers. The government can not expect to create basic innovations in isolation and have these readily built upon by industry. Academic research can be adopted painfully slowly for similar reasons.[11] Private firms without basic research programs of their own will be unsuited to developing other's basic discoveries. Moreover, basic discoveries themselves are often a result of incremental research. Thus, while governments can play a powerful role in encouraging innovation, they should not do so in isolation but rather in close long-term cooperation with the private sector. Drawing academic researchers into closer contact with industry could prove advantageous to both; technological inquiry often fuels scientific insight. The optimal institutional structure has varied across time, country, and industry. A certain flexibility in approach is therefore advisable.[12]

To the extent possible, the government should include small business within its research strategy. Kimmel (1939) argued there was a role for the government in extending credit to small firms during the Depression. One has to be very careful, however, in identifying where exactly private sector banks are failing, in order to avoid the possibility that governments only lend to businesses justifiably viewed by banks as bad risks. If the government becomes a clearing house for technological information, it may be able to dispense both information and credit to innovative firms. The success of agricultural extension indicates that small firms can benefit from government research. As small firms have played a large role in basic innovation, such a strategy could have a major impact.

Growth Is the Only Answer?

There are many who feel not only that the days of rapid growth are numbered but that they should be. Fear that the capacity of the earth to support economic activity is being strained, in terms of both resource supplies and environment, has led to suggestions that we make do with what we have, and resist further growth. The proponents of "limits to growth" arguments in the 1970s made dire predictions of the future. Freeman (1984) has noted that these predictions did not allow for the possibility that radically new technology could totally transform the situation. He notes as well that to stand in the way of growth-enhancing innovation while not prohibiting labor-saving technology is to invite future employment problems. As Mokyr (1991) has argued, technology is the ultimate free lunch. Innovations as yet unimagined can give us growth while at the same time actually reducing resource use and environmental damage. We can, by appropriate policy measures, ensure that research effort is focussed in directions which satisfy these different, but not necessarily incompatible, goals.

The analysis of this book points to a more optimistic view of the effect of pollution control on employment and output. Installation of scrubbers or searching for cleaner raw materials is like new product innovation: it should, at least in the short term, increase employment (if one does not like treating it thus, it could instead be termed negative process innovation; the result is the same). Unlike government pump-priming the cost of pollution control would be borne privately. While some plants or industries may decline as a result of increased costs, if demand curves are inelastic the net general equilibrium effect should be to increase output (by value) and employment.

Some may still be disturbed by the suggestion that only a constant stream of new products can prevent economic malaise. Surely, rampant consumerism is not the only solution? I must confess to being a believer in growth; the pie which grows can more easily be divided equitably. While it might seem that an adequate range of consumer goods already exists, we can all be relieved that our parents or grandparents had not decided that no new products were needed.

Flexibility in working time is likely to become of increasing importance in the future. We have seen that the average working week fell substantially during the early twentieth century. Despite rapid postwar productivity growth, it has declined little since. I am doubtful that, as some claim, the recent stability of the forty-hour week reflects some ideal division of the week between toil and leisure. Moreover, there is certainly much scope for the extension of annual vacations. Much of the productivity increases likely to occur over the next decades could be absorbed in this fashion. The question is whether society would prefer leisure to increased consumption. Certainly there are many who would if given the opportunity. The revision of government policies which penalize part-time work would allow these segments of society to choose to work less. This proportion might be considerably enhanced by improvements in income distribution (or in greater social appreciation of self-actualization).

While hours reductions can substitute for product creation (and thus provide another mechanism by which individuals can bring their income

Unemployment

> "Unemployment is *the* central problem of our age. Unemployment involves a tragic waste of scarce human resources. It is demeaning, debilitating, grossly unfair and massively expensive. Ultimately it threatens the fabric of society. It is all too obvious that it will not cure itself. Active adventurous policies are urgently required" (Sinclair, 1987: 293).

At some point in the future, society will decide that involuntary unemployment is simply unacceptable. The efforts of some economists to deny its existence may delay, but not indefinitely, this resolution. The only question is when. Huge cross-country differentials in unemployment prove that it is not inevitable. While we in North America have since the 1960s become accustomed to unemployment rates of 8% or more (how many even remember that in the 1960s rates over 4% were considered too high?), some countries, such as Switzerland and Sweden (until recently), have averaged rates less than half of ours through the postwar period.

Sinclair (1987) is not alone in arguing that unemployment rates can become self-perpetuating. Workers and employers can adjust their behavior in a number of ways, and society as a whole can become complacent. The fact that some countries have consistently had much lower unemployment rates than others should shake us from such complacency (and the fact that others have experienced high unemployment for decades should decrease our concern over lags in policy implementation). Cross-country comparison points to many reasons for lower unemployment; long-term employment contracts in Japan, guaranteed jobs for the long-term unemployed in Sweden, more rapid rates of economic growth (Sinclair, 1987: 6).

Enhancing economic growth by publicly supporting research is a good starting point. We should not expect this alone to take us to the promised land. We should, given the failure of the existing policy mix, be willing to consider other significant departures from present public policy.[13] Sweden already guarantees jobs to those who would otherwise be the long-term unemployed. This will likely be a more widely acceptable policy in the future (jobs need not all be in the public sector; subsidies and/or penalties could be applied to the private sector). The cost and effectiveness of such a policy will, however, depend on the range of other employment policies in place.

Structural unemployment

Relying for prosperity on continual product innovation must mean that flexibility becomes a guiding principle of economic management. New computer automated production processes may reduce the difficulty of "retooling" to produce a new product. Even this technology only facilitates transfers among relatively similar products, though. The replacement of old products with new is likely to still involve transitional unemployment at least for the foreseeable future. Scandinavian co-operation between business and government has reduced the scope of structural unemployment, but apparently at the cost of sacrificing some flexibility.

Policies which might alleviate cyclical fluctuations are a blunt tool when employed against structural unemployment. Workers with the wrong skills or location will not be affected by increasing aggregate demand, unless dying industries are temporarily revived. Though government training and relocation programs to date leave much to be desired, public policy which focuses on change must also seek means to increase flexibility. This is especially the case for older workers who find it extremely difficult to find re-employment and can, as we have seen, become a major component of chronic unemployment. Nor must the humanitarian motive stand alone; countries with a better environment for worker retraining — Japan, West Germany, Sweden — have also experienced more rapid diffusion of new technology (Mowery, 1988: 499).

Planning?

As Stolper (1991) has noted, an evolutionary approach, which recognizes that the system will endogenously generate periods of high unemployment, leaves a greater role for government stabilization than even Keynesian macroeconomics. It also warns us, though, that it is folly to stand in the way of structural change by, for example, trying to prevent industrial decline and job losses. Gowdy (1992) reaches a similar conclusion: the recognition that natural selection among firms is an important determinant of output and productivity growth warns us to be careful in intervening; the recognition that natural selection is not the only determinant of firm success or failure — firms might, for example, just find themselves in the wrong place at the wrong time — gives us cause to suspect that government intervention might be able to do some good. What, though, should governments do?

Bernstein (1987) after discussing the downturn of the 1970s, makes no explicit policy recommendation but does hint at the need for a planning mechanism to hasten structural change. Many authors in the 1930s, both conservatives and liberals, in part impressed by the achievements of wartime (and Soviet) government intervention, also advocated some sort of planning to deal with the structural imbalance they saw about them.[14] We have certainly seen that economic stability depends on the timing of innovation; the equilibrating mechanisms of the economy operate two slowly to prevent lengthy bouts of unemployment.

To date, though, the practice of economic planning has been hardly compatible with technological innovation. We would not wish to gain stability at the price of sacrificing innovation. Many have, with justification, pointed out that even New Deal policy often stood in the way of the transfer of resources between sectors. Government regulation was often captured by industry leaders and used to pursue their own narrow interest. Flexibility and openness to changes must remain among the primary goals of government policy.

We shall not enter the contentious debate on industrial policy here. We have noted the difficulty in predicting the future, and might wish to be extremely cautious in advocating a role for government in picking industrial winners. However, government is itself a large sector of the economy. There is thus much scope for the proper temporal management of its own expenditures to reduce economic hardship. This, in combination with research support, and structural adjustment policies, could potentially make long bouts of involuntary unemployment a thing of the past. We need only be bold in our thinking.

Yet the argument here does not necessarily imply that government should grow larger. Hughes has pointed out that much of present government is a historical accident. Programs set up to deal with a particular crisis were not reduced when the problem was overcome. Nor was government policy every conceived as a unified whole; gradual accretions need not always be complementary. "Piece by piece our controls were triumphs; in aggregate they are simply a bureaucratic congeries — a heap" (1977: 239). Recognizing that much government effort is counterproductive should not blind us to the fact that it can perform valuable functions. Hughes notes that the steady growth of government must reflect a popular distrust of pure capitalism. We have seen that the public has good cause not to rely on the untrammelled market to guarantee full employment. The post offices and nature trails of the WPA and CCC not only alleviated the unemployment problem somewhat, but provided valuable artifacts for future generations. The challenge is to reorganize government so that useful public projects are always at hand to absorb those unemployed in the private sector.[15] The ability of government to do so will still depend on the generation of new ideas by and for industry. Since this can not provide

Conclusion 329

stability, though, we should make sure that, if government's role is to be cut back, the baby not be thrown out with the bath water.

Business Cycles

One of the problems with Schumpeter's approach was that he tried to explain too much. Both regular business cycles and longer swings in economic fortunes were attributed to technological change. We have focussed on the latter here. We will not join adherents of Real Business Cycles and argue that year to year fluctuations in technology generation explain the business cycle (though see Cheng, 1992). We will join them so far as to support the idea that there should be no cycles in a world without change, and so the forces of change must somehow be tied to the generation of cycles (likely at least in part through the inability of the capital goods sector to smoothly service the needs of the consumer goods sector, and to the tendency for agents to react simultaneously to the same stimuli and thus collectively over-react).

Not all business cycles are the same (a point only rarely recognized in standard theorizing). The 1950s and 1960s witnessed mild recessions because the forces driving industry-level growth were so strong; the 1970s experienced severe recessions because they were so weak. New products do not eliminate cycles but can greatly limit their severity. There is still scope, then, for government countercyclical spending to further iron out the cycle (though modern eyes must view the recessions of the 1950s as almost heavenly). The most efficacious public policy would be one that could guarantee steady product innovation.

New products, of course, can have a similar impact to some forms of government spending (without, however, causing concern about where the government gets the money from). In an economy characterized by market saturation, spending designed merely to increase aggregate demand, either through transfers or lower taxes, may have much less than the desired effect. The result depends to a large extent on which elements of the population receive the cash. Effects on output and investment will depend further on inventory holdings and capacity constraints. By spending money on public works itself, the government can be seen as creating demand for such works.[16] The public buildings and nature trails of the WPA and CCC are fine examples of this. Schumpeter felt that fascist regimes in the 1930s owed much of their success to the fact that government spending focussed not on increasing purchasing power but on producing specific physical outputs (1939: 975). Moreover, to the extent that such spending serves long-felt community needs, and so is likely to replace future spending, much of the concern of rational expectationists that pri-

What Should Hoover and Roosevelt Have Done?

Support of research must necessarily be a long-term strategy. Once the Depression had caught them by surprise, American governments looked for quicker remedies (and thus did not follow through on plans to increase public support of research).[17] Automatic stabilizers could only deal with the next Depression; companies like Eastman Kodak which introduced unemployment insurance in 1931 pursued a laudable but future-oriented goal. It has been common to criticize Hoover for not pursuing stimulative fiscal policy. Yet Roosevelt did not believe in pump-priming per se; he viewed relief expenditure as a direct transfer to those in the greatest need, not as the key to prosperity. Phelps (1993: 162) feels that so-called Keynesian policies only appeared to work in the 1930s precisely because governments directly hired the unemployed rather than relying on demand management.

We have suggested that the marginal propensity to save of large parts of the population was high at the onset of the Depression. Tax cuts or broadly-based spending may thus have had a limited effect on aggregate consumption demand (the multiplier was small). Leven, from expenditure surveys, attempted to estimate the *average* propensity to save in 1929 by income groups. It was negative for those with incomes below $2,000, 5-10% for those between $2,000-$5,000, 17% for $5,000-$10,000, 13% $10,000-$20,000, 12% $20,000-$50,000, 8% $50,000-$100,000 and 34% for those with incomes in excess of $100,000 (1934: 93). The irregularity of the correspondence between income and saving indicates that these estimates are rough indeed. Still, we could wonder if the tax increases Hoover implemented to balance the budget would have had much effect on consumption, the marginal propensity to save being much higher than average.

Roosevelt's New Deal seems a drop in the bucket to modern eyes. (Still, in the three years before July 1, 1938, the WPA alone built 17,562 public buildings, 279,000 miles of highway, 29,000 bridges, 357 airports, 15,000 parks, 6000 water mains, and almost 9000 miles of sewer — Webb, 1939.) The easiest complaint to lodge is that it was not enough. The levels of debt incurred were minuscule compared with those the postwar populace would realize a government could safely accumulate in order to deal with a temporary crisis. Moreover, it would appear that higher taxes and spending (the balanced budget multiplier) in the circumstances of the 1930s would have largely transferred funds from those who saved (income taxes only started at an income of $5,000), to those who desperately wished to consume. While we can quibble about the scope of spending, the general approach was broadly correct (and by the late 1930s as much as half the unemployed were involved in relief work). Specific public projects were chosen that were of considerable public utility and would provide employment to those who needed it. This fed and clothed the needy, and prevented the deterioration of work habits; local multiplier effects were felt as well. One major failing of the New Deal program was that retraining and relocation programs which might have minimized structural unemployment were virtually absent (Brown 1941: 145); the government might, more generally, have done more to limit the decline in education.[18] The elements of gender and ethnic discrimination which infiltrated the relief program were objectionable. Requiring local contributions limited the extent of the program in the hardest hit areas. But the approach (and Roosevelt himself) recognized that the economy was undergoing a structural transformation and replaced flagging private consumption demand with collective demand. We can only hope that if there is to be another crisis, governments will be as farsighted. Unfortunately, the present debt situation may limit their efforts just as the balanced budget ideal limited Hoover and Roosevelt.

Trade Theory

International trade theory has traditionally been an area in which economists have achieved a great degree of consensus. Simple yet elegant theories establish that, with few exceptions, tariffs limit economic efficiency and thus stand in the way of income growth. There have been some who have maintained that the static basis of neoclassical theory leads to an incorrect conclusion.[19] History provides numerous examples of countries prospering behind high tariff barriers. In the 1930s itself, the autarkic Soviet and German regimes fared best. Economic historians and development economists have proven somewhat more willing than the mainstream to suggest that the dynamic benefits of tariffs may outweigh the static disadvantages. Certainly the historical record shows that dynamic considerations determine trade flows; in the late 1980s the industries in which the United States had a trade surplus were those in which it was a technological leader such as aircraft, chemicals, plastics, drugs, and scientific instruments.

As an avowed internationalist, it is only grudgingly that I abandon wholehearted support for my favorite piece of economic orthodoxy. If industry level forces drive economic growth, then protecting certain industries may not be sheer folly from a national perspective. If Canada had proven a lower cost producer, might the United States not have been well advised to protect domestic car makers in the 1920s? Trade theory tells us that the outflow of funds would have been matched by an inflow (eventually), but does not, at least at this stage, guarantee that growth potential exported will be balanced on the import side. Moreover, if the goal is to increase domestic productivity, it appears that possessing industries with much potential for process innovation is a good idea (though, of course, tariffs which protect local firms may remove much of the incentive to lower costs). Again, though, one shudders at the thought of politicians being entrusted with the role of picking winners.

Further Research

The crying need is for historians of particular industries to explicitly address the question of how much of the changes in employment, output, and investment in the 1920s and 1930s were due to technological change and market saturation. We have been forced to answer questions about the Depression from a literature not focussed on the causes of Depression.

We know that officials of research labs often visited those in other industries. Wise, for example, lists a host of visitors to General Electric's lab. We could know more about how the philosophy of lab management evolved, and how this in turn affected the course of lab output.

The postwar era deserves a study in the same vein as this. We are, hopefully, far enough from the early 1970s to be able to apply some historical detachment. A valuable service would be performed in establishing the importance of structural forces in both the unprecedented growth of the 1950s and 1960s and the disappointing performance thereafter. Can we expect the performance of the 1950s again, or was it a one-time fluke?

I have elsewhere been a strong advocate of comparative research. Were it possible, I would have greatly extended the international comparisons contained herein. Countries which lagged in both technology and income could experience a similar crisis at a somewhat different level of development. Research of this type performed on other developed countries could refine this approach while providing important confirmation. It would also serve to answer the question of how so many countries experienced a contemporaneous Depression when international links were weak.

Though I would welcome theoretical insights into dynamic behavior, it is likely that the tradeoff between static losses and dynamic gains in tariff policy is best established by historical research. The role of trade policy in supporting industries which then become a major force in economic growth (or not) deserves inspection. The American chemical industry would provide one interesting case study.

Cross-country comparisons at the industry level would seem, for every time period, the most likely method of arriving at an understanding of why some countries fare better than others. The role of leading sectors in

Conclusion 333

economic development has been too casually dismissed by the profession. While the linkages are not always obvious, it is clear that modern economic growth has depended on the evolution of three interrelated technologies. Scherer (1984) has observed that most of the effect of new technology is felt outside the industry involved. Those nations, such as Japan and Italy, which were able to quickly borrow the elements of the Second Industrial Revolution, were as a result well placed to experience rapid economic growth in this century. Just as industry level analysis can provide a better guide to understanding temporal differentials in economic growth than highly aggregated studies, comprehending why these technologies were adopted in some countries faster than in others could well be the key to understanding spatial differentials in growth.

Economic historians must tread a narrow line; looking for the big picture but not ignoring historical detail. It is especially important not to follow Schumpeter in assuming that the form technology takes is irrelevant. When we see successful technology transmission, for example, are we simply seeing a general receptivity on the part of the receiving country, or were they rather simply ready at the right time for the right technology?

Research along lines such as the above will serve to inform — indeed generate — a new theory of economic growth with technology at its core.

Notes

1. Classical economists such as Ricardo were most interested in dynamic problems, and took for granted the static analysis which has come to dominate economics. Pasinetti feels the marginalist revolution of the 1870s represented a desire to have a new theory to respond to Marx (who had himself started from the same premises as classical economists), even if this meant focussing on trivialities. Economics, then, became more interested in logical consistency than relevance. Pasinetti thus sets out to develop a theory which will have technological change at its core (1981: ch.1).

2. We will argue below that trade theory needs to be reworked in a dynamic context, and have seen before reason to suspect that public finance issues must be placed in historical perspective. If most (all?) fields of economics need a long-term perspective, then a properly constituted course on economic growth might well deserve a place in the core of the economies program (a place economic history held decades ago).

3. Temin (1990: 73) both underlines the importance of the matter and evinces some scepticism: "We should hope that leaders in future crises will take into account the possibility of totalitarian takeover when they make their economic choices".

4. Even at that, some sectors still suffered. In agriculture, productivity increases from mechanization and chemical fertilizers, insecticides, and herbicides caused

rapid output growth and employment decline. There was now some place for these workers to move.

5. This in turn raises the possibility that the phenomenal growth of the next couple of decades does not represent the "natural" state of the economy but was instead a one-time occurrence. Perhaps only with greater societal commitment to research can that record be repeated. There is much evidence that profits fell through the postwar period (Mandel, 1982: 22-3, Freeman, 1982: 151). Mandel speaks of the decline in "super-profits" which had been earned by the technological leaders in fields such as synthetics, computers, and airplanes.

6. In the EEC, output and employment both expanded to 1960, output continued to expand while employment stagnated 1961-1974, and thereafter output growth slowed while employment fell (Freeman, 1982: 153).

7. While forecasters in 1941 correctly recognized the growth potential of pharmaceuticals, nuclear power, and television, they also spoke of long-term weather forecasting, photosynthesis, and prefabricated housing (Temporary National Economic Committee, 1941: 184-9). At present, microelectronics and biotechnology must appear as good bets for the future.

8. Grubel also argues that market saturation and productivity increase have sequentially affected the primary, secondary, and service sector; in the first two cases displaced workers had some place to go (1990: 11). Disaggregation beyond three sectors would seem to remove the potential for imminent disaster.

9. Nor was the argument unknown in the 1920s. Scientists seeking business support in the 1920s noted that the existing prosperity was built on past discoveries and that more scientific discoveries were needed in the present to ensure future prosperity (Wise, 1988: 260-1).

10. A handful of government task forces examined the question of technological unemployment during the New Deal. While the inquiry was undertaken with rhetoric about controlling technology for society's benefit, none ever advocated polices to limit labor-saving innovation. One could not prevent the bad effects of technology without losing the good as well (Pursell, 1979).

11. Adams (1990) has recently found that fundamental knowledge as expressed in scholarly papers is an important source of productivity growth, with on average a twenty year lag. While the results are interesting, one could question the highly aggregative nature of his research.

12. Mowery and Rosenburg also note the beneficial effects of low interest rates, and increased educational spending. Elsewhere, Mowery has suggested the use of standards and tax policies to encourage diffusion (1988: 500). Dasgupta (1989), following M Polanyi, has suggested that government rewards might be superior to patents for basic research (both appropriability and diffusion might be aided) This would mean extending the reward system we have for science into the technological domain.

13. Sinclair notes that while profit-sharing, as in the Japanese bonus system, may have good effects on employment, productivity, and strikes, it may not maximize expected utility for risk-averse workers (1987: ch. 14). He neglects the possibility that by changing expectations the business cycle might be ironed out. He

does note (ch. 15) that reduction in the variance of unemployment should reduce the mean level. His main recommendation is for economy-wide employment subsidies; the subsidy varying with the unemployment rate [the problem of crowding out might be alleviated by tailoring funding to environmental cleaning etc]. Among other policies, he supports affirmative action and the lowering of marginal tax rates for the poor. Phelps (1993: 366) also argues in favor of employment subsidies.

14. "Technological advance brings a great gain in the potential amount of production and leisure. It brings also a serious loss in economic security and a tendency for fluctuations to become wider and more destructive. A concomitant of a highly advanced technology is a demand and a desperately felt need for some form of social (that is political) regulation that will overrule economic mischance to the individual and guarantee continuous employment with at least a minimum remuneration" (Hopkins, 1941: 180). He anticipated a lengthy period of experiment before the best means of doing so were found. Temin (1989) argues that both Soviet and German experience in the 1930s was due to non-economic coercion.

15. Keynes himself suggested an even more extreme solution in the General Theory; "I conceive, therefore, that a somewhat comprehensive socialisation of investment will prove the only means of securing an approximation to full employment..." Hughes (1977: 156) credits wartime and postwar government spending with a major role in fostering employment growth, and feels the New Deal, if more extensive, would have done the same.

16. Sinclair (1987: 292) argues that in modern cost-benefit studies we tend to over-value the cost of capital, and improperly shadow price for unemployment. We therefore invest less in public projects than would be socially optimal. Moreover, demand for public goods likely rises with income.

17. Clark (1937) recognized that, "In the field of education many thousands of additional persons could well be employed...". In health, there was room for hundreds of thousands or even millions. The entire unemployment problem could have been solved by a program to provide "adequate" housing for the entire population. Clark was correct in suggesting that there was much useful work for the unemployed. He also showed a good deal of prescience in unwittingly predicting some postwar growth sectors.

18. Schumpeter believed that the German move toward autarky, while having a negative welfare impact, stimulated investment and "was an important factor in stimulating prosperity" (1939: 1973).

19. Even in a static framework, some argue that protectionism may increase employment. Early trade theorizing focussed on a world where full employment was assumed. More recent general equilibrium models which attempt to uncover employment effects are highly questionable. Still, Sinclair (1987: 268-9) has argued persuasively that tariffs are a poor employment-enhancing tool. Beyond the dangers of retaliation and exchange rate appreciation, he notes the traditional argument that tariffs are a subsidy to certain industries: if we wish to subsidize employment we should avoid distortion and do it economy-wide.

Bibliography

Abernathy, William J. 1978. *The Productivity Dilemma.* Baltimore: Johns Hopkins University Press.
Adams, James. 1990. "Fundamental Stocks of Knowledge and Productivity Growth." *Journal of Political Economy* 98:4, 673-707.
Aitken, Hugh G. 1989. *The Continuous Wave.* Princeton, N.J.: Princeton University Press.
Albion, Mark, and Paul Farris. 1981. *The Advertising Controversy.* Boston: Auburn House.
Aldcroft, Derek. 1977. *From Versailles to Wall Street 1919-1929.* London: Allen Lane.
Allen, Frederick L. 1952. *The Big Change.* New York: Harper.
Amendola, M., and J. Gifford. 1988. *The Innovation Choice.* Cambridge, UK: Blackwell.
Angley, E. 1931. *Oh Yeah?* New York: Viking.
Backman, Jules, and M.R. Gainsbrugh. 1946. *Economics of the Cotton Textile Industry.* New York: National Industrial Conference Board.
Baily, M. N. 1983. "The Labor Market in the 1930's" in J. Tobin, ed., *Macroeconomics, Prices and Quantities.* Washington: Brookings.
Balke, Nathan, and Robert J. Gordon. 1989. "The Estimation of Prewar GNP: Methodology and New Evidence." *Journal of Political Economy* 97:1, 38-92.
Barber, Clarence. 1978. "On the Origin of the Great Depression." *Southern Economic Journal* 44:3, 432-56.
Barger, H. 1951. *The Transportation Industries 1889-1946: A Study of Employment, Output, and Productivity.* New York: National Bureau of Economic Research.
―――― 1955. *Distribution's Place in the American Economy Since 1869.* Princeton: Princeton University Press.
Barger, H., and H.H. Landsberg. 1942. *American Agriculture 1899-1939: A Study of Output, Employment, and Productivity.* New York: National Bureau of Economic Research.
Barger, H., and S.H. Schurr. 1944. *The Mining Industries 1899-1939: A Study of Output, Employment, and Productivity.* New York: National Bureau of Economic Research.
Barker, T.C. 1982. "The Spread of Motor Vehicles Before 1914," in Charles Kindleberger and G. di Tella, eds., *Economics in the Long View.* New York: New York University Press.
Basalla, G. 1988. *The Evolution of Technology.* Cambridge: Cambridge University Press.
Batra, Ravi. 1987. *The Great Depression of 1990.* New York: Simon and Shuster.
Beard, Charles. 1928. *Wither Mankind?* Freeport, NY: Books for Libraries Press.

Bell, Spurgeon. 1940. *Productivity, Wages, and National Income.* Washington: Brookings.

Benjamin, D., and L. Kochin. 1979. "Searching for an Explanation of Unemployment in Interwar Britain." *Journal of Political Economy* 87, 441-78.

Bernanke, Ben. 1983. "Nonmonetary Effects of the Financial Crisis in the Propagation of the Great Depression." *American Economic Review* 73:3, 257-76.

――――1986. "Employment, Hours, and Earnings in the Depression." *American Economic Review* 76:1, 82-109.

Bernanke, Ben, and M. Gertler. 1989. "Agency Costs, Net Worth, and Business Fluctuations." *American Economic Review* 79:1, 14-31.

Bernanke, Ben, and M. Parkinson. 1989. "Unemployment, Inflation, and Wages in the American Depression: Are There Lessons For Europe?" *American Economic Review* 79:2, 210-4.

Bernstein, I. 1960. *The Lean Years.* Boston: Houghton Mifflin.

Bernstein, Michael. 1987. *The Great Depression.* Cambridge: Cambridge University Press.

――――1987b. "The Response of American Manufacturing Industries to the Great Depression." *History and Technology* 3:3, 225-48.

Berry, Brian L. 1991. *Long Wave Rhythms in Economic Development and Political Behavior.* Baltimore: Johns Hopkins University Press.

Bhaskar, R. 1989. *Reclaiming Reality.* New York: Verso.

Bilstein, Roger. 1984. *Flight in America 1900-1983.* Baltimore: The Johns Hopkins University Press.

Bleaney, Michael. 1976. *Underconsumption Theories.* New York: International Publishers.

Blinder, A. 1988. "The Challenge of High Unemployment." *American Economic Review* 78:2, 1-15.

Blinder, A., and L. Maccini. 1991. "Taking Stock: A Critical Assessment of Recent Research on Inventories." *Journal of Economic Perspectives*, 5:1, 73-96.

Bolch, Ben, Rendigs Fels, and M. McMahon. 1970. "Housing Surplus in the 1920s?" *Explorations in Economic History* 7, 259-83.

Bowden, Sue. 1988. "The Market for Domestic Electric Cookers in England 1932-1938: A Regional Analysis." *Explorations in Economic History* 25, 42-59.

Bowden, Sue, and Avner Offer. 1994. "Household Appliances and the use of time: the United States and Britain since the 1920s." *Economic History Review* 47:4, 725-48.

Bowers, Brian. 1982. *A History of Electric Light and Power.* London: Peter Peregrinus.

Braun, Ernest, and S. MacDonald. 1982. *Revolution in Miniature.* 2nd ed. Cambridge University Press.

Bresnahan, T., and D. Raff. 1991. "Intra-Industry Heterogeneity and the Great Depression: The American Motor Vehicle Industry 1929-1935." *Journal of Economic History* 51:2, 317-31.

Brown, Malcolm, and John N. Webb. 1971. *Seven Stranded Coal Towns.* WPA Research Monograph XXIII, 1941.

Brown, Nelson. 1947. *Lumber.* New York: John Wiley and Sons.

Bryant, Keith L., and Henry C. Dethloff. 1990. *A History of American Business.* 2nd ed. Englewood Cliffs N.J.: Prentice-Hall.

Bryant, Lynwood. 1978. "The Internal Combustion Engine," in Trevor Williams, ed., *A History of Technology* v. 7, 997-1024, Oxford: Clarendon.
Burns, R.W. 1991. "The Contributions of the Bell Telephone Laboratories to the Early Development of Television." *History of Technology* 13, 181-213.
Burnside, C., et al. 1993. "Labor Hoarding and the Business Cycle." *Journal of Political Economy* 101, 223-44.
Burstall, Aubrey. 1963. *A History of Mechanical Engineering*. London: Faber and Faber.
Buxton, Neil K. 1979. "Introduction," in Neil K. Buxton and Derek H. Aldcroft, eds., *British Industry Between the Wars*. London: Scolar Press.
Calkins, E.E. 1932. "What Consumer Engineering Really is" in Roy Sheldon and Egmont Arens, eds., *Consumer Engineering*. New York: Harper and Brothers.
Calomiris, Charles W. 1993. "Financial Factors in the Great Depression." *Journal of Economic Perspectives* 7:2, 61-86.
Cardwell, D.S.L. 1972. *Turning Points in Western Technology*. New York: Neale Watson Academic Publications.
Casson, Mark. 1983. *Economics of Unemployment*. Oxford: Robertson.
Catterall, R.E. 1979. "Electrical Engineering," in Neil K. Buxton and Derek H. Aldcroft, eds., *British Industry Between the Wars*. London: Scolar Press.
Cechetti, S. 1992. "Prices During the Great Depression: Was the Deflation of 1930-1932 Really Unanticipated?" *American Economic Review* 82:1, 141-56.
Chandler, A.D. 1963. *Strategy and Structure: Chapters in the History of the American Corporation*. Cambridge, MA: MIT Press.
Chandler, L.V. 1970. *America's Greatest Depression 1929-1941*. New York: Harper and Row.
Chase, Stuart. 1932. "The Heart of American Industry," in Fred Ringel, ed., *America as Americans see it*. New York: Literary Guild.
Cheng, L., and E. Dinopolous. 1992. "Schumpeterian Growth and International Business Cycles." *American Economic Review* 82:2, 409-14.
Church, Roy. 1981. "The Marketing of Automobiles in Britain and the United States Before 1939," in Akio Okochi, ed., *Development of Mass Marketing*. Tokyo: University of Tokyo Press.
Clark, Harold. 1937. *Recent Economic Changes and Their Meaning*. Columbus, OH: Educational Printing House.
Clewett, J. 1951. "Mass Marketing," in H. Williamson, ed., *Growth of the American Economy*. New York: Prentice Hall.
Coen, Robert. 1973. "Labor Force and Unemployment in the 1920's and 1930's: A Re-examination Based on Postwar Experience." *Review of Economics and Statistics* 40:1, 46-55.
Cohen, Avi J. 1984. "Technological Change as Historical Process: The Case of the U. S. Pulp and Paper Industry 1915 - 1940." *Journal of Economic History* 44:3, 775-99.
Constant, Edward. 1980. *The Origins of the Turbojet Revolution*. Baltimore: Johns Hopkins University Press.
Cooper, R., and J. Haltiwanger. 1990. "Inventories and the Propagation of Sectoral Shocks." *American Economic Review* 80:1, 170-90.

Copeland, Morris. 1929. "The National Income and Its Distribution," in *Recent Economic Changes in the United States*. President's Conference on Unemployment.

Coplon, W. 1932. "Lumber: An Old Industry and the New Competition." *Harvard Business Review* 9:2, 161-9.

Cornwall, John. 1977. *Growth and Stability in a Mature Economy*. London: Martin Robertson.

Costrell, Robert M. 1988. "The Effect of Technical Progress on Productivity, Wages, and Distribution of Employment," in Richard M. Cyert and David C. Mowery, eds., *The Impact of Technological Change on Employment and Economic Growth*. Cambridge, MA: Harper and Row.

Cowan, Ruth S. 1983. *More Work for Mother*. New York: Basic Books.

Cox, Charles. 1981. "Monopoly Explanations of the Great Depression and Public Policies Toward Business," in K. Brunner, ed., *The Great Depression Revisited*. Boston: Martinus Nijhoff.

Crabb, A. Richard. 1969. *Birth of a Giant*. Philadelphia: Chilton Book Co.

Cross, Gary, and Rick Szostak. 1995. *Technology and American Society: A History*. Englewood Cliffs, N.J.: Prentice-Hall.

Cunningham, W. 1929. "Railroads," in *Recent Economic Changes in the United States*. President's Conference on Unemployment.

Cyert, Richard M., and David C. Mowery. 1988. "Introduction," in *The Impact of Technological Change on Employment and Economic Growth*. Cambridge, MA: Harper and Row.

Darby, M. 1976. "Three and a Half Million U.S. Employees Have Been Mislaid: Or, An Explanation of Unemployment 1934-1941." *Journal of Political Economy* 84:1, 1-16.

Dasgupta, Partha. 1989. "Economic Organization and Technological Change," in Aubrey Silberston, ed., *Technology and Economic Progress*. Hampshire: Macmillan Press.

Davis, Donald. 1988. *Conspicuous Production: Automobiles and Elites 1899-1933*. Philadelphia: Temple University Press.

Davis, Pearce. 1970. *The Development of the American Glass Industry*. New York: Russell and Russell, (1949).

Dearden, J., et al. 1990. "To Innovate or Not to Innovate: Incentives and Innovations in Hierarchies." *American Economic Review* 80:5, 1105-24.

DeBresson, C. 1991. "Technological Innovation and Long Wave Theory: two pieces of the puzzle." *Journal of Evolutionary Economics* 1, 241-72.

DeLong, J.B., and A. Schleifer, 1991. "The Stock Market Bubble of 1929: Evidence From Closed-End Mutual Funds." *Journal of Economic History* 51:3, 675-700.

Dennison, Henry S. 1929. "Management," in *Recent Economic Changes in the United States*. President's Conference on Unemployment.

Dimand, Robert. 1991. "Cranks, Heretics, and Macroeconomics in the 1930s." *History of Economics Review* 161-72.

Dixit, Avinash. 1992. "Investment and Hysteresis." *Journal of Economic Perspectives* 6:1, 107-32.

Dominguez, K., et al. 1988. "Forecasting the Depression; Harvard versus Yale." *American Economic Review* 78:4, 595-612.

Domowitz, Ian, R. Glenn Hubbard, and Bruce Peterson. 1988. "Market Structure and Cyclical Fluctuations in United States Manufacturing." *Review of Economics and Statistics* 70:1, 55-66.
Dosi, Giovanni. 1988. "Sources, Procedures, and Microeconomic Effects of Innovation." *Journal of Economic Literature* 26, 1120-1171.
Douglas, Paul. 1930. *Real Wages in the United States 1890-1926.* Boston: Houghton-Mifflin.
Douglas, Paul, and A. Director. 1931. *The Problem of Unemployment.* New York: Macmillan.
Douglas, Susan J. 1987. *Inventing American Broadcasting 1899-1922.* Baltimore: Johns Hopkins University Press.
Drummond, I. 1972. "The British Empire Economies in the Great Depression," in H. Van der Wee, ed., *The Great Depression Revisited.* The Hague: Nijhoff.
Du Boff, R. 1979. *Electric Power in American Manufacturing 1889-1958.* New York: Arno Press.
Dummer, G.W.A. 1978. *Electronic Inventions and Discoveries.* 2nd ed. Oxford: Pergamon Press.
Eichengreen, Barry. 1987. "Macroeconomics and History," in A. Field, ed., *The Future of Economic History.* Boston: Kluwer-Nijhoff.
_____1988. "Did International Economic Forces Cause the Great Depression?" *Contemporary Policy Issues* 6:2, 90-114.
_____1990. "Introduction," in E. Aerts and B. Eichengreen, eds., *Unemployment and Underemployment in Historical Perspective.* Proceedings of the Tenth International Economic History Congress, Leuven University Press.
_____1992. *Golden Fetters.* New York: Oxford University Press.
Eichengreen, Barry, and T. J. Hatton. 1988. "Interwar Unemployment in International Perspective: An Overview" in *Interwar Unemployment in International Perspective.* Norwell MA: Kluwer Academic Publishers.
Eichler, Ned. 1982. *The Merchant Builders.* Boston: MIT University Press.
Enos, John L. 1962. "Invention and Innovation in the Petroleum Refining Industry," in *The Rate and Direction of Economic Activity: Economic and Social Factors.* Pp. 299-321. National Bureau of Economic Research.
Epstein, Ralph. 1934. *Industrial Profits in the United States.* New York: National Bureau of Economic Research.
Ewen, Stuart. 1976. *Captains of Consciousness: Advertising and the Social Roots of the Consumer Culture.* New York: McGraw Hill.
Ewen, Stuart, and Elizabeth Ewen. 1982. *Channels of Desire.* New York: McGraw-Hill.
Fabricant, Solomon. 1945. *Factory Employment and Output Since 1899.* National Bureau of Economic Research Occasional Paper No. 4.
_____1945b. *Labor Savings in American Industry 1899-1939.* National Bureau of Economic Research Occasional Paper No. 22.
Fair, R.C., and K. Dominguez. 1991. "Effects of the Changing U.S. Age Distribution on Macroeconomic Equations." *American Economic Review* 81:5, 1276-94.
Fano, Ester. 1987. "Technological Progress as a Destabilizing Factor and as an Agent of Recovery in the United States Between the Two World Wars." *History and Technology* 3:3, 249-74.

_____1991. "A Wastage of Men: Technological Progress and Unemployment in the United States." *Technology and Culture*, 264-92.
Fearon, Peter. 1987. *War, Prosperity, and Depression*. Oxford: Philip Allan.
Field, Alexander J. 1984. "A New Interpretation of the Onset of the Great Depression." *Journal of Economic History* 44:3, 489-98.
_____1984b. "Asset Exchanges and the Transactions Demand for Money 1919-29." *American Economic Review* 74:1, 43-59.
_____,ed. 1987. *The Future of Economic History*. Boston: Kluwer-Nijhoff.
_____1992. "Uncontrolled Land Development and the Duration of the Depression in the United States." *Journal of Economic History* 52:4, 785-805.
Field, D.C. 1958. "Internal Combustion Engines," in Charles Singer, et al., eds., *A History of Technology*. Oxford: Oxford University Press.
Final Report and Recommendation of the Temporary National Economic Committee. 1941. Washington: United States Government Printing Office.
Finegan, T. Aldrich, and Robert A. Margo. 1994. "Work Relief and the Labor Force Participation of Married Women in 1940." *Journal of Economic History* 54:1, 64-84.
Fischer, Claude. 1988. "Touch Someone: The Telephone Industry Discovers Sociability." *Technology and Culture* 29:1.
Fisher, Franklin, J. McKie, and R. Mancke. 1983. *IBM and the U.S. Data Processing Industry: An Economic History*. New York: Praeger Publishers.
Fisher, Irving. 1933. "The Debt-Deflation Theory of Great Depressions." *Econometrica* 1:4, 337-57.
Fleisig, H. 1972. "The United States and the Non-European Periphery During the Early Years of the Great Depression," in H. Van der Wee, ed., *The Great Depression Revisited*. The Hague: Nijhoff.
Folk, George E. 1942. *Patents and Industrial Progress*. New York: Harper and Brothers.
Freeman, Christopher, J. Clark, and L. Soete. 1982. *Unemployment and Technical Innovation*. Westport, CT: Greenwood Press.
Freeman, Christopher. 1984. "Prometheus Unbound." *Futures* 17, 494-506.
French, Michael. 1991. *The United States Tire Industry: A History*. Boston: Twayne.
Freudenthal, E.E. 1968. "Depression, Industrial Concentration, and the Growing Export Market," in Gene Simonson, ed., *The History of the American Aircraft Industry: An Anthology*. Cambridge, MA: MIT Press.
Friedman, Milton, and Anna J. Schwartz. 1963. *A Monetary History of the United States 1867-1960*. Princeton, N.J.: Princeton University Press.
Garraty, John. 1986. *The Great Depression: An Inquiry Into the Causes, Course, and Consequences of the Worldwide Depression of the Nineteen-Thirties, as Seen By Contemporaries and In the Light of History*. San Diego: Harcourt Brace Jovanovich.
Gennsheim, H. 1978. "Photography," in Trevor Williams, ed., *A History of Technology* vol. 7, Oxford: Clarendon.
Gershuny, J.I., and I.D. Miles. 1983. *The New Service Economy*. London: Frances Printer.
Gibbs-Smith, C.H. 1970. *Aviation: An Historical Survey*. London: Her Majesty's Stationery Office.
Giedion, S. 1948. *Mechanization Takes Command*. New York: Oxford University Press.

Gilfillan, S.C. 1935. *The Sociology of Invention*. Cambridge, MA: MIT Press.
Gill, C. 1940. *Unemployment and Technological Change*, Works Progress Administration National Research Project G7.
Gill, C., and E. Ross. 1939. "Price Dispersion and Industrial Action 1928-1938," Works Progress Administration.
Ginzberg, Eli. 1939. *The Illusion of Economic Stability*. New York: Harper and Brothers.
Goldin, Claudia Dale. 1990. *Understanding the Gender Gap: An economic History of American Women*. New York: Oxford University Press.
Goldsmith, Raymond. 1955. *A Study of Saving in the United States*. Princeton N.J.: Princeton University Press.
Goldstein, J. 1988. *Long Cycles: Prosperity and War in the Modern Age*. New Haven: Yale University Press.
Gomulka, Stanislaw. 1990. *The Theory of Technological and Economic Growth*. London: Routledge.
Goodman, J., and K. Honeyman. 1988. *Gainful Pursuits: The Making of Industrial Europe, 1600-1914*. London: Edward Arnold.
Gordon, David, et al. 1982. *Segmented Work, Divided Workers: The Historical Transformation of Labor in the United States*. New York: Cambridge University Press.
Gordon, R.A. 1949. "Business Cycles in the Interwar Period: The Quantitative - Historical Approach." *American Economic Review* 39:2, 47-63.
_____ 1961. *Business Fluctuations*. New York: Harper and Row.
Gordon R.J., and James A. Willcox. 1981. "Monetarist Interpretations of the Great Depression," in K. Brunner, ed., *The Great Depression Revisited*. Boston: Martinus Nijhoff.
Gordon, R.J. 1983. "A Century of Evidence on Wage and Price Stickiness in the United States, the United Kingdom, and Japan," in J. Tobin, ed., *Macroeconomics, Prices, and Quantities*. Washington: Brookings.
Gort, Michael. 1962. *Diversification and Integration in American Industry 1929-54*. Princeton: Princeton University Press.
Gould, J.M. 1946. *Output and Productivity in the Electric and Gas Utilities 1899-1942*. Cambridge, MA: National Bureau of Economic Research.
Gourvitch, Alexander. 1966. *Survey of Economic Theory on Technological Change and Employment*. Works Progress Administration, 1940, reprinted NY: Augustus M. Kelley.
Gowdy, J.M. 1992. "Higher Selection Processes in Evolutionary Economic Change." *Journal of Evolutionary Economics* 2:1, 1-16.
Gramlich, Edward. 1985. "Government Services," in Robert P. Inman, ed., *Managing the Service Economy*. Cambridge: Cambridge University Press.
Grebler, Leo, et al. 1956. *Capital Formation in Residential Real Estate*. Princeton, N.J.: National Bureau of Economic Research.
Green, Alan, and Mary MacKinnon. 1990. "Regional Employment in the Depression: A Northern View," in E. Aerts and B. Eichengreen, eds., *Unemployment and Underemployment in Historical Perspective*. Proceedings of the Tenth International Economic History Congress, Leuven University Press.
Gries, John M. 1929. "Construction," in *Recent Economic Changes in the United States*. President's Conference on Unemployment.

Griliches, Z. 1990. "Patent Statistics as Economic Indicators: A Survey." *Journal of Economic Literature* 28:4, 1661-1707.

Grubel, Herbert G., and Michael A. Walker. 1989. *Service industry growth: causes and effects*. Vancouver: Fraser Institute.

Haber, L.F. 1971. *The Chemical Industry 1900-1930*. Oxford: Clarendon.

Haberler, Gottfried. 1957. *Prosperity and Depression*. Cambridge, MA: Harvard University Press.

Hamermesh, Daniel S. 1989. "Labor Demand and the Structure of Adjustment Costs." *American Economic Review* 79:4, 674-89.

Hamilton, James D. 1987. "Monetary Factors in the Great Depression." *Journal of Monetary Economics* 19, 145-69.

_____1992. "Was the Deflation During the Great Depression Anticipated? Evidence From the Commodity Futures Market." *American Economic Review* 82:1, 157-78.

Hampe, E., and M. Wittenberg. 1964. *The Lifeline of America*. New York: McGraw-Hill.

Hansen, Alvin. 1938. *Full Recovery or Stagnation?* New York: Norton.

Harberger, A. 1960. "Introduction," in *The Demand for Durable Goods*. Chicago: University of Chicago Press.

Hatton, T.J. 1985. "The British Labor Market in the 1920s: A Test of the Search-Turnover Approach." *Explorations in Economic History* 22, 257-70.

_____1986. "Structural Aspects of Unemployment in Britain Between the World Wars." *Research in Economic History* 10, 55-92.

Hayes, Douglas. 1951. *Business Confidence and Business Activity: A Case Study of the Recession of 1937*. Ann Arbor: University of Michigan Press.

Haynes, Williams. 1948. *American Chemical Industry*. 5 vols. D. Van Nostrand.

Heim, Carol E. 1984. "Structural Transformation and the Demand for New Labor in Advanced Economies: Interwar Britain." *Journal of Economic History* 44:2, 585-95.

Henderson, John. 1961. *Changes in the Industrial Distribution of Employment 1919-1959*. University of Illinois Press.

Hickman, B. 1974. "What Became of the Building Cycle?," in P. David and M. Reder, eds., *Nations and Households in Economic Growth*. New York: Academic Press.

Higgs, R. 1992. "Wartime Prosperity? A Reassessment of the U.S. Economy in the 1940s." *Journal of Economic History* 52:1, 41-60.

Hirschhorn, Larry. 1988. "Computers and Jobs: Services and the New Mode of Production," in Richard M. Cyert and David C. Mowery, eds., *The Impact of Technological Change on Employment and Economic Growth*. Cambridge, MA: Harper and Row.

Hogan, William T. 1950. *Productivity in the Blast Furnace and Open Hearth Segments of the Steel Industry 1920-1946*. New York: Fordham University Press.

Hogan, William T. 1971. *Economic History of the Iron and Steel Industry in the United States*. Lexington, MA: Lexington Books.

Hollander, Samuel. 1965. *The Sources of Increased Efficiency*. Cambridge, MA: MIT Press.

Holt, C. F. 1977. "Who Benefitted From the Prosperity of the Twenties?" *Explorations in Economic History* 14, 277-89.
Holthausen, D. 1940. *The Volume of Consumer Installment Credit 1929-1938.* New York: National Bureau of Economic Research.
Hopkins, John. 1973. *Changing Technology and Employment in Agriculture.* New York: Da Capo Press. (Government Printing Office 1941).
Hosios, Arthur J. 1994. "Unemployment and Vacancies with Sectoral Shifts." *American Economic Review* 84:1, 124-44.
Hounshell, David., and J.K. Smith Jr. 1988. *Science and Corporate Strategy: Du Pont R. and D. 1902-1980.* New York: Cambridge University Press.
Hughes, Jonathan R.T. 1977. *The Governmental Habit.* New York: Basic Books.
_____1987. *American Economic History.* 2nd. ed. Glenview, IL: Scott, Foresman, and Co.
Hughes, Thomas P. 1983. *Networks of Power: Electrification in Western Society 1880-1930.* Baltimore: Johns Hopkins University Press.
Hunnicut, B.K. 1988. *Work Without End: Abandoning Shorter Hours For the Right to Work.* Philadelphia: Temple University Press.
Inkster, Ian. 1991. *Science and Technology in History.* London: Macmillan.
Jacoby, Sanford. 1985. *Employing Bureaucracy: Managers, Unions, and the Transformation of Work in American Industry 1900-1945.* New York: Columbia University Press.
Jaeger, Hans. 1972. "Business in the Great Depression," in H. Van der Wee, ed., *The Great Depression Revisited: Essays on the Economics of the Thirties.* The Hague: Nijhoff.
Jenkins, Reese V. 1975. *Images and Enterprise: Technology and the American Photographic Industry.* Baltimore: Johns Hopkins University Press.
Jensen, Vernon H. 1945. *Lumber and Labor.* New York: Farrar and Rinehart.
Jerome, Harry. 1934. *Mechanization in Industry.* New York: National Bureau of Economic Research.
Jewkes J., et al. 1968. *The Sources of Invention.* London: Macmillan.
Jones, Eric. 1988. *Growth Recurring; Economic Change in World History.* Cambridge: Clarendon.
Katon, H., and G. Saeger. 1939. *Changes in Technology and Labor Requirements in the Crushed Stone Industry.* Philadelphia: Works Progress Administration.
Katsoulocas, Y. 1986. *The Employment Effect of Technical Change.* Lincoln: University of Nebraska Press.
_____1991. "Technical Change and Employment under Imperfect Competition with Perfect and Imperfect Information." *Journal of Evolutionary Economics* 1:3, 207-18.
Katz, B., and A. Philips. 1982. "Government, Technological Opportunities, and the Emergence of the Computer Industry," in H. Giersch, ed., *Emerging Technologies: Consequences For Growth, Structural Change, and Employment.* Tubingen: J.C.B. Mohr.
Keller, R.R. 1973. "Factor Income Distribution in the United States During the 1920's: A Reexamination of Fact and Theory." *Journal of Economic History* 33:2, 252-73.

Kendrick, John. 1985. "Measurement of Output and Productivity in the Service Sector," in Robert P. Inman, ed., *Managing the Service Economy*. Cambridge: Cambridge University Press.

Kennedy, E.D. 1939. *Dividends to Pay*. New York: A.M. Kelley.

———1941. *The Automobile Industry*. New York: Reynal and Hitchcock.

Kesselman, J.R., and N.E. Savin. 1978. "Three and a Half Million Workers Were Never Lost." *Economic Inquiry* 16:2, 205-25.

Keynes, J.M. 1930. "Economic Possibilities For Our Grandchildren," in *Essays in Persuasion*. New York: Norton (1963).

———1930b. *A Treatise on Money*. New York: Harcourt, Brace and company.

———1936. *The General Theory of Employment, Interest and Money*. London: Macmillan.

Kimball, D. 1929. "Changes in New and Old Industries," in *Recent Economic Changes in the United States*. President's Conference on Unemployment.

Kimball, Miles S. 1994. "Labor Market Dynamics when Employment is a Worker Discipline Device." *American Economic Review* 84:4, 1045-59.

Kimmel, Lewis. 1939. *The Availability of Bank Credit 1933-1938*. National Industrial Conference Board.

Kindleberger, Charles P. 1973. *The World in Depression 1929-1939*. Berkeley: University of California Press.

Klamer, Arjo. 1984. *Conversations with Economists*. Totowa, NJ: Rowman and Allenheld.

Klein, B. 1977. *Dynamic Economics*. Cambridge, MA: Harvard University Press.

Klein, J. 1928. "Business," in Charles Beard, ed., *Whither Mankind?* Freeport, NY: Books for Libraries Press.

Kleinknecht, Alfred. 1987. *Innovation Patterns in Crisis and Prosperity: Schumpeter's Long Cycle Reconsidered*. London: Macmillan.

Koenigsberger, F. 1978. "Production Engineering," in Trevor Williams, ed., *A History of Technology* vol. 7., 1046-59. Oxford: Clarendon.

Kraft, James P. 1994. "Musicians in Hollywood: Work and Technological Change in Entertainment Industries 1926-1940." *Technology and Culture* 35:2, 289-314.

Krooss, H., and C. Gilbert. 1972. *American Business History*. Englewood Cliffs, NJ: Prentice-Hall.

Kuznets, Simon. 1966. *Modern Economic Growth: Rate, Structure, and Spread*. New Haven: Yale University Press.

Laidler, David. 1989. "Dow and Saville's Critique of Monetary Policy: A Review Essay." *Journal of Economic Literature* 27:3, 1147-59.

Larson, H., E. Knowlton, and C. Popple. 1971. *New Horizons 1927-1950: History of Standard Oil Company*. New York: Harper and Row.

Layton, David, 1978. "Education in Industrialized Societies," in T. Williams, ed., *A History of Technology*. Oxford: Clarendon.

Leonard, Jonathan, S. 1988. "Technological Change and the Extent of Frictional and Structural Unemployment," in Richard M. Cyert and David C. Mowery, eds., *The Impact of Technological Change on Employment and Economic Growth*. Cambridge, MA: Harper and Row.

Leuchtenberg, William E. 1958. *The Perils of Prosperity 1914-32*. Chicago: University of Chicago Press.

Leven, M., et al. 1934. *America's Capacity to Consume*. Washington: Brookings.
Leveson, Irving. 1985. "Services in the U.S. Economy," in Robert P. Inman, ed., *Managing the Service Economy*. Cambridge: Cambridge University Press.
Liebenau, Jonathan. 1987. *Medical Science and Medical Industry*. Baltimore: Johns Hopkins University Press.
Lilien, David. 1982. "Sectoral Shifts and Cyclical Unemployment." *Journal of Political Economy* 90:41, 777-93.
Lindbeck, Assar, and D.J. Snower. 1988. *The Insider-Outsider Theory of Employment and Unemployment*. Cambridge, MA: MIT Press.
Linder, Staffan. 1970. *The Harried Leisure Class*. New York: Columbia University Press.
Lindert, Peter. 1981. "Comments on Understanding 1929-1933," in K. Brunner, ed., *The Great Depression Revisited*. Boston: Martinus Nijhoff.
Lodewijks, J.K. 1990. "Market Structure and Industrial Innovation." *Prometheus* 8.
Lorant, J.H. 1966. *The Role of Capital-Improving Innovations in American Manufacturing During the 1920's*. Columbia University thesis; reprinted 1973 by University Microfilms, Ann Arbor, MI.
Lough, William. 1935. *High-Level Consumption*. New York: McGraw-Hill.
Lubin, I. 1929. *The Absorption of the Unemployed by American Industry*. Washington: Brookings.
Lucas, R., and L. Rapping. 1972. "Unemployment in the Great Depression: Is There a Full Explanation?" *Journal of Political Economy* 80:1, 186-191.
Luckoff, H. 1979. *From Dits to Bits: A Personal History of the Computer*. Portland: Robotics Press.
MacKinnon, Mary. 1990. "Relief not Insurance: Canadian Unemployment Relief in the 1930s." *Explorations in Economic History* 27:1, 46-83.
MacLaurin, W.R. 1949. *Invention and Innovation in the Radio Industry*. New York: Macmillan.
Maidique, M.A., and S. Zirger. 1985. "The New Product Learning Cycle." *Research Policy* 14, 299-313.
Mandel, Ernst. 1980. *The Second Slump*. London: Verso.
Mankiw, N.G. 1989. "Real Business Cycles: A New Keynesian Perspective." *Journal of Economic Perspectives* 3:3, 79-90.
_____ 1990. "A Quick Refresher Course in Macroeconomics." *Journal of Economic Literature* 28:4, 1645-60.
Mansfield, E. 1968. *The Economics of Technological Change*. New York: Norton.
Margo, Robert. 1988. "Interwar Unemployment in the United States: Evidence From the 1940 Census Sample," in B. Eichengreen and T.J. Hatton, eds., *Interwar Unemployment in International Perspective*. Norwell, MA: Kluwer Academic Publishers.
Margo, Robert. 1993. "Employment and Unemployment in the 1930s." *Journal of Economic Perspectives* 7:2, 41-60.
Matthews, R.C.O. 1968. "Why Has Britain Had Full Employment Since the War?" *Economic Journal* 78, 555-69.
Mayer, Thomas. 1978. "Consumption in the Great Depression." *Journal of Political Economy* 86:1, 139-45.

McCraw, T., and F. Reinhardt. 1989. "Losing to Win: US Steel's Pricing, Investment Decisions, and Market Share 1901-1938." *Journal of Economic History* 49:3, 593-620.

McKie, James W. 1993. *Tin Cans and Tin Plate*. Cambridge MA: Harvard University Press.

Mensch, G. 1979. *Stalemate in Technology*. Cambridge, MA: Ballinger.

Mercer, Lloyd J., and W.D. Morgan. 1972. "Alternative Interpretations of Market Saturation: Evaluation For the Automobile Market in the Late 1920s." *Explorations in Economic History* 9:2, 269-90.

Mercer, Lloyd J., and W.D. Morgan. 1973. "Housing Surplus in the 1920s: Another Evaluation." *Explorations in Economic History* 10:3, 295-303.

Mieszkowski, Peter, and Edwin S. Mills. 1993. "The Causes of Metropolitan Suburbanization." *Journal of Economic Perspectives* 7:3, 135-47.

Miller, M., and R.A. Church. 1979. "Motor Manufacturing," in Neil K. Buxton and Derek H. Aldcroft, eds., *British Industry Between the Wars*. London: Scolar Press.

Mills, Frederick C. 1932. *Economic Tendencies in the United States*. New York: National Bureau of Economic Research.

Mills, Frederick C. 1938. *Employment Opportunities in Manufacturing Industries in the United States*. New York: National Bureau of Economic Research.

Mingos, Howard. 1968. "Birth of an Industry," in Gene Simonson, ed., *The History of the American Aircraft Industry*, 23-69. Cambridge, MA: MIT Press.

Miron, Jeffrey. 1986. "Financial Panics, The Seasonality of the Nominal Interest Rate, and the Founding of the Fed." *American Economic Review* 76:1, 125-40.

Mishkin, F.S. 1978. "The Household Balance Sheet and the Great Depression." *Journal of Economic History* 38:4, 918-37.

Mitchell, W.C. 1929. "Conclusion," in *Recent Economic Changes in the United States*. President's Conference on Unemployment.

Mokyr, Joel. 1990. "Punctuated Equilibria and Technological Progress." *American Economic Review* 80:2, 350-4.

———1991. *The Lever of Riches*. New York: Oxford University Press.

Mowery, David C. 1983. "Industrial Research and Firm Size, Survival, and Growth in American Manufacturing 1921-1946: An Assessment." *Journal of Economic History* 43:4, 953-80.

———1983b. "The Relation Between Intrafirm and Contract Forms of Industrial Research in American Manufacturing 1900-1940." *Explorations in Economic History* 20, 351-74.

———1988. "Diffusion of New Manufacturing Technology," in Richard M.Cyert and David C. Mowery, eds., *The Impact of Technological Change on Employment and Economic Growth*. Cambridge, MA: Harper and Row.

Mowery, David C., and Nathan Rosenberg. 1989. *Technology and the Pursuit of Economic Growth*. Cambridge: Cambridge University Press.

Mueller, Willard. 1962. "The Origins of the Basic Inventions Underlying Du Pont's Major Product and Process Innovations 1920-1950," in *The Rate and Direction of Inventive Activity*. National Bureau of Economic Research.

National Industrial Conference Board. 1935. *Machinery, Employment, and Purchasing Power*. New York.

Neisser, Hans. 1942. "Permanent Technological Unemployment." *American Economic Review* 32:1, 50-71.
Nell, Edward. 1988. *Prosperity and Public Spending*. Boston: Unwin Hyman.
Nelson, Richard. 1959. "The Simple Economics of Basic Research," *Journal of Political Economy* 67, 297-306.
_____1962. "The Link Between Science and Invention: The Case of the Transistor," in *The Rate and Direction of Inventive Activity*. National Bureau of Economic Research.
Nelson, R., and S. Winter. 1982. *An Evolutionary Theory of Economic Change*. Cambridge, MA: Harvard University Press.
Noble, David. 1977. *America by Design*. New York: Knopf.
Norton, R.D. 1986."Industrial Policy in America." *Journal of Economic Literature* 24:1, 1-40.
Nye, David. 1985. *Image Worlds*. Cambridge, MA: MIT Press.
O'Brien, Anthony Patrick. 1989. "The ICC, Freight Rates, and the Great Depression." *Explorations in Economic History* 26:1, 73-98.
_____1990. "A Behavioural Explanation for Nominal Wage Rigidity During the Great Depression." *Quarterly Journal of Economics* 104:4, 719-36.
Oi, Walter Y. 1988. "The Indirect Effect of Technology on Retail Trade," in Richard M. Cyert and David C. Mowery, eds., *The Impact of Technological Change on Employment and Economic Growth*. Cambridge, MA: Harper and Row.
Olney, Martha L. 1989. "Credit as a Production-smoothing Device: The Case of Automobiles 1913-1938." *Journal of Economic History* 49:2, 377-92.
_____1989b. "Consumer Durables in the Interwar Years." *Research in Economic History* 12, 119-50.
_____1990. "Demand for Consumer Durable Goods in Twentieth Century America." *Explorations in Economic History* 27, 322-49.
_____1991. *Buy Now, Pay Later*. Chapel Hill: University of North Carolina Press.
Olson, Mancur. 1982. *The Rise and Decline of Nations*. New Haven: Yale University Press.
Pasinetti, Luigi. 1981. *Structural Change and Economic Growth*. New York: Cambridge University Press.
Patinkin, Don. 1956. *Money, interest, and price: an integration of monetary and value theory*. Evanston, IL: Row, Peterson.
Peppers, Larry C. 1973. "Full Employment Surplus Analysis and Structural Change: The 1930s." *Explorations in Economic History* 10, 197-210.
Perazick, G., and P. Field. 1940. *Industrial Research and Changing Technology*. Works Progress Administration, National Research Project M-1, Philadelphia.
Phelps, Edmond S. 1993. *Structural Slumps*. Cambridge, MA: Harvard University Press.
Philips, A. 1971. *Technology and Market Structure: A Study of the Aircraft Industry*. Lexington, MA: Heath Lexington Books.
Platt, H.L. 1988. "City Lights: The Electrification of the Chicago Region, 1880-1930," in J. Tarr and G. Dupuy, eds., *Technology and the Rise of the Networked City in Europe and America*. Philadelphia: Temple University Press.
Plosser, C.I. 1989. "Understanding Real Business Cycles." *Journal of Economic Perspectives* 3:3, 51-78.

Podgursky, Michael. 1988. "Job Displacement and Labor Market Adjustment: Evidence From the Displaced Worker Surveys," in Richard M. Cyert and David C. Mowery, eds., *The Impact of Technological Change on Employment and Economic Growth*. Cambridge, MA: Harper and Row.

Pursell, Carroll W. 1979. "Government and Technology in the Great Depression." *Technology and Culture* 20:1, 165-78.

Rae, John. 1968. *Climb to Greatness: The American Aircraft Industry 1920-1960*. Cambridge, MA: MIT Press.

———1984. *The American Automobile Industry*. Boston: Twayne Publishers.

Ramey, Valerie. 1989. "Inventories as Factors of Production and Economic Fluctuations." *American Economic Review* 79:3, 338-54.

———1991. "Nonconvex Costs and the Behavior of Inventories." *Journal of Political Economy* 99:2, 306-34.

Rasmussen, W., and J. Porter. 1972. "Agriculture in the Industrial Economies of the West During the Great Depression, with Specific Reference to the United States," in H. Van der Wee, ed., *The Great Depression Revisited*. The Hague: Nijhoff.

Raupach, H. 1972. "The Impact of the Great Depression on Eastern Europe," in H. Van der Wee, ed., *The Great Depression Revisited*. The Hague: Nijhoff.

Reekie, W.D. 1975. *The Economics of the Pharmaceutical Industry*. London: MacMillan.

Reich, Leonard. 1985. *The Making of American Industrial Research*. Cambridge MA: Cambridge University Press.

Robertson, D.H. 1948. "New Light on an Old Story." *Economica* 15, 294-300.

Rolt, L.T.C. 1986. *Tools For the Job*. London: Her Majesty's Stationery Office (1965).

Romer, Christina. 1992. "The Great Crash and the Onset of the Great Depression." *Quarterly Journal of Economics* 107, 597-623.

———1992b. "What Ended the Great Depression?" *Journal of Economic History* 52:3, 757-84.

———1993. "The Nation in Depression." *Journal of Economic Perspectives* 7:2, 19-40.

Rose, Mark. 1988. "Urban Gas and Electric Systems and Social Change 1900-1940," in J. Tarr and G. Dupuy, eds., *Technology and the Rise of the Networked City in Europe and America*. Philadelphia: Temple University Press.

Rosenberg, Nathan. 1963. "Technological Change in the Machine Tool Industry 1840-1910." *Journal of Economic History* 23:2, 414-46.

Rostow, W.W. 1971. "The Strategic Role of Theory: A Commentary." *Journal of Economic History* 31:1, 76-86.

———1978. *The World Economy: History and Prospect*. Austin: University of Texas Press.

Rotella, E. 1981. *From Home to Office*. Ann Arbor: UMI Research Press.

Rowthorn, R., and A. Glyn. 1990. "The Diversity of Unemployment Experience Since 1973." *Structural Change and Economic Dynamics* 1:1, 57-89.

Sachs, Jeffrey. 1980. "The Changing Cyclical Behavior of Wages and Prices 1890-1976." *American Economic Review* 70:1, 78-90.

Safarian, A.E. 1958. *The Canadian Economy in the Great Depression*. Toronto: University of Toronto Press.

Sahal, D. 1985. "Technological Guideposts and Innovation Avenues." *Research Policy* 14, 61-82.
Saint-Etienne, C. 1984. *The Great Depression 1929-1938.* Stanford: Hoover Institution Press.
Sardoni, Claudio. 1987. *Marx and Keynes on Economic Recessions.* New York: New York University Press.
Saul, S.B. 1985. *The Myth of the Great Depression 1873-1896.* 2nd ed. London: MacMillan.
Sawers, David. 1978. "The Sources of Innovation," in Trevor Williams, ed., *A History of Technology.* vol. 6. Oxford: Clarendon Press.
Scherer, F.M. 1984. *Innovation and Growth: Schumpeterian Perspectives.* Cambridge, MA: MIT Press.
Schlebecker, John T. 1975. *Whereby We Thrive: A History of American Farming 1607-1972.* Ames, Iowa: Iowa State University Press.
Schroeder, Fred. 1986. "More 'Small Things Forgotten': Domestic Electrical Plugs and Receptacles 1881-1931." *Technology and Culture* 27:3, 525-43.
Schultz, Theodore. 1990. *Restoring Economic Equilibrium: Human Capital in the Modernizing Economy.* Cambridge, MA: Blackwell.
Schumpeter, Joseph. 1939. *Business Cycles.* New York: McGraw-Hill.
Schurr, S.H., and B.C. Netschert. 1960. *Energy in the American Economy, 1850-1975.* Baltimore: Johns Hopkins University Press.
Schwartzmann, David. 1976. *Innovation in the Pharmaceutical Industry.* Baltimore: Johns Hopkins University Press.
_____1985. "The Growth of Sales Per Manhour in Retail Trade," in Robert P. Inman, ed., *Managing the Service Economy.* Cambridge: Cambridge University Press.
Scitovsky, Tibor. 1986. *Human Desire and Economic Satisfaction.* New York: New York University Press.
Scoville, John, and Noel Sargent. 1942. *Fact and Fancy in the TNEC Monographs.* New York: National Association of Manufacturers.
Seltzer, Lawrence. 1928. *A Financial History of the American Automobile Industry.* Boston: H. Mifflin.
Shannon, David. 1960. *The Great Depression.* Englewood Cliffs, N.J.: Prentice-Hall.
Sharlin, Harold. 1963. *The Making of the Electrical Age.* London: Abelard-Schuman.
Sharpe, Steven A. 1994. "Financial Market Imperfections, Firm Leverage, and the Cyclicality of Employment." *American Economic Review* 84:4, 1060-74.
Sheldon, Roy, and Egmont Arens, eds. 1932. *Consumer Engineering.* New York: Harper and Brothers.
Simon, Curtis J., and Clark Nardinelli. 1992. "Does Industrial Diversity Always Reduce Unemployment? Evidence from the Great Depression and After." *Economic Inquiry* 16:2, 384-97.
Sinclair, Peter. 1981. "When Will Technical Progress Destroy Jobs?" *Oxford Economic Papers* 33:1, 1-19.
_____1987. *Unemployment: Economic Theory and Evidence.* Oxford: Basil Blackwell.
Slichter, Sumner. 1929. "The Current Labor Policies of American Industries." *Quarterly Journal of Economics* 64, 393-435.

Smiley, Gene. 1983. "Did Incomes For Most of the Population Fall From 1923 Through 1929?" *Journal of Economic History* 43:1, 209-16.

―――― 1983b. "Recent Unemployment Rate Estimates For the 1920s and 1930s." *Journal of Economic History* 43:2, 487-93.

Smith, Philip. 1970. *Wheels Within Wheels.* New York: Funk and Wagnalls.

Solomou, S. 1987. *Phases of Economic Growth, 1850-1973: Kondratieff Waves and Kuznets Swings.* New York: Cambridge University Press.

Soule, George H. 1947. *Prosperity Decade; from war to depression: 1917-1929.* New York: Rinehart.

Stanback, Thomas M. Jr., et al. 1981. *Services: The New Economy.* Totowa, NJ: Allanheld, Osmon & Co.

Staudenmaier, John. 1989. *Technology's Storytellers.* Cambridge, MA: MIT Press.

Stigler, George. 1956. *Trends in Employment in the Service Industries.* Princeton, NJ: Princeton University Press.

Stolper, W.F. 1991. "The Theoretical Bases of Economic Policy: the Schumpeterian Perspective." *Journal of Evolutionary Economics* 1:3, 189-205.

Stricker, Frank. 1983. "Affluence For Whom? - Another Look at Prosperity and the Working Classes in the 1920s." *Labor History* 24:1, 5-33.

Sturmey, S.G. 1958. *The Economic Development of Radio.* London: Gerald Duckworth.

Sundstrom, William. 1990. "Was There a Golden Age of Flexible Wages: Evidence From Ohio Manufacturing 1892-1910." *Journal of Economic History* 50:2, 309-20.

―――― 1992. "Last Hired, First Fired? Unemployment and Urban Black Workers During the Great Depression." *Journal of Economic History* 52:2, 415-30.

―――― 1994. "The Color Line: Racial Norms and Discrimination in Urban Labor Markets 1910-1950." *Journal of Economic History* 54:2, 382-96.

Svennilson, Ingvar. 1954. *Growth and Stagnation in the European Economy.* Geneva: United Nations Commission For Europe.

Swann, John. 1988. *Academic Scientists and the Pharmaceutical Industry.* Baltimore: Johns Hopkins University Press.

Sylos-Labini, Paolo. 1984. *The Forces of Economic Growth and Decline.* Cambridge, MA: MIT Press.

Szostak, Rick. 1991. *The Role of Transportation in the Industrial Revolution: A Comparison of England and France.* Montreal: McGill-Queen's University Press.

Taylor, F.S. 1957. *A History of Industrial Chemistry.* New York: Abelard-Schuman.

Taylor, Morris P. 1933. *Common Sense About Machines and Unemployment.* Philadelphia: Winston.

Temin, Peter. 1976. *Did Monetary Forces Cause the Great Depression?* New York: Norton.

―――― 1981. "Notes on the Causes of the Great Depression," in K. Brunner, ed., *The Great Depression Revisited.* Boston: Martinus Nijhoff.

―――― 1989. *Lessons From the Great Depression.* Cambridge, MA: MIT Press.

―――― 1991. "Soviet and Nazi Economic Planning in the 1930s." *Economic History Review* 44:4, 573-93.

―――― 1993. "Transmission of the Great Depression." *Journal of Economic Perspectives* 7:2, 87-102.

Temporary National Economic Committee. 1941. *Investigation of Concentration of Economic Power.* Monograph no. 22: "Technology in our Economy." United States Government Printing Office.

Terbogh, George. 1945. *The Bogey of Economic Maturity.* Chicago: Machinery and Allied Products Institute.

Thomas, David. "Cinematography," in T. Williams, ed., *A History of Technology.* vol. 7, 1302-16. Oxford: Clarendon.

Thomas, Mark. 1988. "Labour Market Structure and the Nature of Unemployment in Interwar Britain," in B. Eichengreen and T.J. Hatton, eds., *Interwar Unemployment in International Perspective.* Norwell, MA: Kluwer Academic Publishers.

Thorne, S. 1984. *The History of Food Processing.* Kirkby Lonsdale, Cumbria: Parthenon Publishing.

Thorp, Rosemary. 1984. "Introduction," in R. Thorp, ed., *Latin America in the 1930s.* London: Macmillan.

Thorp, Willard. 1929. "The Changing Structure of Industry," in *Recent Economic Changes in the United States.* President's Conference on Unemployment.

Tillman, J.R., and D.G. Tucker. 1978. "Electronic Engineering," in T. Williams, ed., *A History of Technology.* vol. 7, 1091-1125. Oxford: Clarendon.

Tilton, John. 1971. *International Diffusion of Technology: The Case of Semiconductors.* Washington, DC: Brookings.

Tobin, James. 1975. "Keynesian Models of Recession and Depression." *American Economic Review* 65:2, 195-200.

Tracey, Michael. 1972. "Agriculture in the Great Depression; World Market Developments and European Protectionism," in H. Van der Wee, ed., *The Great Depression Revisited.* The Hague: Nijhoff.

Udelson, Joseph. 1982. *The Great Television Race.* University of Alabama Press.

Usher, A.P. 1954. *A History of Mechanical Inventions.* Cambridge, MA: Harvard University Press.

Van Duijn, J.J. 1983. *The Long Wave In Economic Life.* London: Allen and Unwin.

Venables, A.J. 1985. "The Economic Implications of a Discrete Technical Change." *Oxford Economic Papers* 37:2, 230-48.

Ville, Simon. 1990. *Transport and the Development of the European Economy 1750-1918.* New York: St. Martin's.

Von Tunzelmann, G.N. 1982. "Structural Change and Leading Sectors in British Manufacturing," in Charles Kindleberger and G. di Tella, eds., *Economics in the Long View.* New York: New York University Press.

Wagoner, Harless D. 1968. *The U.S. Machine Tool Industry From 1900 to 1950.* Cambridge, MA: MIT Press.

Wallace, Henry. 1937. *Technology, Corporations, and the General Welfare.* Chapel Hill: University of North Carolina Press.

Wallis, John. 1989. "Employment in the Great Depression; New Data and Hypotheses." *Explorations in Economic History,* 45-72.

Wallis, John, and D. North. 1986. "Measuring the Transactions Sector in the American Economy 1870-1970," in S. Engerman and R. Gallman, eds., *Longterm Factors in American Economic Growth.* Chicago: University of Chicago Press.

Walsh, M. 1990. "'See This Amazing America' The Long Distance Bus Industry's Use of Advertising in its First Quarter Century," *Journal of Transport History* 3rd Series, 11:1, 61-89.

Walton, G.M., and R.M. Robertson. 1983. *History of the American Economy*. 5th ed. San Diego: Harcourt Brace Jovanovich Publishers.

Wandersee, Winifred D. 1981. *Women's Work and Family Values: 1920-1940*. Cambridge, MA: Harvard University Press.

Warren, Kenneth. 1973. *The American Steel Industry 1850-1970*. Oxford: Clarendon.

Warsh, David. 1984. *The Idea of Economic Complexity*. New York: Penguin.

Waters, Joseph. 1977. *Technological Acceleration and the Great Depression*. New York: Arno.

Webb, John, and Joseph Bevis. 1939. *Facts about Unemployment*. United States Government Printing Office.

Weil, Phillipe. 1989. "Increasing Returns and Animal Spirits." *American Economic Review* 79:4, 889-94.

Weinstein, Michael. 1980. *Recovery and Redistribution under the NIRA*. Amsterdam: North Holland.

Weintraub, D. 1937. *Unemployment and Increasing Productivity*. Philadelphia: Works Progress Administration, National Research Project.

Weir, Ron. 1989. "Rationalization and Diversification in the Scotch Whiskey Industry 1900-1939: Another Look at Old and New Industries." *Economic History Review* 42:3, 375-95.

Whatley, Warren. 1990. "Getting a Foot in the Door: 'Learning', State Dependence, and the Racial Integration of Firms." *Journal of Economic History* 50:1, 43-66.

Wheelock, David C. 1989. "The Strategy, Effectiveness, and Consistency of Federal Reserve Monetary Policy 1924-1933." *Explorations in Economic History* 26:4, 453-76.

———1991. *The Strategy and Consistency of Federal Reserve Monetary Policy 1924-1933*. Cambridge: Cambridge University Press.

White, Eugene. 1984. "A Reinterpretation of the Banking Crisis of 1930." *Journal of Economic History* 44, 119-38.

———1990. "The Stock Market Boom and Crash of 1929 Revisited." *Journal of Economic Perspectives* 4:2, 67-84.

White, Eugene, and P. Rappoport. 1991. "Was There a Bubble in the 1929 Stock Market?" National Bureau of Economic Research Discussion Paper 3612.

Wicker, Elmer. 1980. "A Reconsideration of the Causes of the Banking Panic of 1930." *Journal of Economic History* 40:1, 55-66.

———1982. "Interest Rate and Expenditure Effects of the Banking Panic of 1930." *Explorations in Economic History* 19, 435-45.

Wied-Nebbeling, S. 1993. "Technical Change and Market Structure." *Journal of Evolutionary Economics*, 3:1, 43-61.

Williams, Trevor. 1982. *A Short History of Twentieth-Century Technology c. 1900-c. 1950*. Oxford: Clarendon Press.

Williamson, H., et al. 1963. *The American Petroleum Industry: The Age of Energy 1899-1959*. Evanston, IL: Northwestern University Press.

Williamson, Jeffrey, and Peter Lindert. 1980. *American Inequality: A Macroeconomic History.* New York: Academic Press.

Wise, George. 1985. *Willis R. Whitney, General Electric, and the Origins of Industrial Research.* New York: Columbia University Press.

Wolff, Edward. 1987. *Growth, Accumulation, and Unproductive Activity.* New York: Cambridge University Press.

Wolman, Leo. 1929. "Consumption and the Standard of Living," in *Recent Economic Changes in the United States.* President's Conference on Unemployment.

Woodbury, Robert S. 1978. "Machine Tools," in Trevor Williams, ed., *A History of Technology.* vol. 7, 1035-1045, Oxford: Clarendon.

Woolf, Arthur. 1984. "Electrification, Productivity, and Labor Saving in American Manufacturing 1900-1929." *Explorations in Economic History* 21, 176-91.

Wright. Gavin. 1987. "Labor History and Labor Economics," in A. Field, ed., *The Future of Economic History.* Boston: Kluwer-Nijhoff.

Zevin, R.B. 1982. "The Economics of Normalcy." *Journal of Economic History* 42:1, 43-52.

Index

Abernathy, William, 208, 210, 227-8
Accelerator mechanism, 35, 124
Adams, James, 334
Advertizing, 45, 53, 67, 80-1, 86, 127, 170, 183, 186, 216, 240-1, 245-6, 273-4, 280, 282
    See also Style changes
Agriculture, 18, 41, 70, 83-4, 104, 112, 116, 149, 152, 237, 243, 264-9, 294, 296, 300, 306-7, 333-4
Airplane, 4-5, 7, 36, 69, 91-2, 115, 164, 193-4, 196, 221-5, 294-5, 307
    See also Jet engine
Aitken, Hugh, 180-2, 184
Albion, M., 81
Aldcroft, Derek, 13, 30, 47, 54, 69, 82, 104, 195, 301-2, 305-6
Allen, Frederick, 14, 73-4, 86, 91, 94, 218, 226
Alloys, 104-5, 158, 160, 193-4, 196, 207-8, 220, 223, 225, 236, 250
Aluminium, 159-60, 166, 196, 207, 223, 227
Amendola, M., 109
Appliances, 72, 177, 188-92, 195, 218-9, 240, 277, 292, 296, 319
    refrigeration, 4, 9, 10, 191-2, 242, 280, 282, 292
    vacuum, 79, 188, 190, 192
    washing machine, 79, 190, 192, 285
Argentina, 265, 300
Assembly line, 6, 7, 103, 105, 111-3, 132, 149, 207-9, 260, 288-9
Australia, 23, 83, 208, 300
Automobile, 4-8, 24, 36, 69, 72, 76-80, 85, 87, 90-1, 96, 98, 102-3, 112, 114, 134, 151-2, 158, 162, 164, 189, 193-6, 205-18, 220, 233-6, 242-3, 255-9, 262, 268, 276, 280-2, 291, 296, 302-6, 317, 319

Backman, Jules, 244-7
Baily, Martin, 15, 21, 40-1, 43, 45, 105
Balke, Nathan, 45
Banking crises, 9, 13-14, 28, 30-2, 34, 46, 58-9, 226, 301, 311
Bankruptcy, 24, 32, 38-9, 92, 137, 214, 264
Barber, Clarence, 32-3, 83-4, 254
Barger, H., 229, 232, 236-8, 259-63, 275, 285
Barker, T.C., 206, 208
Basalla, G., 108, 113
Batra, Ravi, 70, 96
Belgium, 23, 217
Bell, Alexander Graham, 178-9
Bell, Spurgeon, 73
Bell Telephone (AT&T), 108, 121, 177, 182, 198-9, 201, 290
Benjamin, D., 63
Bernanke, Ben, 27, 29, 31-2, 38, 40-1, 43, 47, 94
Bernstein, I., 15, 18, 20, 23, 244, 247
Bernstein, M., 10, 32, 35, 44, 46, 58, 81, 90-4, 102, 104-5, 143-4, 216, 227-8, 240-1, 244-7, 297, 303, 308, 320, 327
Berry, Brian, 55
Bhaskar, R., 27
Bicycle, 85, 193, 206-8, 216
Bilstein, Roger, 222, 224, 228
Bleaney, Michael, 52-5, 90
Blinder, A., 40, 92, 109
Bolch, Ben, 77, 254-6
Boulding, Kenneth, 8
Bowden, Sue, 192, 201

357

Braun, Ernest, 126, 198-9, 314
Bresnahan, T., 58, 137, 154, 214
Britain, 8, 12, 23, 78, 91, 94, 115-6, 126, 132, 135, 153, 162, 164, 166-7, 172, 181, 193, 206, 211-2, 215, 248, 257, 301-7, 319
Brown, Malcolm, 249-50, 331
Brown, Nelson, 243-4
Brunner, K., 33, 143
Bryant, Lynwood, 207, 221, 244, 259
Burns, Arthur, 44, 86, 253
Burns, R.W., 185-6, 201
Burnside, C., 140
Burstall, Aubrey, 126, 193-4, 207, 221
Buxton, Neil, 126, 153

Calkins, E.E., 75, 89
Calomiris, Charles, 29, 34
Canada, 23, 32-3, 95, 208, 265-6, 283, 300, 302, 306
Canning, food, 217-9, 221, 239-42
Capital goods sector, 56-7, 83, 87, 125, 192-6, 296, 329
Cardwell, D.S.L., 206
Carothers, Wallace, 165
Casson, Mark, 62, 92, 133
Catterall, R.E., 305
Cechetti, S., 29, 46
Cellophane, 158, 163, 239, 244
Chandler, A.D., 87-8, 113, 122, 278
Chandler, L.V., 23, 34, 38, 267, 318
Chase, Stuart, 80, 105
Chemical industry, 5-7, 69, 85, 92, 98, 105, 111-2, 114-6, 127, 135, 151, 155-67, 170-1, 173, 175, 177, 191, 193, 210, 221, 223, 233-4, 249-50, 306
    See also Pharmaceuticals, Plastic, Synthetic fibers
Chemistry, advances in, 115-6, 155-6, 159, 162-5, 218
Cheng, L., 329
Church, Roy, 305
Cigar manufacture, 131
Civilian Conservation Corps (CCC), 23, 328-9
Clark, Harold, 335

Clerical work, 196, 261, 274-5, 309
Clewett, J., 75, 89, 278-9
Coal, 71, 104, 116, 133, 161, 190, 232-3, 236-8, 263, 293
Coen, Robert, 15
Cohen, Avi, 106, 126, 246
Computers, 91, 196-9
Constant, Edward, 114, 225
Construction, 5-6, 36, 56, 69, 77, 84, 87, 91, 96, 116, 152, 201, 217-9, 242, 253-8, 292, 304, 306
    roads, 69, 151, 208-9, 215, 255, 257, 273, 319
Construction materials, 226, 237, 242, 257-9, 292
Consumption, fall in, 5-6, 8, 13, 34-7, 47-8, 52, 60, 67-82, 84-6, 102-3, 131, 136, 265
    See also Market saturation, Propensity to consume
Continuous processing, 6, 7, 103, 105, 113, 149, 156, 158, 160, 216-8, 234-5, 240-1, 249, 288-9
Coolidge, Calvin, 129, 291
Cooper, R., 64
Copeland, Morris, 69, 73, 79-80, 201, 223, 232, 245, 285
Coplon, W., 242-3
Cornwall, John, 82, 85, 87, 140, 253-4
Costrell, Robert, 271
Cowan, Ruth, 74, 188-91, 201-2, 248, 277, 281
Cox, Charles, 45-6
Crabb, Richard, 207
Credit, availability of, 10, 31-3, 37-9, 54, 56, 58, 79-81, 87, 91-2, 116, 223, 228-9, 264, 320, 324
    See also Banking, Instalment buying, Interest rates
Cross, Gary, 111, 127
Cunningham, W., 69, 282
Cyert, Richard, 322

Darby, M., 15
Dasgupta, Partha, 334
Davis, Lance, 153
Davis, Pearce, 217-8

Index

Dearden, J., 107
DeBresson, C., 57, 110-1
Deflation, 9, 13-14, 24, 29, 31, 34, 37-41, 46, 54-5, 68
DeForest, Lee, 181-2
DeLong, J.B., 48
Dimand, Robert, 52
Distribution, 84, 92, 112, 133, 152, 163, 170-1, 218, 239-42, 276, 278-82, 295
Dixit, A., 139
Dominguez, K., 61
Domowitz, I., 64
Dosi, Giovanni, 67, 117, 125, 127, 147
Douglas, P., 15, 47, 73, 86
Douglas, Susan, 185
Drummond, I., 300
DuBoff, R., 176-7
Dummer, G.W.A., 178, 198
DuPont Corp., 108, 121, 157-9, 163-5, 191, 210, 217, 239, 289-90, 314
Durable goods, 5, 37, 40, 47, 68-9, 74-7, 80, 84-8, 91, 137, 192, 195, 213, 278, 296
See also Appliances, Automobiles
Dyes, 116, 155-6, 158, 160, 167, 170-1

Eastman, George, and co., 108, 121, 167, 330
Economic growth, 48-50, 53, 60, 82, 88-90, 101, 315-7, 324-6, 333
Economic history, 153, 315-8, 331
Economic planning, 74, 80, 327-9
Edison, Thomas, 108, 121, 168, 175-7, 181
Education, 19, 73, 115, 126-7, 273, 331, 335
Efficiency wages, 9, 41, 72
Eichengreen, Barry, 21-3, 27, 33, 40, 44, 47, 63, 153, 301, 304, 308
Eichler, Ned, 257
Einstein, Albert, 185, 239
Electric generators, 159, 175-6, 180, 237
Electric motors, 104, 175-7, 188, 220, 228, 247
Electrical transmission, 176-7

Electrification, 6-7, 72, 74, 85, 90, 103-4, 111, 116, 151, 159, 175-203, 220, 228, 231-3, 237-8, 247, 255, 258, 262, 268, 288-9, 296, 305, 307
Electrochemicals, 114, 156, 159-61
Energy
  See Coal, Electric generators, Nuclear power, Petroleum
Engineering, 115, 156, 158, 160-1, 165, 179, 185, 193, 199, 202, 228, 237
Enos, John, 147, 234-5
Entrepreneur, role of, 6, 7, 10, 35, 52-3, 55, 89-90, 101, 122-3, 129, 134-5, 215, 228
Epstein, Ralph, 86, 151
Europe, 13, 17, 78, 83, 86, 95, 177, 186, 199, 208, 221-2, 225, 301-4, 334
  See also Belgium, Britain, France, Germany
Ewen, Stuart, 246, 250
Exchange rates, 28, 33, 78, 95, 212, 301
Expectations, 6, 8-10, 29, 35, 37, 46-7, 54, 59, 92, 94, 123, 134-5, 137, 296, 316
  of deflation, 9, 29, 38, 40-1, 54
Exports, 33, 67, 78-9, 93-5, 211-2, 228, 233, 243, 250, 263-6, 269, 299-302, 306, 331-2

Fabricant, Solomon, 46, 103, 130
Fair, R.C., 83
Fano, E., 73, 103, 110, 158
Faraday, Michael, 175-6, 180
Farnsworth, Philo, 187
Fearon, Peter, 15, 21, 97, 247, 250, 265, 267, 307
Federal Communications Commission, 187
Federal Reserve System, 9, 15, 28-33, 37, 47
Federal Trade Commission, 44
Fertilizers, chemical, 112, 160-1, 283, 333
Fessenden, Reginald, 180-1
Field, A., 30, 35, 256-7
Field, D.C., 205

Firm size, 32, 44-6, 71, 107-9, 120, 147, 162, 170, 219, 234, 323-4
Fiscal policy, 67-8, 93-4, 296
Fischer, Claude, 179
Fisher, Franklin, 198
Fisher, Irving, 8, 27, 38-9, 59, 136
Fleisig, H., 8
Fleming, John, 172, 181
Folk, George, 120
Food processing, 84, 112, 152, 207, 217-9, 221, 239-42, 294, 307
Ford, Henry, and co., 41, 76, 78, 80, 103, 207-8, 210, 212-4, 217, 227
Foreign trade
  See Exchange rates, Exports, Tariffs
Foreign investment, 33, 78, 300-2, 306
Freeman, Christopher, 36, 82, 90, 106-8, 111, 114, 130, 163, 166, 203, 319, 322, 324, 334
France, 23, 86, 162-3, 171-2, 206, 208, 212, 235, 283, 301-2, 305
French, Michael, 216
Freudenthal, E.E., 223, 228
Friedman, Milton, 28-31
Furniture, 79, 250, 292

Galbraith, J.K., 68, 80-1, 94, 108
Garraty, John, 17, 41, 46, 105, 283, 301, 308
General Electric (GE), 108-10, 121, 177, 181-3, 188, 191, 201, 314, 332
General Motors (GM), 76, 211-3, 227, 261
Gennsheim, H., 167
Germany, 94, 116, 156-7, 162-5, 167, 171, 196, 212, 220, 290, 301-3, 305, 310, 315, 327, 331, 335
Gershuny, J.I., 271, 276, 321
Gibbs-Smith, C.H., 222-3
Giedion, S., 188
Gilfillan, S.C., 113
Gill, C., 14, 105, 108, 133, 195-6, 215, 237, 267
Ginzberg, Eli, 35, 42, 46, 87, 101, 214
Glass, 69, 91, 158, 160, 216-8, 292-3

Goldin, Claudia, 308
Goldsmith, Raymond, 96
Goldstein, J., 55
Gomulka, S., 101, 321, 323
Goodman, J., 206-7
Gordon, R.A., 8, 30, 36, 46, 48, 82, 90, 143, 148, 190, 213, 216, 255, 266, 309, 319
Gordon, R.J., 29, 31, 33-4, 36-7, 46, 56, 253-4
Gort, Michael, 108-9
Gould, J.M., 178
Gourvitch, Alexander, 131
Government, 9, 42, 45, 92, 182, 197, 208, 222-5, 243, 256-7, 264-5, 271-3, 285, 290, 296, 306, 319, 326-31
  research policy, 12, 106, 322-4, 326, 329-30, 334
  See also Education, Fiscal policy, Military, Regulation, Relief, Various government bodies
Gowdy, J.M., 327
Gramlich, Edward, 271
"Great Depression" of the late nineteenth century, 14, 111, 115-6
Grebler, Leo, 256, 258
Green, Alan, 17, 284, 306
Gries, John, 255
Griliches, Z., 113, 125
Grubel, Herbert, 270, 273, 284, 321, 334

Haber, L.F., 156, 158-9, 161, 283
Haberler, G., 65
Hamermesh, 39, 150
Hamilton, James, 29-30
Hampe, E., 240-1, 279-80
Hansen, Alvin, 24, 45, 52, 54, 62, 82-3, 90, 125, 253
Harberger, A., 96
Hatton, T.J., 23, 63, 307
Hayes, Douglas, 14, 32
Haynes, William, 98, 158, 160-2, 164, 170-2, 249
Health, 18-9, 170-2, 273, 335

Heim, Carol, 132, 304
Henderson, John, 145, 253, 256, 266, 270, 274
Hertz, H., 180, 185
Hickman, B., 253, 255-6
Hicks, John R., 48, 58
Higgs, R., 93, 272
Hirschhorn, Larry, 138
Hobson, John, 51
Hogan, William, 189-90, 195-6, 208, 211, 219-21, 226, 235, 248, 250, 257-61, 263, 268, 283
Hollander, Samuel, 163
Holt, C.F., 70-1
Holthausen, D., 79-80, 98, 189
Hoover, Herbert, 15, 42, 51, 93-4, 121, 318, 330-1
Hopkins, John, 266-8, 285, 307, 335
Hounshell, David, 45, 121, 157-8, 163, 165-6, 191, 207, 212-3, 273
Hours of labor, 9, 13, 15-9, 43, 76, 105, 130-1, 237, 247, 268, 325
Hughes, J.R.T., 13, 44, 73, 92, 94, 272, 328, 335
Hughes, Thomas, 176
Hunnicut, B.K., 17

Immigration, 41, 43, 83-4, 95, 168, 222, 254-5, 277
Implicit contracts, 41-2, 93
Income distribution, 13, 51, 53, 67, 70-4, 94, 190, 209, 211, 227, 245-6, 256, 265, 277, 287, 300, 305, 325
Industrial Revolution (British), 6, 24, 55, 84-5, 90, 101, 115-6, 130, 155, 218, 315
Inkster, Ian, 233
Innovation, timing of, 6-7, 106-22, 149, 151-2, 289-91, 314
Insider-outsider theory, 9, 43
Instalment buying, 37, 67, 79-81, 189, 192, 201, 210
Interchangeable parts, 104, 206-7
Interest rates, 14, 28-32, 34-6, 47, 59, 71, 256, 311, 316
Internal combusion engine, 6, 111, 116, 151, 193-4, 205-6, 209, 221-5, 232, 244, 248-9, 260-2

International trasmission of the Depression, 12, 28, 33, 78, 94-5, 299-301
Inventory investment, 69-70, 86, 92-3, 96, 150
Investment, decline of, 5-9, 13-14, 34-7, 39, 51-4, 56, 60, 67, 69, 72, 77, 81-93, 105-6, 131, 134, 150-1, 184, 195-6, 213, 216, 219, 235-6, 256-7, 263, 287-8, 297
Investment opportunities, 9, 32-3, 35, 47, 81-93, 134-5, 242, 247, 254, 288, 294-5, 326
Iron and steel, 69, 84-5, 115-6, 210, 214-5, 218-21, 223, 237, 258-60, 262, 292-3, 319
See also Alloys

Jacoby, Sanford, 18, 21, 23, 25, 41-2, 104, 130
Jaeger, Hans, 14, 61, 69, 228, 259, 303
Japan, 122, 250, 303, 314, 317, 327
Jenkins, Reese, 166-8
Jensen, Vernon, 25, 41-2, 139, 243
Jerome, Harry, 103-5, 110, 130-1, 139, 148, 169, 179, 195, 200-2, 216-8, 238, 244, 247-8, 261, 266
Jet engine, 180, 198, 206, 225
Jones, Eric, 24

Kalecki, M., 44, 62, 90
Katon, H., 258
Katsoulacas, Y., 58, 131-2, 139, 320, 322
Katz, B., 197
Keller, R.R., 71
Kendrick, J., 103, 148, 284
Kennedy, E.D., 154, 227
Kesselman, J.R., 16-7, 132
Keynes, J.M., 35-6, 38, 75, 130, 265, 304, 335
Keynesian theory, 6, 8-9, 28, 33-8, 45, 48, 50, 52-4, 58, 67-8, 93-4, 138, 151, 255, 265, 318, 327, 330
Kimball, D., 41, 80, 83, 108
Kimmel, Lewis, 98, 324
Kindleberger, Charles, 265, 294, 301
Klein, Burton, 82, 107-8, 224, 227

Klein, J., 35, 271
Kleinknecht, Alfred, 55, 109
Koenigsberger, F., 194
Kondratieff or Kuznets cycles
See Long Waves
Kraft, James, 169
Krooss, H., 45, 80, 83-4, 86
Kuznets, Simon, 58, 70, 86, 88, 122-3, 134, 271

Labor dishoarding, 6, 8, 17, 49, 93, 129, 136-7, 149-50, 237, 243, 247, 261, 288, 292-5, 320
Laboratories, industrial research, 7, 89, 102, 108-10, 112, 117-22, 147, 157-9, 164-5, 170, 177, 181-3, 186-8, 193, 199, 223, 225, 234-5, 273, 290, 313-4, 323
Labor-saving technology, 6, 50, 54-5, 59, 71-2, 75, 91, 103-6, 110, 112-3, 126, 130-5, 148-52, 156-7, 160, 163-4, 169, 178-9, 195-6, 209-10, 214, 216-20, 235, 237-8, 243-7, 258, 261-3, 267-8, 275-6, 278, 288, 291-5, 320-2
Laidler, David, 62
Larson, H., 235-6
Layton, David, 115, 127
Lebergott, S., 15, 17
Leonard, Jonathan, 24, 65, 136
Leuchtenberg, William, 45, 71
Leven, M., 72-3, 330
Leveson, Irving, 283
Liebenau, Jonathan, 170-1
Lightbulb, 162, 175-8, 181, 189-90, 196, 217-8, 233
Lilien, David, 50
Lindbeck, Assar, 43
Linder, S., 75
Lindert, Peter, 33-4
Lodewijks, J.K., 107
Long Waves, 9, 10, 55-7, 90, 111, 116, 122, 131, 253, 300
Lorant, J.H., 124-5
Lough, William, 85-6, 89
Lucas, Robert, 40, 63
Luckoff, H., 197-8

Lumber industry, 152, 164, 190, 216, 231-2, 242-4, 292, 307

Machine tools, 6, 57, 87, 104-5, 112, 149, 151, 160-1, 176-7, 192-6, 218, 232, 241, 248, 260, 274, 292, 296, 306
MacLaurin, W.R., 181-5, 187-8, 322
Macroeconomics, 8-9, 27-40, 44-50, 52-3, 57-9, 67-8, 93-4, 131, 137-8, 150-1, 255, 265, 296, 317-20, 327, 329-30
Maidique, M.A., 118, 125
Management, 9, 23, 25, 42, 84, 87-8, 104-5, 107, 112, 134, 137, 247
Mandel, E., 54-5, 256, 319-20, 334
Mankiw, N.G., 44-5, 49-50
Mansfield, E., 125
Marconi, G., and co., 180-2
Margo, Robert, 16-7, 21-3, 41-2, 58, 125, 143, 149
Market saturation, 4-5, 10, 69-70, 72, 76-80, 84-6, 103, 123, 131, 135, 138, 148, 150-2, 184, 189-91, 210-5, 245, 255-6, 287, 291-5, 302-3, 305, 319, 329
Marxian theory, 9, 53-5, 131, 151, 333
Mass production, effect on innovation, 89, 113, 126, 170, 227
Matthews, R.C.O., 135
Maxwell, James Clerk, 115, 175, 180
Mayer, Thomas, 36, 64
McCraw, T., 219
McKie, James, 240
Means, Gardiner, 46
Mellon, A., 42, 71
Mendel, Gregor, 115, 268
Mensch, G., 55, 57, 106, 109, 111, 114, 116, 188, 261, 290
Mercer, Lloyd, 77, 213, 256
Mieszkowski, Peter, 255
Military, 110, 160, 178, 185, 197, 199-200, 222-5, 272-3, 290
Miller, M., 212, 215, 305
Mills, Frederick C., 15, 105, 139-40, 195
Mingos, Howard, 221

Mining, 131, 152, 237-8, 307
Miron, J., 30
Mishkin, F.S., 37, 46
Mitchell, W.C., 24, 63, 73, 85, 88-9, 95-6, 103-4, 131, 266
Mokyr, Joel, 106, 325
Monetary explanation of Depression, 8-9, 15, 28-34, 38, 47, 59
Motion pictures, 24, 36, 158, 162, 166-70, 186, 293, 307
Mowery, David, 108, 110, 112, 115, 121, 225, 239, 315, 324, 327, 334
Mueller, Willard, 163
Multiplier mechanism, 5-6, 12, 17, 74, 98, 136-8, 150, 295-6, 330

National Industrial Conference Board, 42-3, 98, 139
National Industrial Recovery Act (NIRA), 41-2, 105, 296
National Recovery Administration (NRA), 92, 247
Neisser, Hans, 139
Nell, Edward, 62, 81-2, 321
Nelson, Richard, 64, 109, 180, 199
New Deal, 44, 94, 106, 257, 296, 328, 330-1, 335
Noble, David, 117, 119, 125, 323
Norton, R.D., 88, 114
Nuclear Power, 158, 238-9
Nye, David, 167, 175-6, 189
Nylon, 158-9, 165

O'Brien, A.P., 41, 43, 63, 260
Oi, Walter, 280, 282, 285
Olney, Martha, 37, 85, 96, 189-90, 192, 210
Olson, Mancur, 45
Output, changes in, 13-14, 17, 19, 31, 33, 55, 103, 150-2, 184, 190, 195, 213, 218, 235, 243-4, 264, 282

Paper industry, 106, 112, 126, 157, 160, 246, 266, 294, 306
Parke-Davis, 108, 170
Pasinetti, Luigi, 84, 89, 333
Pasteur, Louis, 115, 172, 239-40

Patents, 7, 106-7, 109-10, 120, 157, 162, 165, 168-70, 177, 179, 182-4, 186-7, 191, 221, 234, 249
Patinkin, Don, 44
Peppers, Larry C., 93-4
Perazick, G., 109-10, 119
Petroleum, 36, 71, 87, 112, 135, 147, 158, 160-1, 190, 205, 207-9, 215, 219, 223, 225, 232-6, 262-3, 279, 293-4
Pharmaceuticals, 4-5, 7, 91, 156, 158, 160, 170-2, 269, 294
Phelps, Edmund, 9, 59-60, 83, 131, 330, 335
Philips, A., 223-4, 315
Photography, 158, 166-8, 222, 293
Pipelines, 218-9, 258, 262-3
Plastic, 4, 7, 91, 156, 158-9, 161-6, 178
Platt, H.L., 124, 177, 189, 255
Plosser, C.I., 49
Podgursky, Michael, 25, 138
Population, 5, 9, 17, 29, 53, 56, 59, 67, 72, 74, 83-4, 253-4, 256, 264, 304, 319-20
  See also Immigration
Price flexibility, 13, 29, 44-6, 54, 73, 129, 131-2, 137-8, 219, 237, 245, 260
Process versus product innovation, 6-7, 10, 39, 59, 87, 90-1, 101-3, 106-7, 112-3, 120-2, 129-32, 227, 287-8, 303, 305, 313-4, 320, 323, 325
Product cycle, 78-95
Productivity, 72, 88, 103, 264, 270-1
  capital, 13, 55, 103, 231
  labor, 6, 13-14, 17, 19, 41-4, 49, 55, 91, 96, 103-6, 116, 130, 137, 147-52, 195-6, 209-10, 216-20, 229, 231, 235-8, 244, 246-7, 258-61, 266-7, 279-82, 288, 291-5, 305, 321
Profits, 14, 35, 47-8, 53-4, 58, 71, 92, 109, 125, 151, 163, 169, 185, 188, 227-8, 235, 244-5, 279, 334
Propensity to consume, 5, 17, 34, 48, 51, 53, 67-79, 131, 136, 330
Pursell, Carroll W., 106, 334

Radar, 178, 187, 202
Radio, 4, 7, 36, 79-80, 85, 120, 158, 162, 164, 168-9, 177-86, 188, 192, 197-9, 292, 306
Radio Corporation of America (RCA), 182-4, 186-7, 201
Rae, John, 205, 211, 215, 217, 220, 222-4, 310-1
Railroads, 69, 83, 85, 90, 116, 131, 206, 219, 232-3, 236-7, 243, 250, 258-63, 293
Ramey, Valerie, 93, 96
Rasmussen, W., 283
Raupach, H., 309-10
Rayon (cellulose), 36, 86, 158, 160, 162-3, 244, 246, 266, 294, 307
Real Business Cycles, 9, 48-50, 53, 150, 329
Reconstruction Finance Corporation, 94
Recovery from Depression, 6, 8, 14, 29, 31, 37-8, 58-9, 94, 195, 253, 296-7, 300
Reekie, W.D., 171-2
Regulation, 92, 185-7, 208, 243, 256, 260, 274
Reich, Leonard, 108, 110, 112, 121, 126-7, 177
Relief measures, 10, 15-8, 21, 93-4, 131, 330-1
Research
  See Laboratory, Science, Universities
Roads, 69, 151, 208-9, 215, 255, 257, 273
Rolt, L.T.C., 119, 192-4
Romer, C., 31, 47, 69
Roosevelt, Franklin, 39, 44, 93-4, 330-1
Rose, Mark, 201
Rosenberg, Nathan, 206
Rostow, W.W., 54, 82, 86, 97, 104, 114, 255, 257, 310
Rotella, E., 275, 309
Rowland, Henry, 119
Rowthorn, R., 97
Rubber, 36, 69, 72, 95, 105, 157, 159, 161, 163-5, 216-7, 292-3

Sachs, J., 45
Safarian, A.E., 306
Sahal, D., 114, 117
Saint-Etienne, C., 33, 35-6, 94, 304
Sardoni, Claudio, 38, 53-4, 62, 131
Saul, S.B., 14, 55, 116
Sawers, David, 108, 110, 126, 221, 224
Scherer, F.M., 126-7, 333
Schlebecker, John, 265, 268
Schmookler, J., 109, 113
Schroeder, Fred, 189
Schultz, Theodore, 139
Schumpeter, Joseph, 3, 7, 10, 49, 56, 90, 102, 104, 112, 116, 122-4, 126, 148, 183, 253, 264, 290, 307, 329, 333, 335
Schurr, S.H., 231, 233, 236, 242
Schwartz, Anna J., 28-9, 34
Schwartzmann, David, 172, 281-2
Science, 111, 115, 119-20, 126, 156, 158, 162-3, 175, 180-2, 185, 194, 199, 218, 228, 238-40
Scitovsky, T., 4
Scoville, John, 73
Second Industrial Revolution, 6-7, 102, 111-5, 152, 218, 313-4
Secular forces and the Great Depression, 8, 9, 17, 41, 52-3, 56, 147-8, 254, 270, 274
Seltzer, Lawrence, 98, 207, 215, 227
Servants, 74, 188-9, 276-7, 295
Services, 112, 116, 152, 177, 269-82, 289, 296-7, 308, 321
  See also Advertizing, Clerical, Distribution, Servants
Shannon, David, 18-9, 24
Sharlin, Harold, 175, 179
Sharpe, Steven, 58, 137
Sheldon, Roy, 69-70, 85, 89, 167
Shipbuilding, 222, 232, 243, 261, 263, 293, 319
Sinclair, Peter, 43, 63, 140, 278, 311, 326, 334-5
Slichter, Sumner, 21, 41, 84
Smiley, Gene, 15, 71
Smith, Philip, 210, 212-3
Solomou, S., 55
Solow, Robert, 49

Index

Soule, George, 47, 247, 253-4, 266
Soviet Union, 265, 306, 316-7, 331, 335
Stanback, Thomas, 75, 271, 274, 284, 320, 322
Standard of living, 37, 42, 70, 74, 208, 265
Staudenmeier, John, 114, 117, 317
Stigler, G., 95, 270, 273, 276-8, 280, 285
Stock market, 9, 10, 13, 29-30, 33, 35, 37, 46-8, 58, 60, 78, 92, 96-7, 227, 253-4, 274, 295, 320
Stolper, W., 327
Stricker, Frank, 18, 72, 265
Structural unemployment, 10, 129-30, 132-3, 169, 247, 249-50, 288, 304, 327, 331
Sturmey, S.G., 181, 184
Style changes, 80-1, 89, 127, 189, 201, 212-3, 227
Suburbanization, 24, 69, 87, 97, 112, 192, 226, 255
Sundstom, W., 63, 309
Svennilson, Ingvar, 17, 83, 97, 303-4
Swann, John, 170
Sylos-Labini, Paolo, 83-96
Synthetic fibers, 4, 7, 91, 158-9, 161-6, 265
See also Rayon

Tariffs, 45, 78, 95, 157, 212, 265, 300, 331
Taylor, F.S., 155-6, 162, 170
Taylor, Frederick, 104, 193-4, 207
Taylor, Melvin, 76
Taylor, Morris, 27, 73, 130
Technological unemployment, 6, 11-12, 106, 129-32, 148-9, 267, 274, 334
See also Consumption, Investment, Labor-saving technology
Technology clusters, 106-11
Technology diffusion, 164, 200, 327, 334
Technology transfer, 157, 162-6, 170, 196, 202, 235, 316, 322
Telephone, 177-80, 182-3, 185, 192, 197

Television, 4-5, 7, 91, 120, 169-70, 184-8, 192, 319
Temin, Peter, 28-34, 36, 51, 93-4, 254, 265, 301-2, 333, 335
Temporary National Economic Committee (TNEC), 45, 51, 73, 85, 97, 105, 120, 131, 156, 243, 334
Terbogh, George, 51-3, 97
Tesla, N., 176
Textile manufacture, 83-4, 95, 112, 116, 131, 133, 137, 152, 158, 160, 163, 244-8, 294
Thomas, David, 168
Thomas, Mark, 21, 23, 63
Thorne, S., 241-2
Thorp, Rosemary, 300
Thorp, Willard, 45
Tillman, J.R., 178, 202
Tilton, John, 182, 200, 202
Tobin, James, 57, 315
Tracey, Michael, 265
Tractors, 104, 112, 267-8
Transistor, 198-200, 319
Transportation, 69, 73, 152, 206, 239, 258-63, 293
Trucks, 206, 236, 255, 259-61, 268-9, 282
Tugwell, Rexford, 32

Udelson, Joseph, 184, 186-8, 201
Underconsumption theory, 9, 51-4
Unemployment
  aggregate measures of, 13-9
  by sector, 144-7
  duration of, 20-2
  age distribution of, 17, 23
  regional distribution of, 17, 307
  race and, 17, 309
  gender and, 308-9
  psychological effects, 18-21
Unions, 18, 42-3, 71, 104, 139, 169, 307
Universities, 108, 119, 127, 159, 170, 172, 184, 290, 324, 334
Usher, A.P., 113

Vacuum tube, 168, 178, 181-4, 197-9, 217
VanDuijn, J.J., 109
Venables, A.J., 140
Ville, Simon, 207, 226
Vitamins, 4-5, 158, 172
VonTunzelmann, G.N., 91, 297, 304, 310

Wages, 9, 14, 39-45, 47, 54, 60, 71, 76, 78, 92, 96, 110, 131-6, 237, 246, 264, 307
Wagoner, Harless, 124, 195, 321
Wallace, Henry, 84, 98-9
Wallis, John, 278, 284, 307
Walsh, M., 283
Walton, G.M., 264-5
Wandersee, Winifred, 37, 74, 82, 244, 308
Warren, Kenneth, 311
Warsh, David, 143
Waters, Joseph, 47, 73, 84
Webb, John, 18, 24, 330
Weil, Phillipe, 61
Weintraub, David, 131, 133, 148, 154, 238, 261
Weir, Ron, 32, 304
Westinghouse Co., 177, 182-3, 186, 202
Whatley, Warren, 309
Wheelock, David, 29-30
White, Eugene, 30, 47-8
Wicker, Elmer, 30-1
Wied-Nebbeling, S., 140
Williams, Trevor, 106, 121, 168, 179, 187-8, 193, 221, 228, 237, 239, 247-8, 258
Williamson, Harold, 120, 234-6, 248-9, 262-3, 279
Williamson, Jeffrey, 72
Wise, George, 88, 109, 119-21, 191, 332, 334
Wolff, Edward, 284
Wolman, Leo, 85, 179, 209, 245
Women, labor force participation rate, 17, 239, 275, 278, 308
Woodbury, Robert, 193-6
Woodruff, R.W., 35
Woolf, Arthur, 71, 103-4, 148
Works Progress Administration (WPA), 17-8, 94, 105, 131, 215, 257, 292, 328-30
World War One, 24, 42, 55, 72, 79, 82-3, 110, 126, 157, 160-1, 163, 170, 182, 188, 217, 221-3, 238, 255, 263, 290, 301
World War Two, 5, 14, 23, 79, 94, 99, 126, 158, 165-6, 172, 187, 202, 214, 217, 225, 239, 241, 267, 290, 319
Wright, Gavin, 41, 43, 72, 74

X-rays, 155, 177

Zevin, R.B., 47, 253
Zworykin, V., 186

About the Book and Author

In this unique volume, Rick Szostak takes an innovative approach toward analyzing the Great Depression of the 1930s. Most of the literature focuses on the movement in aggregate variables, but Szostak provides evidence primarily at the sectoral level, being careful to show that this argument is consistent with aggregate data.

Combining a fresh theoretical viewpoint and industry-level analysis, Szostak contends that an abundance of process technology made it possible for industry to produce the existing range of products with a much smaller labor input, while a shortage of new product technology severely limited the introduction of new products. Pinpointing how the timing of the Second Industrial Revolution affected the evolution of the workplace and how the industrial research laboratories that emerged in the United States in the twentieth century initially emphasized process over product innovation, he explains why this conjunction of technological forces caused both consumption and investment to fall so precipitously in early 1929.

In addition to exploring the technological and employment experience of specific sectors, Szostak looks at trends in income distribution and population and other factors that created the ultimate economic depression.

Rick Szostak is associate professor of economics at the University of Alberta.

DATE DUE			
			OCT 1 1 1996
MAY 04 1996			
	SEP 01 1996		
	NOV 3 0 1998		
	NOV 2 3 1996		
			Printed in USA

HIGHSMITH #45230